NEUROLOGICAL CONTROL SYSTEMS
Studies in Bioengineering

NEUROLOGICAL CONTROL SYSTEMS
Studies in Bioengineering

by Lawrence Stark, M. D.
Professor of Physiological Optics
University of California, Berkeley

Foreword by Warren McCulloch
Research Laboratory for Electronics
Massachusetts Institute of Technology
Cambridge, Massachusetts

PLENUM PRESS · NEW YORK · 1968

First Printing — October 1968
Second Printing — April 1972

Library of Congress Catalog Card Number 68-14855
ISBN   0-306-30325-6

© 1968 Plenum Press
A Division of Plenum Publishing Corporation
227 West 17 Street, New York, N. Y. 10011

Printed in the United States of America

To my parents,
Edward and Frieda Stark

Foreword

To anyone who worked long on the functional organization of living systems, it seems obvious that the central problems arise from a multiplicity of closed loops simultaneously active in the control of every act. Severally, they may be stable but, combined, they may crack up or break into schizogenetic oscillation. Whatever they are, linear or not, is beside the point, for in either case the same difficulty of analysis arises. Nor does it help our argument that we deal only with formal neurons or those having greater similitude to real ones. This limited Pitts and me, in 1943, to three theorems of Part III of our " Logical Calculus of Idea Immanent in Nervous Activity." We still seek a transparent terminology in which to discern their properties. In the face of such difficulties it takes great courage to attempt a servosystem analysis of the control mechanisms mediating any reflex if one would do this as a good engineer.

Walt Whitman says of the "Beginners," "How dear and dreadful they are to the earth — appearing at intervals — How all times mischoose their objects of adulation and reward — And how the same price must still be paid for the same great services!"

Larry Stark is a beginner. His book must not be mistaken for physiology — and there is nothing wrong with his physiology. Nor may it be mistaken for mere physics, for physics deals only with things that simply happen or do not happen and has no truck with their utility. No! He, a physician, is an engineer in his attack and is entitled to distinguish work from energy and signal from noise.

Because his hypotheses are explicit, his data reliable, and his aims patent, it would be dangerous for anyone to attack his work and its presentation. Its claims are explicit. Woe to him who says it is not all one could ask of physiology or of physics; for they simply missed the point, and time will catch up with them.

But it is difficult to understand, from the vantage point of an antiquated discipline of science, what is actually happening without a bit of history.

Larry Stark was graduated from Albany Medical School in 1948, the very year he had come to know Shannon's mathematical theory of communication. Two years later he was in the go of these problems, and in 1951 he walked into my laboratory in the Department of Psychiatry at the University of Illinois in Chicago. Since this time I have been well informed of his doings and thinkings.

The next year (1952) he was in the U.S. Navy concerned with the mechanisms of fire control and soaked himself in Wiener's Cybernetics. When he returned to medicine in 1954, it was as Assistant Professor of Neurology at Yale, where he studied in the graduate courses of electrical engineering concerned with control, and the very next year he was off on the theory of pupillary mechanisms.

What tempts an ambitious spirit into a novel field is usually the most difficult, because the most important, kind of problem. So it was with him as with the rest of us who dare to dream. Sooner or later we discover our limitations and are willing to solve simpler things in simpler ways. So it was with him. But, like every good teacher, he does not take his student first through the places where he is still struggling, but through simpler things he has learned to handle. This has determined the book, its structure, and its references.

No one could ask a better beginning in this mode of analysis, and those who are not content could do no better than to get to know him in his working days.

Warren S. McCulloch

Introduction

Engineering as science has a rather short history. Until the second quarter of the present century, engineers looked to chemists and physicists for the basic scientific information upon which they could base their engineering applications — bridge construction, prime movers, power machinery. With the inventions of the telephone, the triode, feedback amplifiers, radio systems, and automatic pilots, the development of control and communication engineering systems proceeded. When engineers again turned to physicists and chemists for basic scientific information concerning these complexly organized systems, they found none available: when they turned to biology, they found that biology had not yet been transformed into an advanced quantitative science. Being intelligent, well-trained, and practical men, they developed the science of engineering systems themselves. People like Vannevar Bush, Norbert Wiener, Claude Shannon, Warren McCulloch, John von Neumann, Julian Bigelow, and Hendrick Bode pushed ahead and created what we now know as modern electrical engineering and computer science. Even those aspects based closely upon the works of the nineteenth-century French school of mathematical analysis — Laplace, Fourier, Lagrange, Cauchy — have been elucidated by constant use in engineering design.

These pioneering individuals were interested in biology as an example of a complexly organized set of systems; and cybernetics was well named by Norbert Wiener as "control and communication in animals and machines." Their influence on biology, starting with those engineering scientists who were originally biologists, has grown considerably. A landmark is a special issue of the Proceedings of the Institute of Radio Engineers published in November 1959 with papers on the frog's eye and on the pupillary servomechanism, which give some early examples of careful application of cybernetics to neurophysiology and neurological organization. The material in this book further represents the intellectual impact of engineering science on neurology.

As a young neurologist, trying to make sense of disorders of motor coordination so often seen in a variety of neurological syndromes, I found

classical neurophysiology without adequate explanations to offer, and turned
to engineering science or cybernetics. Our approach to neurological systems was to define, measure, quantify, analyze, and form concepts having
that same hard mathematical structure found in the physical sciences.
These quantitative descriptions do not merely summarize in a trivial fashion the verbal and intuitive qualitative feelings that neurologists have already understood. Science proceeds by alternate steps: formulation of an
intuitive concept for a new phenomenon, incorporation of the new and other
related phenomena into a mathematical structure, and finally, again, development of intuitive concepts of further phenomena utilizing the added
perspective obtained with the formal elegance of the mathematical model
or theory. As an example, we offer the comparison between the first scientific appreciation of gravity as both the force causing the apple to hit Newton's head and the force acting between planetary bodies, as contrasted with
the mathematical formulation of gravitational laws that permit detailed
calculations and predictions to be made which are the basis of the current
technological ability to make a soft landing on the moon.

Our work has dealt mainly with the engineering science approach to
four neurological motor feedback systems — the pupil, the lens, eyeball
rotation, and hand movement. Neurological control systems appear to fall
into two general classes: (1) The eye—hand tracking systems have intermittent characteristics enabling them to be represented by sampled data
models, show prediction operators giving them input adaptive properties,
and are otherwise fairly linear with low noise levels. (2) The regulators,
of which the pupil is a classical example, are generally continuous and
without prediction operators. Many of their essential design characteristics lie in nonlinear features, such as strong scale-compressions or saturations, and important asymmetries not eliminated by small signal approaches. They are often multi-input systems, present high levels of noise,
and "biological" adaptation is commonly found.

Other ancillary areas include study of the underlying neurophysiological mechanisms, the elements from which these neurological servomechanisms are constructed. These elements with their neuronal organ-
izations and functions must be analyzed in detail; here, classical neurophysiological techniques of microelectrode recording and stereotaxic stimulation are embedded within the engineering approach to specify information rates, nerve codes, transfer functions, nonlinear operations, and discontinuous transformations of signal flow. This experimental approach
turns away from classical decerebrate and anesthetized animals and seeks
new and ethical methods to deal with the awake, intact brain. Finally, our
research also utilizes digital and analog computers to make mathematical
models of these neurological systems so that explicit quantitative formulations can be integrated and understood.

This volume is an attempt to bring together an integrated collection of laboratory experiments which illustrate the variety of situations to which quantitative methods may be applied. No attempt has been made to explain the fundamentals of systems analysis as these have already been admirably presented in such textbooks as those by Trimmer, Truxal, and Kuo. It goes without saying that there are many difficulties in applying quantitative experimental techniques to the intact nervous system, and it has in fact been necessary to search carefully in order to find the appropriate part of the nervous system that best illustrates each characteristic design principle. Selection of this type is nothing new in biology where the particular experimental material must often be used to demonstrate the general principle. Without the use of the squid axon in neurophysiology or of bacteria in genetics, the modern development of these subjects would have been severely handicapped. The sections are not arranged chronologically; they simply represent an attempt to develop the subject systematically, starting with the simple and proceeding to the complex.

Section I on the crayfish introduces the idea of input—output analysis in its simplest forms, defining an element of a larger system. Students appreciate having an early look at a "wet dissection" experiment which gives them a better feeling for the neurophysiological elements of these neurological systems. The trains of nerve impulses which carry the messages to the central nervous system of invertebrates and vertebrates alike is dealt with and shown to be a simple average frequency code.

In Section II on the pupil, the important concept of *feedback* is introduced. It is illustrated in terms of a linear transfer function approach which enables quantitative predictions to be made concerning such crucial features of control systems as stability conditions and the development of oscillatory behavior. Because "plant dynamics" dominate much of the behavior of the pupil, it is possible to present this classical linear control theory application to a system which is essentially nonlinear. The nonlinearities are approached utilizing the interesting Wiener G-functional formulation. *Noise*, an essential feature of modern control and communication theory, is significantly present in the pupillary system, and further is shown to be not additive, but multiplicative noise. In this experimental approach, "dry dissection" utilizes the multi-input and multi-output topology of the pupil control system in order to break up a single "black-box" into a multiple subsystems block model.

In Section III on the lens, nonlinearities are shown to dominate accommodative behavior. The describing function approach, an extension of transfer function analysis, serves to clarify the role of nonlinearities in shaping both noise and dynamic behavior. An interesting "even-error" signal feature of accommodation is also treated.

Section IV on the eye introduces us to class I systems with their important prediction operators, that is, learning of repetitive input patterns and compensation for attenuations and phase lags observed with unpredictable inputs. Intermittency in the response leads us to employ a sampled-data model which is justified by accuracy in predicting behavior under variable feedback conditions. It is impressive how the low inertia of the eyeball permits the neurological computations to be reflected so directly in the output of the system.

Finally, in Section V on the hand, we come to the most complex control system studied. Not only does it have those operators — prediction and intermittency — that characterize the eye movement control system, but also the output adaptive mechanisms to adjust for varying loads. The neurophysiological subsystems underlying the neurological performance of the hand involve muscle spindles and other receptors, alpha and gamma motor neurons, nonlinear muscle dynamics, spinal cord reflexes, and higher brain stem, cerebellar, and cerebral nervous activity. In order to integrate all this physiological material into a behavioral model, we utilized extensive computer modeling.

The computer early played an essential role in bioengineering studies. Simulation, using hybrid analog-digital computers, proves valuable for understanding models whose features are too complex for either analytical computation or intuitive recognition. A typical experimental approach to a neurological control system employs programs in user-oriented languages for on-line digital computer control of (1) such preliminary details of experimental design as Latin square sequencing of stimulus variables, (2) the ongoing experimental excitation function generation and response reception, (3) storage and readout for real time or clean-up analysis, following preediting of data, and finally, (4) such man-oriented man-machine communications as graphical displays of experimental results. Rather than have a separate section devoted to computer techniques, we have elected to introduce this material throughout the volume in the context of its uses in obtaining specific research results.

Bioengineering is a two-way street — biologists obtain scientific descriptions and deeper understanding of these elegant neurological control systems. Engineers profit because of the stimulus the complex biological systems provide for encouraging the development, elaboration, and refinement of system theoretic approaches. Because of its exciting scientific discoveries and because of its role in advancement of engineering theory, the area of biological control, cybernetics is playing an essential intellectual role in promoting the interaction between biology and engineering.

Finally, I would willingly like to acknowledge my debt to colleagues and students who have worked with me over the years, many of whose particular roles in our research on neurological servomechanisms are discussed in the respective sections of this volume.

Most of the research described in this volume was carried out from 1954–1965 at Yale University and Massachusetts Institute of Technology where I was an Assistant Professor of Neurology and Head of the Neurology Section, respectively; recently this research has been continued and extended at the University of Illinois at Chicago Circle where I have been Professor of Bioengineering and Chairman of the Biomedical Engineering Department of the Presbyterian-St. Luke's Hospital and College of Engineering. Our work through the years has been supported by research grants and contracts from the National Institutes of Health (NB–3055, 3090, 06197, and 06487; MH–06175 and GM–01436), the Office of Naval Research [609(39), 1841(70), 0014], the Air Force [AF–33(616)–7282 and 7588, AFOSR–49(638)1313], and the Army (DA–18–108–405) as well as university and private sources (Clement V. Stone Foundation, F. N. Bard Fund, Yale Neurology Fund, Massachusetts Institute of Technology, University of Illinois at Chicago Circle, and Smith, Kline & French Foundation).

February, 1968   Lawrence Stark
Chicago, Illionois

Contents

SECTION I: THE CRAYFISH

SECTION II: THE PUPIL

Chapter 1

Chapter 2

Chapter 3

Chapter 4

SECTION III: THE LENS

Chapter 1

Chapter 2

Chapter 3

SECTION IV: THE EYE

Chapter 1

Chapter 2

Chapter 3

SECTION V: THE HAND

Chapter 4

Chapter 5

THE CRAYFISH

INTRODUCTION

Biological transducers such as sensory organs or sensory elements provide an opportunity for full description of the limits of their signal transfer, since their stimulating input energy can be both manipulated and monitored while their output nerve impulses are recorded. It is believed that the signal which represents the input stimulus, and which is transmitted to other centers of the central nervous system, is the average frequency of the receptor's output of impulses. By correlating the instantaneous values of the stimulus and the averaged frequency of receptor signal output, it is possible to make a rigorous statement regarding the transfer characteristics of the sensory receptor. The task, then, is to find the equation (generally an integro-differential equation) which characterizes the functional relationship between input and output [16, 52–54].

Before beginning the work on the transfer function of the crayfish photoreceptor described in this section, my colleagues and I had already studied transfer functions of the human pupil and motor coordination systems (see Sections II and V). However, the transfer functions we had obtained from these black box studies of human servomechanisms were but a beginning to our approach to these complex and interesting systems. What we really wanted was a system of nonlinear equations where the structure or "topology" of the equations related to the "topology" of the system in question, and the parameters of various portions of the equations related to the individual physical or physiological elements within the system. A number of "dry dissection" techniques such as multi-input and multi-output analysis, environmental clamping, and various pharmacological procedures were utilized to try to analyze the system without physically dissecting the human subject. However, we were inevitably driven to wet dissection methods. Justification for physical or "wet" dissection into a human nervous system, except where used as a therapeutic measure for neurological disease, is questionable. Wet dissection in animals raises other problems, since the anesthesia ethically required for the elimination of pain also impairs the very neural sensory and motor aspects under study. We therefore turned toward invertebrate animals where certain measurements could be made directly without interfering with their primitive nervous systems. Here we combined systems theory with more classical neurophysiological techniques. The subtle but essential difference between our combined studies and the classical neurophysiological approach alone is that our neurophysiological studies were done while the system was embedded in an input−output systems analysis framework. Thus, we knew that our inputs and outputs were physiological and represented the real function of the system within the behavioral range. This section is presented first because of its essential simplicity of approach. However, it historically followed our system studies and was really an attempt to study the neurophysiological mechanisms underlying the neurological control systems treated later in this book.

In 1948, when I was still a medical student, I had obtained Professor Shannon's paper, "Mathematical Theory of Communication" [49] and for the last fifteen years I have been stimulated by Prof. Warren McCulloch's work in the exciting field of bioengineering and cybernetics [29, 60]. Having contact with them and Prof. Norbert Weiner at M.I.T. contributed to my desire to undertake this investigation into the coding characteristics of signal transmission along nerve fibers.

My own background in neurophysiology, of course, led me to perform these studies, and I was influenced by the attempt of Pringle and Wilson [39] to obtain a transfer function in the cockroach. Although they were unable to collect enough data from their preparation, their study suggested a highly nonlinear switchlike system, which thus was not very amenable to transfer function analysis. I was looking for a simple preparation which would give a rather continuous response to a continuous input, and therefore chose the crayfish tail receptor, which is a simple paucicellular system that had been studied by Prosser [40−42], Wiersma [62], Kennedy [22−25], and by a colleague at Yale, Prof. Alexander Mauro. Together with Dr. Valentino Braitenberg (now at the Cybernetic Section of the Institute of Theoretic Physics, University of Naples, then visiting at Yale University), and a student, Mr. (now Dr.) Roger Atwood, we began to do transfer function studies in the laboratory of Prof. Harold Saxon Burr.

The transfer function study went through several successive refinements. The first paper written by myself was a report to Wright-Patterson Air Force Base [52]. (A section of this paper, "Plans for Future Research" is included as an appendix following Chapter 1, for historical interest.) After I moved to M.I.T., Dr. Howard Hermann and Mr. Paul A. Willis took up this problem with me, and we developed some more precise techniques [17], and obtained further refinements [53, 54], including a single unit study [16].

The second chapter deals with studies on the probabilistic random walk system itself [14], nonlinearities in the response [15, 54], inhibitory effects [15, 36], and the artificial photosensitivity of the ventral nerve cord [30]. This latter phase was contributed by Mr. Ronald Mellicia, who has since gone on to do graduate work with Prof. Mauro, now at the Rockefeller Institute.

The third chapter devotes itself to the question of the nerve impulse code [34, 5, 37, 57]. A more detailed definition of the neuroanatomy and neurophysiology of the crayfish light-to-walking movement reflex was necessary. Next, dissection experiments established that either one of the two photoreceptor fibers alone could drive the response bilaterally. It was found that the detailed patterns of nerve impulse intervals of the two nerve fibers were not correlated, even when identical stimuli were impinging upon them. Thus although approximately 100 bits/sec would be transmitted in this manner, we could exclude the complex pattern code. The simpler, average frequency code appeared more likely even though it carries only 3 bits/sec of information. I was able to develop this exciting phase of the crayfish research when two biophysicists from the University of Mexico, Prof. Jose Negrete and his wife, Prof. Guillermina Yankelevich Negrete, spent a year as visiting fellows at M.I.T. Associated with us was Dr. George Theodoridis, a former student of mine and a nuclear engineer deeply interested in biology.

Chapter 1

Transfer Function of a Photoreceptor Ganglion

INTRODUCTION

The terminal abdominal ganglion of the crayfish, shown by Prosser [40−42] and Welsh [61] to be a photoreceptor, is particularly suitable for studies of receptor organ transfer functions [16, 52−54] (see Fig. 1). It is an extremely primitive system, with no special optical or photochemical organization yet discovered. However, it has been shown to affect the light-avoidance reflex behavior of the crayfish [27]. Since the system shows little or no physiological adaptation, fairly stationary stochastic time series experiments may be performed. Furthermore, the crayfish is small, easily kept in laboratory conditions, and surgically requires neither anesthesia nor elaborate dissection techniques.

The terminal ganglion, a 2−mm−diameter translucent collection of neural cells and neuropil, receives fibers of several sensory modes from the uropods [25, 66]. In Fig. 2, an anatomical drawing of the crayfish photoreceptor, it can be seen that there are a number of nerve fiber groups which spread outward to supply the telsun and the last abdominal segment. (The abdominal ganglion develops embryologically through fusion of the two last abdominal ganglia.)

The large cell bodies, neurons, are located around the central mass of neuropil, an area composed exclusively of synaptic connections between the various neurons. (The neuropil can be identified as the small speckled region.) In the top portion of the drawing the central nerve cord can be seen going to the next abdominal ganglion. It should be noted that both the ganglion and the nerve cord are bilaterally symmetrical. Recording electrodes are placed on the nerve cord a suitable distance (1−10 mm) from the abdominal ganglion. The determination of which ganglionic element is photosensitive, yet to be made, cannot be done with the technique in use,

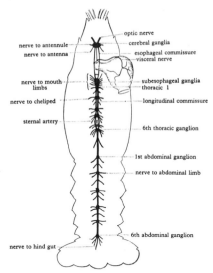

Fig. 1. The central nervous system of
Astacus. [From the Invertebrata, by
L. A. Borradaile and F. A. Potts. Third
ed., revised by G. A. Kerkut. Cambridge
University Press (1959).]

since the stimulating light beam, 1 mm in diameter, is of the same order
of magnitude as the entire ganglion. However, it is known that a small
number (ca. 10) of neurons transduce photic energy in the ganglion. Lack
of optical apparatus, absence of apparent spatial arrays in its structure,
and sluggishness of behavioral response suggest that the ganglion serves
to detect average levels of background illumination. In Fig. 3 the sixth

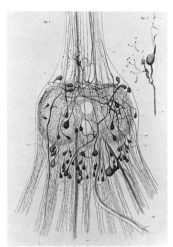

Fig. 2. An anatomical
drawing of crayfish photoreceptor ganglion (Retzius,
1890).

Fig. 3. Sixth abdominal ganglion
of crayfish (1 mm).

abdominal ganglion is shown as it appears when microphotographed after
silver staining.

METHODS

Preparation

A moderate-sized specimen of the common crayfish was fixed to a
cork board with soft rubber tubing in a supine position. [One group of our
crayfish was identified as belonging to *Orconectes virilis* (Hagen) although
it is probable that other related species were used over the 8 years of
study.] The animal was kept in a recording chamber providing control of
light, vibration, and temperature (19 ± 0.5°C). Humidity control (100%
saturation) prevented drying or drop accumulation, and so permitted recording of nerve pulses without recourse to oil layer and perfusion fluid
techniques [43, 54, 59]. Maintenance of adequate blood pressure and minimization of blood loss proved to be critical in the preparation. The hydro-
static level of the tail ganglion was approximated to a level several milli-
meters above the ventral coelum. With careful attention to homeostasis,
steady-state recording for 6−8 hours was possible.

In the case of gross recording from the sixth abdominal ganglion, a
window was cut in the ventral abdominal exoskeleton and gross electrodes
were applied to the ventral nerve cord. For single unit recording, once
access to the cord was gained through a 3 × 12 mm window in the exoskeleton, the ventral artery supplying the ganglion was carefully dissected
free and placed to one side. (Puncture or damage to this artery will weak-
en the preparation.) The tough fibrous sheath surrounding the cord was
first hemisected by means of glass microknives. Under microscopic visual control the cord was first hemisected and then progressively cut with
microscissors and fractionated with the microknives until a single light

unit was isolated (determined by monitoring on a cathode-ray oscillo-scope).

APPARATUS

An electronically controlled light source was used to generate tran-sient steps as well as sinusoidal variations of light. The light source con-tained a monitoring photocell where output could be recorded during the experiment [51]. The light stimulus was focused to an approximate $3\ mm^2$ spot falling on the photosensitive abdominal ganglion, and was calibrated by means of an illuminometer. The ventral nerve cord and the other ab-dominal ganglia gave no response when illuminated.

Signals were led off to conventional alternating current amplifiers by gross platinum electrodes hooked about the ventral abdominal cord in varying positions along the chain of ganglia. Visual monitoring of the nerve impulses was not detectably altered by shifts in location of the elec-trodes; obviously change in cross-sectional location of a given light fiber produced very little change in potential. Correlation of pulse velocity and pulse height supported the visual impression that pulses of the same height (but from different fibers) showed equal velocities wherever recorded in the cord. The arrangement of the experimental apparatus is shown in block diagrammatic form in Fig. 4.

The apparatus as described was used in the first series of experi-ments. For the second and third series, advantage was taken of pulse height constancy through the use of a "pulse height window" which screen-ed out those pulses whose height was greater or less than the population

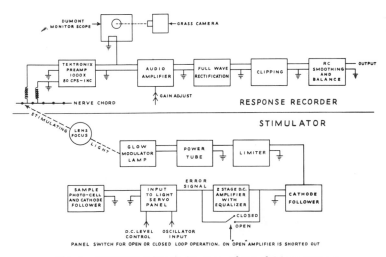

Fig. 4. Functional block diagram of crayfish experiment.

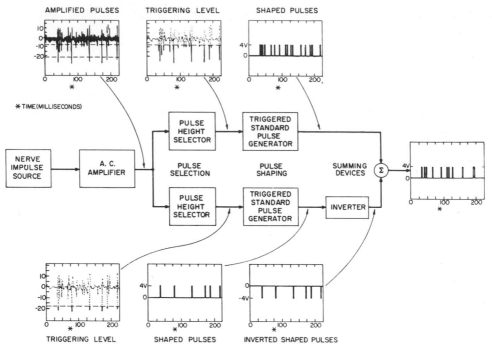

AMPLIFIED PULSES   TRIGGERING LEVEL   SHAPED PULSES

*TIME (MILLISECONDS)

TRIGGERING LEVEL   SHAPED PULSES   INVERTED SHAPED PULSES

Fig. 5. Block diagram of "pulse height window" arrangement of instrumentation.

of light-sensitive neurons (see Fig. 5). The relevant neurons showed a homogeneous pulse height and, of course, velocity distribution [17]. Individual pulses whose height was greater than a preselected level triggered a standard pulse, as may be seen in Fig. 5. The pulse height groups could be separated into three crude divisions: a midrange group "B" ($75-125\,\mu$V) carrying the light signal; a low voltage group "C" ($35-50\,\mu$V) correlating inversely with the light fiber activity; and a high voltage group "A" ($125\,$mμV) which was uncorrelated. The "C" group was screened out by raising the bias level. By taking the difference of the "A" group from

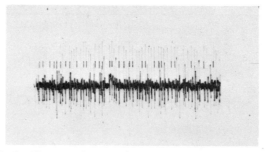

Fig. 6. Oscilloscope display of nerve impulses: Pulses selected by electronic window are tagged by means of Z-modulator.

the remaining "A" and "B" group, a "B" window could be formed. The adjustment of the window, enabling selection of the desired pulses, was facilitated by tagging the selected pulses with the help of the Z-modulator of an oscilloscope (see Fig. 6), which controls the intensity of the cathode-ray image. Our results were confirmed using a conventional pulse height discrimination counter [17]. Pulse selection and shaping resulting from this technique is shown in Fig. 7. Figure 8 illustrates the effect of illuminating the ganglion with sinusoidal inputs, then using the "pulse height window" to select only "B" fiber impulses.

The standard pulses generated were then passed through a low-pass filter whose time constant was long with respect to the pulse repetition rate, but short with respect to the response frequency of the photoreceptor. The output of the low-pass filter (an analog voltage whose level corresponds to the instantaneous output frequency of the neuron population) was displayed on a standard pen-writer.

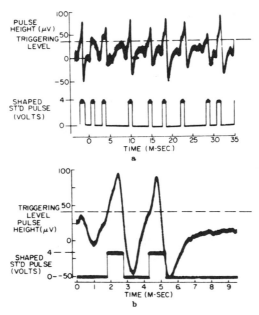

Fig. 7. Pulse selection and pulse shaping. Top trace shows raw nerve impulse train. Bottom trace shows shaped pulses (1 msec, 4 V) triggered at a selected level of nerve impulse which may be either positive or negative with regard to the base line as desired. Same pulse selection and pulse shaping may be seen at greater magnification. The width of standard pulse is set at 0.5 or 1.0 msec to allow for easy averaging but still separate trigger pulses close together in time.

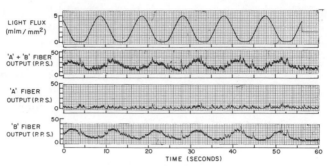

Fig. 8. Sinusoidal variations in light flux (top trace) driving
light-sensitive B-fiber with an admixture of uncorrelated A-
fiber impulses (second trace). When A-fiber impulses are isolated (third trace) and subtracted from combined A+B response,
an´"impulse amplitude window" is formed to pass exclusively
B-fiber impulses (fourth trace).

CONSTANT-LEVEL INPUTS

A representative sample of crayfish nerve impulse data obtained in
the first series of experiments under light-off conditions is shown in Fig. 9.
Because of the propagated nature of the nerve impulses and the separation
of the electrodes (approximately 1 cm), this data is spatially differentiated.
The abscissa is time, with 5 cm (distance between vertical lines, or one
frame) equivalent to 100 msec. The frames numbered 1 to 18 in Fig. 9
correspond to a continuous sample of 1.8 sec. The ordinate is the height
of the nerve action potential in millivolts on a relative scale. (The thin
horizontal line was added to aid in data reduction.)

These nerve impulses arrive from a population of nerve fibers, some
of which transmit sensory data irrelevant to the light intensity. The irregular grouping of the nerve impulses is the subject of Chapter 3. The
light-off operating condition was achieved by using a black screen on the
window and a red safety lamp, since the crayfish spectral sensitivity is
roughly comparable to the human rod; that is, relatively insensitive to red
light.

A representative sample of crayfish nerve impulse data under steady-
state light-on condition is shown in Fig. 10, where it can be seen that the
application of light to the ganglion causes a marked increase in frequency
of nerve impulses as compared to Fig. 9. The "B" impulses described
earlier are the primary contributors to the frequency increase. Since the
height of a nerve impulse is not a function of the stimulus intensity, but

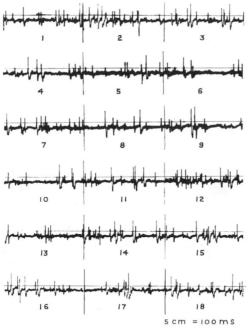

5 cm = 100 ms

Fig. 9. Crayfish nerve impulses with light off.

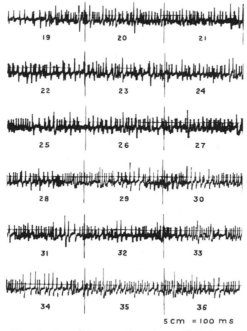

5 cm = 100 ms

Fig. 10. Crayfish nerve impulses with light on.

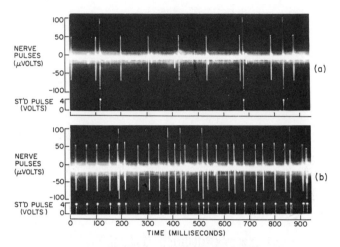

Fig. 11. Nerve impulse activity with low (a) and high (b) illumination. Activity consists of an A-fiber unit (large amplitude
proprioceptor) and one B-fiber unit (medium amplitude light
fiber). Trace pair (a) shows nerve cord activity (upper) and
shaped triggered pulses (lower) corresponding solely to A-
fiber nerve impulses. Trace pair (b) shows nerve cord activity (upper) and shaped triggered pulses (lower) corresponding to both A- and B-fiber nerve impulses. C-fiber
activity and noise are excluded from the standard pulse
generator in both a and b. See text description of window technique for obtaining shaped triggered pulses for
only B-fiber units.

rather of the size of the nerve axon and the recording conditions, information concerning light intensity is obviously transmitted as a function of
nerve impulse frequency. Since no decrease was seen in impulse frequency under steady illumination, the crayfish photoreceptor is considered to
be nonadapting in the physiological sense. Thus, stationarity of the stochastic processes involved seems a reasonable assumption.

In Fig. 11a and b, nerve impulse activity is shown under low and
high levels of illumination. Nerve pulses may be seen to be more frequent
in Fig. 11b, with higher illumination than in Fig. 11a.

EXPERIMENTAL RESULTS

Transient Inputs

When the "pulse height window" is used to screen out impulses from
"A" and "C" fibers, the response of the photosensitive ganglion is measured in terms of the average frequency of firing of the population of "B"
nerve fibers in the ventral nerve cord. The stimulus is recorded as the
calibrated output of the monitoring photocell. A sudden change in the level
of illumination, or "step-change," produces rapidly increased firing rate of
the nerve fiber population, as shown in Fig. 12. The response exhibits

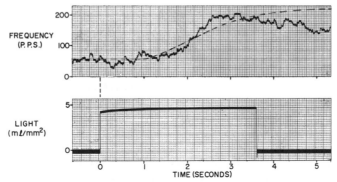

Fig. 12. Response of photosensitive ganglion to
a step change light. Response of photoganglion
in terms of average frequency of nerve impulses
from B fibers, to a step of light flux. Noise
seen comes, in part, from random groupings of
nerve impulses from different fibers. Solid line
represents response of linear model system described by transfer function. Lack of fit is explained by two factors. First, random variations
in a single step function are considerable. Second, small signal approximation is violated in step
experiments and certain nonlinearities produce di-
vergences.

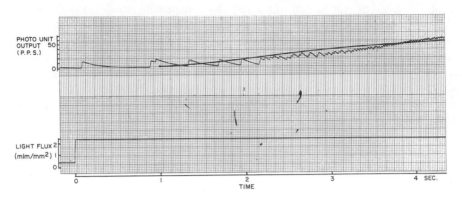

Fig. 13. Response (notched pulse-rate trace) to a small input step change of light is shown together with prediction of response (continuous line) from transfer function.

some characteristic dynamic features: long latent period, inflected rise to maximum with little overshoot or ringing, and maintained d-c response. These features will be measured and examined in detail in the experiments described below. The irregularity of the base line consists of high-frequency drifts. Irregular grouping of the various nerve fibers impulse trains, seen also with steady-state light-on conditions, is probably the cause of much of this variation, and as noted, may be treated conveniently as noise.

Another illustration of receptor response to a step change of light is given in Fig. 13, in which the light input is smaller than that of Fig. 12, and averaged impulses of a single receptor unit rather than the whole ganglion are being recorded.

The time delays shown in the preceding two figures could be due to nerve conduction time or to processes in the photosensitive ganglion. Two pairs of electrodes monitoring the nerve impulse response were located at either end of the abdominal ventral nerve cord. The responses from both electrode pairs exhibit long latent periods of equal duration. Calculation of the nerve conduction velocity confirms what this experiment demonstrates, namely, that conduction time (5 msec) is negligible compared to the latent period (1 sec). It is of interest to see how fine details of the response are transmitted faithfully along the nerve cord, suggesting absence of synaptic relays, with the attendant opportunities for convergence, divergence, and shift in temporal patterns of the impulses.

In this experiment, rather more overshoot is apparent, as the input signal was intentionally made large by dark-adapting the photosensitive ganglion.

SINUSOIDAL INPUTS

In order to quantify carefully the dynamic characteristics of the photoganglion nerve impulse response, we used sinusoidal inputs. By this means we were able to subject the system to a steady-state analysis. In Fig. 14 the response of the crayfish ganglion to sinusoidal variation in light intensity is shown. The nerve impulses shown in Figs. 9 and 10 have been further amplified, rectified, clipped, and integrated using an R-C low-pass filter. The scale of the pulse rate shown in Fig. 14 was determined by direct calibration, which consisted of photographing and counting nerve impulses while displaying pulse rate on the response recorder. The pulse rate fluctuated a good deal about a mean sinusoidal response to a si-nusoidal light intensity input. These fluctuations were treated as noise, and a mean response was empirically obtained and used for data re-duction. The sample shows a nonlinearity of response as well as var-iation in response from cycle to cycle. This is characteristic of bio-logical system responses and deserves further analysis. In an attempt to minimize nonlinearities, a small input signal about a steady "d-c" light level was used. (In order to achieve linearization, the degree of modula-tion with respect to the d-c level of the light is the important parameter to minimize, and not the absolute amplitude of the input sinusoidal func-tion.) From an experiment such as was shown in Fig. 14, average ampli-tude and phase lag of responses were obtained. The raw analysis of the original data record is also shown.

A similar set of experiments is shown in Fig. 15, in which again small stimulating signals were used, and noise and time variations in re-sponsiveness of the ganglion were minimized by averaging responses over

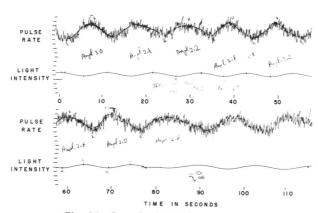

Fig. 14. Sample sinusoidal experiment.

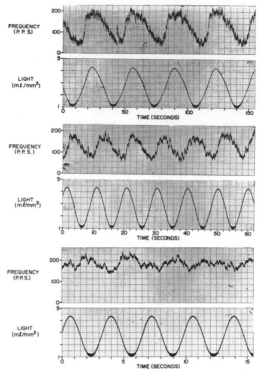

Fig. 15. Sinusoidal response.

a number of cycles. Harmonic distortions, although present, did not contribute much power as compared to the fundamental response.

    Figure 16 shows the response of a single ganglionic "B" fiber to three driving frequencies of sinusoidal light input. These input fluctuations produce regular responses in terms of average pulse repetition rate, as shown in the figure. On subsidence of initial condition transients, response is related to input by two parameters at each frequency: gain (relative amplitude of input and output) and phase shift (amount of lag of response after stimulus). The harmonic distortion seen in this figure is typical, and indicates that little power is present in higher harmonics as compared to fundamental responses. Noise and time variations again were minimized by averaging. Arranging the sequence of stimulating frequencies in random order eliminated possible trends. Clearly, response diminishes at higher frequencies; simultaneously, phase lag increases.

EXPERIMENTAL ANALYSIS

    From the amplitude and phase data derived from the first series of sinusoidal experiments, several displays of the photoreceptive behavior as

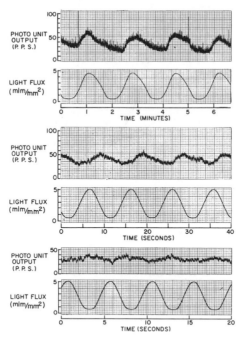

Fig. 16. Sinusoidal response of single unit.

a communications component can be obtained. The first is the Bode plot of frequency-response data, shown in Fig. 17, which enables one to assemble gain and phase data from many frequency-response runs in a compact and understandable form. Gain is plotted on a log–log scale. The points are

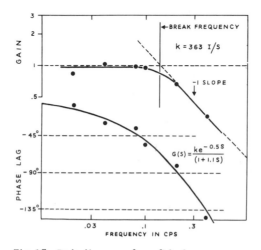

Fig. 17. Bode diagram of crayfish photoreceptor.

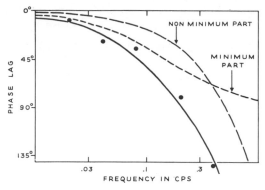

Fig. 18. Phase analysis.

experimental, while the continuous lines are fitted, using as guides the
asymptotes (short dashed line) on either side of the break frequency (the
vertical line) as well as the phase analyses described in Fig. 18. Beyond
the break frequency, the response of the receptor attenuated with a −1
slope (6 db/octave) shown on a scale normalized to k. Attention is called
to the low frequencies shown on the abscissa scale.

In Fig. 18 is shown the phase analysis of the Bode plot of Fig. 17.
Phase analysis consists of division of the phase into minimum and nonminimum phase components. From the −1 attenuation slope of the Bode
plot, a 6 db/octave attenuation, it can be deduced that the maximum value
of the minimum part (short dashed line) of the phase lag will be 90°, with
a 45° phase lag at the break frequency. A transport delay will produce a
nonminimum phase lag which is proportional to frequency and continues to
increase as frequency increases. A transport delay of 0.5 sec would produce the nonminimum phase lag (long dashed line) shown in the figure.
These two lines added together give the solid line, which can be seen to be
a fairly good fit for the experimental points.

Another method of displaying behavior is the Nyquist diagram, a polar plot shown in Fig. 19. This vector plot of gain and phase angle is often
used to show clearly the behavior of important characteristics of a system
— stability of operation with adequate speed, and accuracy of response. The
solid line of Fig. 19 is derived from the continuous line of the Bode diagram. Frequency is an increasing function along the Nyquist curve but is
not a regular function. The curve curls around the origin in a decreasing
spiral because of the transport delay.

In addition to the above qualitative and semiquantitative discussion of
the photoreceptor transducer, it is also desirable to have a full but concise

mathematical description of system behavior. Such a canonical equation is the transfer function, for convenience written as a function of the "complex frequency" operators. The data displayed in the Bode diagram in Fig. 17 will now be used to derive the transfer function $G(s)$. Low-frequency gain is normalized to $k = 1$. This is determined from calibration studies to equal 363 impulses/sec, meaning that with the signal intensity change used, the variation in nerve impulse rate was this amount at low frequencies. The attenuation curve appears to have an asymptotic slope of 6 db/octave beyond the break frequency, which can be represented by one time factor. The actual value of the time constant is difficult to determine exactly from the experimental data, but as rough approximation it is set to 1.1 sec. This time lag accounts for a portion of the phase shift at higher frequencies. By referring to the phase analysis diagram of Fig. 18 it can be seen that the remaining phase shifts can be attributed to a nonminimum phase shift, equivalent to a time delay of 0.5 sec, expressed as $e^{-0.5s}$. These calculations now enable us to write the transfer function in terms of the Laplace transform complex variable, "s" (roughly equivalent to d/dt)

$$G(s) = \frac{ke^{-0.5s}}{(1 + 1.1s)} \qquad k = 363 \text{ impulses/sec} \qquad (1)$$

In the second series of frequency-response experiments, utilizing the "pulse height window" and recording from more experimental points than in the first series, the data obtained supported the initial studies. However, when the displays of the two experimental series and their transfer function equations are compared, certain differences can be noted.

The second series of experiments was designed to eliminate trends. The driving frequencies were randomly ordered inside of each sequence

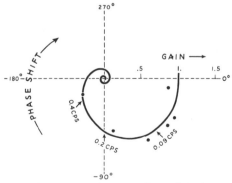

Fig. 19. Nyquist diagram of crayfish photoreceptor.

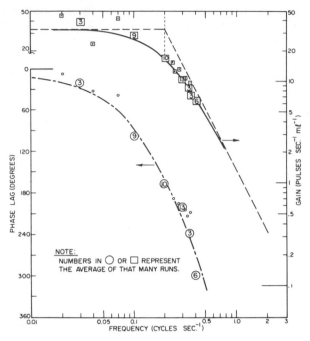

Fig. 20. Bode plot. Gain plotted on a log-log scale,
phase lag on linear-log scale. Points are experimental.
Continuous line of gain plot is guided by asymptotes on
either side of break frequency. Response down 6 db at
break frequency of 0.2 cycles/sec, characteristic of second-order system. Phase lag curve is computed from
sum of minimum and nonminimum phase elements.

and the sequence repeated. No sequence coincided in order of frequency
presentation, but each sequence spanned the entire spectrum.

From the Bode plot shown in Fig. 20 it may be seen that the low-frequency gain is 32 pulses per second per millilumen. The break frequency
was found to be at 0.2 cycles/sec. Above this frequency the response attenuates with a slope of -2 (12 db/octave, 40 db/decade, or 20 decilog/dec-
ade). At the break frequency the gain is down 6 db or 3 decilog from the
low-frequency value. The phase data show a great deal of phase lag near
the low-frequency, high-gain response region. This will be seen to be due
to a large nonminimum phase element. The composite theoretical phase
curve and its elements, seen clearly in Fig. 21, adequately fit the experimental data.

When the transfer function is constructed, it appears as:

$$G(s) = \frac{32e^{-1.0s}}{(1 + 1.2s)^2} \text{ pulses sec}^{-1} \text{ mlm}^{-1} \tag{2}$$

where 32 is the low-frequency gain, $e^{-1.0s}$ represents the nonminimum phase element or delay time, and $(1 + 1.2s)^2$ defines the second-order critically damped lagging elements which attenuate the high-frequency response. It is somewhat more convenient that the equivalent integro-differential equation:

$$32L\ (t-1) = 1.44\frac{d^2P}{dt^2} + 2.4\frac{dP}{dt} + P \tag{3}$$

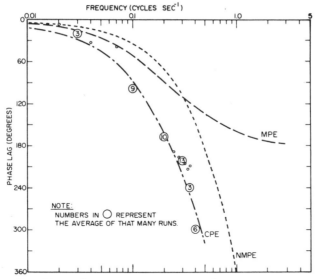

Fig. 21. Bode plot. Decomposition of phase elements. Composite curve CPE is computed from joint contributions of nonminimum phase elements (NMPE) and minimum phase elements (MPE). MPE (lag elements) computed from experimentally obtained equation of the second-order system. NMPE is due to transport delay (1.0 sec) and represents element which, when added to MPE, produces good approximation to experimental data. Note no phase advance appears in frequency range studied. Points are experimental.

where L is the light flux in millilumens, P is the number of pulses per second, and t is time in seconds.

In the polar plot of gain as a function of phase lag, frequency is a monotonically but not regularly increasing function in the clockwise direction. The nonminimum phase element, the time delay, spirals the curve about zero with a phase lag proportional to frequency. The lack of adaptation is clearly demonstrated by the absence both of phase advance and low-frequency attenuation in the range studied.

From the transfer function one can construct the theoretical response of the linear system to a step change of light. This computed response is shown in Fig. 12 as the dashed line superimposed, for comparison, upon the experimentally found response. The striking differences are outside the limits of experimental error. The computed response has been obtained by linearized (small signal) approximation to the real system. The step experiment in Fig. 12 clearly violates the small signal approach utilized in the frequency-response studies. The discrepancies are due to nonlinearities which become significant in the large signal step experiments.

Comparison may now be made with the set of experimental data, derived from recordings from single units within the crayfish photoreceptor. As shown in a Bode plot (Fig. 22), the curves and asymptotes, derived from a best fit transfer function, summarize quantitative features of the response. Low-frequency gain is 12 pulses sec^{-1} min^{-1} mm^2. The break frequency is 0.12 cycles/sec (0.75 radian/sec): equivalent to stating that the time constant of the system is 1.3 sec. At the break frequency, gain is down 6 db. At higher frequencies, gain decreases with a -2 slope (12 db/octave, 40 db/decade, or 20 decilog per decade), equivalent to a second-order critically damped lag element. This can be represented as $1/(1 + 1.3s)^2$. Divergence of low-frequency data suggests the presence of very slow adaptation (equal to a zero on the negative real axis) which is not quantitatively examined in these experiments, nor represented in the transfer function.

Theorems relating phase lag and gain for minimum phase elements (MPE) enable one to assign the phase appropriate to a second-order lag in the plot of phase as a function of frequency. Phase lag due to a nonattenuating, time-delay factor, the nonminimum phase elements (NMPE), is linearly proportional to frequency, and shows an ever-decreasing slope on the logarithmic frequency scale. The two components MPE and NMPE sum together to produce total system phase lag. The 1-sec time-delay factor

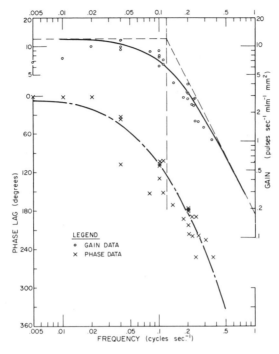

Fig. 22. Bode plot. Gain is plotted on a log-log scale, phase lag on linear-log scale; points are experimental. Continuous line of gain plot is guided by asymptotes on either side of break frequency. Response is down 6 db at break frequency of 0.12 cycles/sec, which is characteristic of a second-order critically damped system. Phase lag curve is computed from sum of the minimum and nonminimum phase elements in transfer function. Points are experimental data from one preparation: quantitatively reproduced results were obtained in even preparations.

has the form $e^{-1.0s}$. Experimental phase points appear to fit the total phase curve (except at low frequencies where the adaptation element produces some phase advance).

We may assemble the transfer function:

$$G(s) = \frac{e^{-1.0s}}{(1 + 1.3s)^2} \text{ pulses sec}^{-1} \text{ mlm}^{-1} \text{ mm}^2 \qquad (4)$$

where components on the right-hand side have been assigned forms and parameters from the above experimental frequency-response results. As an integro-differential equation, it may be written:

$$12I(t-1.0) = 1.69\frac{d^2F}{dt^2} + 2.6\frac{dF}{dt} + F \qquad (5)$$

where t is the time in seconds, F is the number of pulses per second, and I the light flux intensity in millilumens per square millimeter.

The transfer function summarizes the dynamic behavior of the system not only to sinusoidal stimulation but all physically realizable inputs. One important example is the step function input. The predicted step response of the photoreceptor unit is indicated in Fig. 13, together with actual experimental responses. The time delay of 1 sec and the second-order rise (one inflection point) clearly show the same characteristics as derived in the frequency-response analysis. Further quantitative agreement between the transfer function prediction (continuous line) and the experiment (the sawtooth pulse rate trace) should be noted.

The crayfish tail photoreceptor shows unit and population responses to be almost identical. In fact, by comparing the Bode plot of Fig. 23 of frequency-response data from the population and from a single unit in the same animal, it is seen that the phase data are not distinguishable. The gain data, fitted by asymptotes for the same transfer function, are similar save the d-c gain. Population gain is three times unit gain. The population in this experiment apparently consisted of three individual units.

DISCUSSION

The crayfish photoreceptor ganglion clearly acts as a light intensity-to-pulse rate modulation transducer. This is true despite its lack of many components present in more sophisticated visual organs, i.e., systems with accommodative mechanisms, pupil systems for control of light intensity and depth of focus, and spherical surface spatial array of retinal elements. However, the essential features of a photosensitive transducer are present: a photosensitive pigment, an energy amplifying process (probably proportional to the logarithm of stimulation intensity), and a nerve generator mechanism (generating pulse rate proportional to depolarization).

Several interesting features of this photoreceptor's response should be stressed. First, it is a population response, and although much of the noise (not at all germane to the biological system) results from this, certain other experimental advantages accrue, namely, an averaging of response over short and long periods of time. Pringle and Wilson [39], the

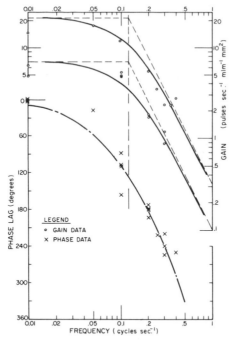

Fig. 23. Bode plot comparing population and single unit response from same preparation. Curves and asymptotes are derived from previous single unit transfer function and applied to single unit data. Same transfer function but with gain increased threefold is applied to population data. Population response is assumed to be composed only of three B-fiber units. Fit of curves to experimental points is strong evidence supporting very limited number of light-responding units.

only other authors working with a transfer function analysis of a biological receptor, were able to obtain only inconsistent data from the cockroach proprioceptor: steps and only three frequency points for sinusoidal data. By contrast, the crayfish ganglion had a maintained response and a slower frequency response with respect to the pulse repetition rate.

The nonadaptive nature of the crayfish photoreceptor, evidenced by its low-frequency gain in the Bode plot, is also of interest. This is in addition to the quite low-frequency range of this transducer's operation. The break frequency and the attenuation slope are clearly not due to passive electrotonic properties of the nerve membrane, and the transport delay is probably not a function of nerve conduction time to any significant degree.

The slow dynamics and long latent period of the ganglion suggest that the ganglion photoresponse drives a tonic reflex, and indeed this is so. The random walk behavior activated by photoreceptor stimulation is low priority activity which shows itself only if no more active behavior is called for. Then a high level produces a photokinetic random walk into a region of subdued illumination [14].

The small signal requirement for linear approximation only partly eliminated persistent nonlinearities such as asymmetrical response, d-c level dependence on driving frequency, and harmonic distortion. When a large signal, as in the step experiment of Fig. 12, is employed, departures from linearity can be marked. The linear response predicted from the transfer function should be symmetrical about the inflection point and critically damped, without overshoot or ringing. In certain amplitude ranges, an inverse relationship holds between damping and input amplitude. Despite these observed deviations from linearity, the experimental analysis has succeeded in defining the system dynamics with considerable validity. The population averaging, the stationarity of the preparation, the fact that the frequency-response range is much slower than the pulse repetition rate, and of course the biological factors involved in careful maintenance of the preparation have all contributed to the success of the approach. Perhaps its most significant part is the use of small sinusoidal inputs which permit the linear characteristics to be extracted with minimum interference from nonlinearities. The ability to average over many cycles and tens of seconds of response increased the precision of the experimental measurements.

APPENDIX A

PLANS FOR FUTURE RESEARCH

As often happens in research, our investigation of the crayfish photoreceptor ganglion has exposed more problems than it has solved. These new avenues of research [52] may be grouped under five headings: noise, nonlinearities, dissection into single units, and structure.

The noise may be considered as a general model of central nervous system noise in all animals. There is good evidence that neurofunction is at least qualitatively similar in many aspects in a wide variety of animals from the invertebrate squid to the vertebrate man. This interest in biological noise parallels the development of noise elimination criteria in engineering servomechanisms and communications systems. The relationship of the noise in a receptor system to the noise in a whole system in which such a receptor is only one component might be studied in such an example of a complete system as the human pupil servosystem. The pupil servomechanism is known to have prominent noise characteristics; these have been studied in our laboratory, and it would be valuable to relate these two studies of noise. Another aspect of the proposed study of noise is a further use of the digital computer in developing relationships between various aspects of the transfer function and the noise characteristics of the crayfish system. An analog of the crayfish transfer function programmed on the digital computer is being developed, as well as one of noise generation in such a system. This should display the relationship between conditional probability, weighting function, and band limitation in frequency response. These should be approached by the use of autocorrelation and power spectroanalysis techniques on the time series generated by the noise program. The probability density spectra of the noise model should be related to the real noise generated by the crayfish.

Nonlinearities in the transfer function such as saturation, dead space, zero offset, and asymmetrical response dependent on the differential stimulus intensity are noticeable to a greater or lesser degree in the already accomplished study on the crayfish. In addition, temperature effect on the transfer function characteristics such as attenuation slope and break frequency should be defined. One of the advantages of undertaking an investigation of these secondary effects upon the linear transfer function would be that various components in the transfer function might well be isolated by certain experimental procedures and thus enable the experimenter to carefully define them in relative experimental isolation.

Many of the aspects of the transfer function now obtained may well be due to the population response of the system as discussed above. Several methods are available for undertaking as analysis of single unit activity. These include finer dissection techniques, and a possible approach by means of pulse height discrimination network. Such an amplitude slicer might enable attainment of single unit activity without necessitating physiological and anatomical breaking-into the crayfish system. This may also be an effective method for eliminating steady-state nerve impulse fluctuations due to nonphotosensitive receptors. Various types of microelectrodes such as glass micropipettes and fine insulated metal electrodes might also be useful in this regard, as well as the possible use of microbeam stimulating technique.

This microbeam technique relates to the fifth grouping of proposed studies, namely, the location in the structural mechanism of the ganglion of the photosensitive elements. Such a microbeam might be two or three orders of magnitude smaller than the present beam and could thus be used to locate the light-sensitive element.

Studies of the structure of the photosensitive ganglion using modern neurological methods such as improved silver staining are also necessary if further progress is to be made in this field.

Chapter 2

The Random Walk System

THE SYSTEM

The crayfish has a pair of ordinary cephalic eyes capable of spatial perception. In addition, it possesses a primitive photosensitive tail ganglion, A_6, of the ventral nerve cord, which even in the absence of normal visual input, permits light avoidance behavior in the form of an aversive walk. This response induces the animal to seek out and remain in dark environments, providing that no overriding stimuli exist [27, 41, 61, 63]. In order to study this response it is useful to scrutinize the entire sequence

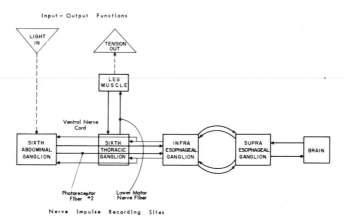

Fig. 24. Experimental arrangement for analysis of photoreceptor walking movement system. Note that classical neurophysiological nerve impulse monitoring techniques are embedded within an input-output systems approach. Photocell, strain gauge, and nerve impulse amplifiers and signal processing units monitored experiment continuously.

of signals: (a) light energy, (b) ascending nerve impulses from photoreceptor to brain, (c) descending nerve impulses from brain to thoracic ganglia,
(d) motor nerve impulses, (e) muscular contraction. A first stage of our
research is to record the nerve signals along the path of the random walk
response [14−16, 30, 36, 55].

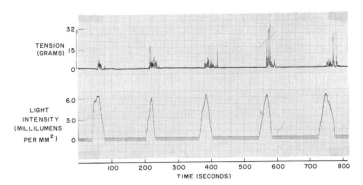

Fig. 25. Random walk response. Note regularity of response. Experimenter turned the light off when response occurred, thus pre-
serving light sensitivity.

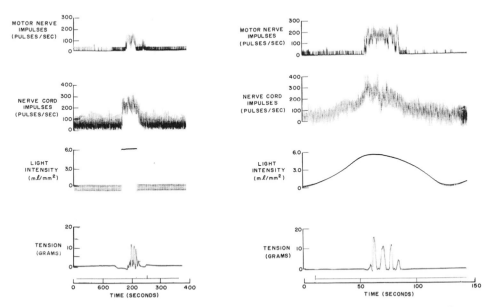

Fig. 26. Random walk response. Experiment displays monitored light input averaged frequency of ventral nerve cord nerve impulse, averaged frequency of motor nerve impulse, and
leg tension.

The ventral nerve cord of the crayfish contains thousands of nerve fibers [63]. However, when the aversive response to light occurs, only a few fibers are activated. These include two ascending "B" fibers from the photosensitive tail ganglion [22, 40, 61] and about double that number of descending "A" fibers conveying nerve impulses from the esophageal ganglia. If the photosensitive ganglion is stimulated with a controlled light source, and a responding leg is fixed to a strain gauge, regular sequences of the aversive or random walk response can be seen. The experimental arrangement for such a study is shown in Fig. 24. Figure 25 shows the random walk response elicited by light flashes applied to the tail ganglion. Simultaneous recordings of the light input signal, ventral cord impulses, motor nerve impulses, and the tension produced by the muscles are shown in Fig. 26. Note that the random walk response seems to last for relatively fixed periods of 80 ± 20 sec. Therefore, any description of a simple dynamic relationship must be attempted with caution. Further evidence for the complexity of this response is the marked response regularity obtained by covering the complex cephalic eyes, and the response blocking caused by strong light stimuli to the eyes.

The interweaving of low-priority background responses such as the random walk response with high-priority behavior such as attention to strong visual stimuli may provide a clue to the mechanisms that permit the crayfish, despite its small brain size and limited repertory of behavioral response, to meet the requirements of everyday living.

CONDUCTION

Using an experimental arrangement similar to that described in the preceding chapter, the conduction velocity of the nerve fibers involved in the random walk response can be calculated. In Fig. 27 the response of the ganglion and ventral nerve cord to a step change of light is shown. The monitored nerve impulses, as photographed from the cathode-ray oscilloscope, were taken during the bracketed segment of the impulse-frequency curve. The relative displacements of the individual identifiable nerve impulses recorded between the fifth and sixth and between the second and third ganglia are due to the 3.8 m/sec velocity of the nerve impulses. The regularity of the displacement indicates that probably no synaptic events occurred in the intervening three ganglia. The impulse-frequency curve also shows the unchanged details of the response at the two sites of electrode placement.

In an additional experiment to determine the cause of the time delay or "nonminimum phase element," two pairs of electrodes monitoring the nerve impulse response were located at either end of the abdominal ventral

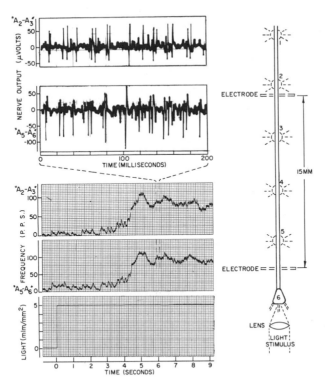

Fig. 27. Experimental arrangement. Schematic diagram on right depicts relationships of ganglia and ventral nerve cord to light stimulus and electrodes. Dotted circles with adjacent numbers indicate paired ganglia (sixth is a fused double ganglion) and their characteristic triple nerve roots. Vertical double lines represent ventral cord with its enclosed B-fibers, which carry light signals from caudal eye to brain centers. Upper inset illustration was photographed from a monitoring cathode-ray oscilloscope during segment of average frequency curves (upper two of the three graphs) demarcated by dashed lines. Letters A_2-A_3 and A_5-A_6 denote recording electrode positions.

nerve cord. The responses from both electrode pairs exhibited long latent periods of equal duration. Calculation of the nerve conduction velocity indicated that conduction time (5 msec) is negligible compared to the latent period (1 sec). It is of interest to see how fine details of the response are transmitted faithfully along the nerve cord, suggesting absence of synaptic relays with the attendant opportunities of convergence, divergence, and shifts in temporal patterns of the impulses. The delay is probably attributable to delays in the transduction processes in photoreception.

NONLINEARITIES

Although as shown in the preceding chapter, 75−80% of the power in the wave form of the response to sinusoidal stimuli to the photoreceptor ganglion is contained in the fundamental frequency, thus implicitly justifying linear analysis, it should be stressed that interesting nonlinearities appear consistently in our data. An asymmetry in rate of change of pulse frequency can be seen in Fig. 28. Since pulse rate rise is more rapid than pulse rate fall, the sawtooth wave form seen in the figure was thus produced. This effect is more prominent with large signal inputs.

Saturation, another important effect also more prominent at large signal input levels, is shown in Fig. 29. Output amplitude is shown as a function of input amplitude at one frequency, 0.08 cycles/sec. Two scales are used, ΔF vs. ΔI, and $\Delta F/\overline{F}$ vs. $\Delta I/\overline{I}$ for normalized units. Here ΔF is change in frequency of nerve pulse and $\overline{F}$ is mean pulse rate and similarly ΔI is change in light flux/unit area and $\overline{I}$ is mean light flux/unit area. In Fig. 29 (bottom), gain is displayed as a function of input amplitude; normalized values and scales are also furnished. In the particular experiment shown, the curves are quite similar. However, if ΔI is kept constant and $\overline{I}$ varied, clear evidence is obtained indicating the necessity for normaliza-

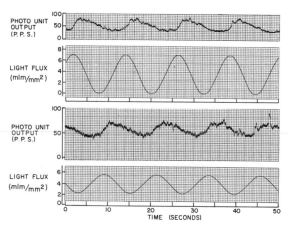

Fig. 28. Sawtooth nonlinearity in single unit response. Upper pair of traces shows larger signal driving function and response; lower pair is smaller signal excitation and response. Scale of response is doubled in lower trace. Less sawtooth asymmetry is apparent in small signal operating condition G = 7.6 and 90° in small signal case, and G = 6.6 and 72° for larger signal case, showing nonlinearity produces apparent relative phase lead and some decrease in gain.

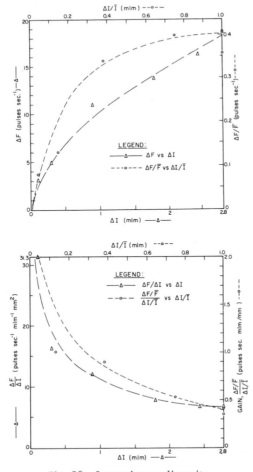

Fig. 29. Saturation nonlinearity.

tion. This is merely an instance of the well-known Weber–Fechner law which states that $\Delta I / \bar{I}$, or the fractional change of stimulus energy, is the effective stimulus. The experiments here reported have been designed with constant $\bar{I}$ and small ΔI, to avoid these complications. The preceding material dealing with asymmetry, saturation, and scale compression non-linearities has been included to provide quantitative estimates of these limitations on the previously described linear analysis leading to the transfer function.

The dynamic shape of a response to a step stimulus should be independent of amplitude in a linear system. Marked changes are easily noted in the crayfish responses shown in Fig. 30 where both time delay and rise time increase with smaller stimuli. This is clear evidence for a "memory-dependent" nonlinearity.

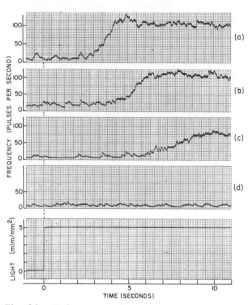

Fig. 30. Light attenuation series. Trace (a) recorded without attenuation of 5 mlm/mm^2 step
change in light stimulus. In traces (b), (c), and
(d), neutral density filters of 0.3, 0.6, and 1.0,
respectively, are inserted in light path. Attenuation to one-half (b), one-quarter (c), and one-
tenth (d) light intensity of trace (a) results in decreasing response and increasing latency of (b)
and (c). In (d), no response was detected even
after 100 sec of stimulation. However, only first
10 sec are shown.

When dealing with absolute calibrations of sensitivities of biological
systems involving nonlinearities, a complex set of experiments is necessary; first to determine the time course of a given response, and second to
demonstrate the nonlinear relationships (over considerable range) between
response and stimulus. If a response characteristic is to be obtained without involving nonlinearities, and thus the use of complex, possibly incomplete calibrations, a null method should be used; that is, the stimuli should
be adjusted to yield a predetermined constant output from the preparation.
In particular, this applies to spectral sensitivity studies reported for the
photoreceptor [24, 55, 58], where it was necessary to demonstrate the linear relationship (over a range of two $\log_{10}$ units) between log response and
log stimulus. (It is not clear at which point in the time response this relation exists, and therefore various curves have been published [24].)

DISSECTION

When analyzing the random walk response of the crayfish to light inputs received by its tail photoreceptor, it is necessary to recall that the
crayfish has an extremely well-developed and sensitive pair of cephalic
eyes capable of complex visual information processing. When extremely
high-intensity light stimuli impinge diffusely on a 3 mm × 15 mm area of
the crayfish abdomen, illuminating the entire chain of ganglia, some response mediated via these cephalic eyes is to be expected due to scatter-
ing in the experimental chamber. The experiment illustrated in Fig. 31
shows a careful control of this factor. The cord is cut just caudal to the

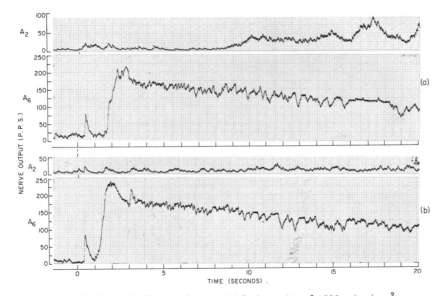

Fig 31. High-density light experiment. Light intensity of 4500 mlm/mm^2 was
employed to stimulate entire nerve cord under study. Upper and lower of each
pair on traces represent cephalad and caudad electrodes. In (a) cord was transected between A_4 and A_5. Small, sharp, early response. detected a few hundred
milliseconds after stimulation is found only in high-intensity stimulation. Early
response resembles in frequency and form rapidly adapting "on" response recorded
from cephalic eye single units of crayfish. The main response rising rapidly 1 sec
after stimulation dwarfs early response. Around 10 sec after onset of light an
anomalous late response appears at the A_2-A_3 recording site. Since cord has been
transected at A_4-A_5, it cannot originate in A_6 ganglion. In (b), same preparation,
cord was again transected, now between A_1 and A_2. Upper trace is cephalad electrode, recording as before at the A_2-A_3 interval; lower trace, caudad electrode
again at A_5-A_6. No response seen in cephalad electrode. It is clear that anomalous late response was due to stimulation of elements cephalad to A_2. Very likely
cephalic eyes, picking up scattered light, are responding.

fifth ganglion, causing light response nerve impulses from the sixth ganglion to reach only the caudal electrodes. However, the cephalad electrode
picks up a long-latency, slowly rising anomalous response. When the nerve
cord is transected cephalad to the second abdominal ganglion, this anomalous response is eliminated. Thus it is attributable to a response to scattered light, mediated by the cephalic eyes.

A further experimental demonstration that the sixth ganglion alone
produces nerve impulses in response to light may be seen in Fig. 32. First

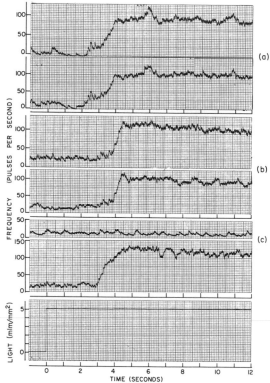

Fig. 32. Transection experiments. Each pair of traces
represents one experiment. Upper trace of each pair
is cephalad electrode; lower is caudad electrode.
Trace (a) serves as control showing typical response to
step change in light as recorded at both electrode sites.
In trace (b) all roots of A_6 have been severed. No effect on light response may be seen. In trace (c) cord
is transected at A_4-A_5; light response disappears at
cephalad electrode, but remains unchanged in caudad
electrode. Light stimulus falls on A_5 and A_6 only.

is shown (a) the normal response; next (b) the response negligibly altered after transection of the nerve roots of the sixth ganglion. Finally, when the ventral nerve cord is transected just below the fifth ganglion, the nerve impulses emanating from the sixth ganglion are noted only in the caudal electrode as shown in Fig. 31c. The light response impulses are seen to be of moderate amplitude as opposed to the small spontaneous impulses (which of course are not excluded from the cephalad electrode).

INHIBITION

Through suitable choice of neutral density filters we were able to attenuate the stimulus flux and, accordingly, to record the changed response of the sixth abdominal ganglion, as shown in Fig. 30 [36]. The photoreceptor response, although reduced and delayed, is still in evidence at one-fourth the level of the original stimulus flux. However, when the stimulus light was focused onto the sixth ganglion nerve cord, containing both afferent and efferent nerve fibers, or onto the other abdominal ganglia, no response was evident.

While engaged in the study of the nonlinearities of the crayfish photosensitive ganglion as a light intensity-to-frequency transducer, we found some inhibitory light effects that can be considered important in the analysis of the nonlinearities of the system.

In *in vivo* experiments, the sixth ganglion and its roots were left attached to the body of the animal while the fifth and fourth ganglia were isolated by cutting the ganglionic roots and the surrounding connective tissue.

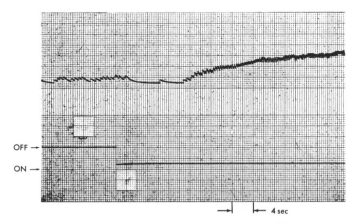

Fig. 33. On-inhibition. Average response to a step of light. Low activity of preparation allows individual pulses to be seen. Notice inhibition of spontaneous activity following onset of the stimulation.

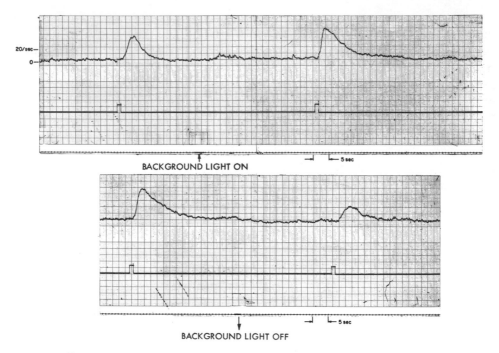

BACKGROUND LIGHT ON

BACKGROUND LIGHT OFF

Fig. 34. Off-inhibition. Average response to light pulses under different background illumination. Two fragments of figure belong to a continuous record, lower following upper one. Notice that when background illumination is removed the response to the light pulse is considerably lower than before or during the background illumination.

The ganglionic chain was cut between the fourth and third ganglia. The
electric signal was then picked up by a pair of platinum electrodes, one of
which was positioned between the fifth and fourth ganglia, thus supporting
the chain in the air, and the second was connected to the animal in the vicinity of the sixth ganglion.

In the *in vitro* experiments the sixth ganglion was detached, and the
whole chain was transferred into a Ringer's solution bath [43, 59]. During
the recording, the sixth ganglion and the second electrode were immersed
in Ringer's solution while the chain was lifted in the air by one of the recording electrodes.

The preparations were maintained in a recording chamber with 100%
humidity and constant temperature (around 20°C).

In some *in vitro* preparations that were kept for as long as 24 hours,
the delay of the light response increased to several seconds. In such cases
an "on-inhibition" of the spontaneous activity was observed a few seconds

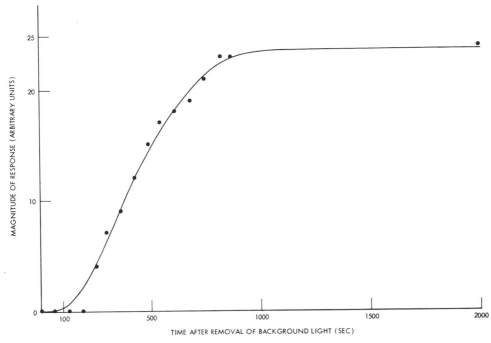

Fig. 35. Time course of the off-inhibitory effect.

after the light stimulation (see Fig. 33). A similar type of "on-inhibition" has been observed in crayfish preparations artificially photosensitized with dyes, to be discussed shortly.

Another type of inhibition, "off-inhibition," was caused by removal of a background light, as shown in Fig. 34. Short-pulse stimulation was applied before, during, and after the background illumination. It can be seen in this figure that the response to the pulse is smallest following the removal of the background light. At higher intensities of the background light the response could be eliminated completely. The off-inhibition caused by the removal of the background light recedes slowly over a period of a few minutes, as seen in Fig. 35.

From the described experiments, it appears that an inhibitory period follows both the onset and the removal of a light stimulation. The fact that the on-inhibition was seen only in long-lasting preparations could be explained by the effect of long usage of the preparation on the timing of competing excitatory and inhibitory mechanisms. The off-inhibition, following the removal of the light, seems to affect the excitability of the system and not its d-c response.

ARTIFICIAL PHOTOSENSITIZATION

As mentioned in the preceding pages, when the stimulus light is focused onto the sixth ganglion nerve roots containing both afferent and efferent nerve fibers, or onto the other abdominal ganglia, no response is evident (see Fig. 36). This was also true for any part of the ventral abdominal nerve cord in the interganglionic regions. Furthermore, even light sources of 300-fold maximum glow modulator intensity produced no effect in abdominal or tail structure, except the sixth ganglion.

However, it is possible to photosensitize the ventral cord artificially by staining it with certain organic dyes, and then to classify and locate the response thus elicited [31].

The ventral nerve cord was stained with methylene blue (1 : 10,000) for 90 min within an electrically insulated humidity chamber (100% saturation, 10°C) which kept the nerve cord active for many hours.

The cord was stimulated with a standard microscope illuminator, the light being focused to a spot 2.55 mm in diameter (0.06 lm/mm^2). Gross platinum hook electrodes were used to record the response nerve pulses, which were fed into an electronic amplifying and windowing system and then displayed on a pen recorder.

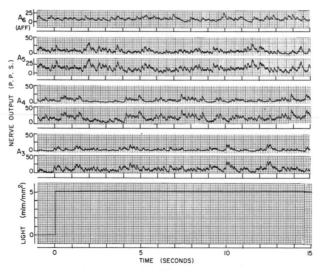

Fig. 36. Response to light stimulation of ventral abdominal nerve cord and ganglia exclusive of A_6. A_6 alone was shielded in these experiments. In top trace, "A_6 aff," nerve roots supplying A_6 were stimulated. All other sites are indicated by ganglion number (A_{2-5}) and include ganglion and adjacent nerve cord. No light response may be found anywhere.

Three characteristic features of the nerve responses were evident. The first was an excitatory response, an increase in the nerve pulse rate when the nerve was stimulated with light. The second was an inhibitory response, a decrease in the pulse rate when stimulated. The third was an "off" response, an increase in the pulse rate at the cessation of the stimulation.

The physiological location of the origin of each of these responses was made by carefully noting the site of stimulation associated with each response. An excitatory response occurred when the nerve cord was stimulated at a ganglion, while an inhibitory response followed by an "off" response occurred when a segment of axons was stimulated. Figure 37 shows a recording of each of these responses and of a composite response obtained by stimulating a ganglion and a segment of axons simultaneously.

Using the electronic windowing technique described in the preceding chapter, only a small number of fibers, the "B" fibers, were found to contribute to the photic response.

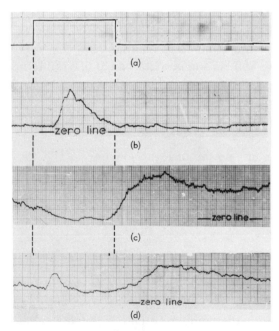

Fig. 37. Response of artificially photosensitized crayfish ventral cord. (a) Light stimulus (0.06 lm/mm^2); (b) excitatory response of ganglion; (c) inhibitory and "off" response of axon; (d) composite response.

The minimum stimulus needed to evoke a response was found to be 0.4 sec of illumination from the microscope illuminator, which, in other terms, is approximately 2.5 photons/dye molecule (assumed to be the response of a single "B" fiber).

The experimental results suggest the existence of two types of artificial photoreceptor, excitatory and inhibitory. The excitatory photoreceptor is localized in the vicinity of the ganglia while the inhibitory photoreceptor is localized in the vicinity of the axons. The results also suggest that the receptors are associated primarily with the "B" fibers that seem to have some generalized photosensitive properties.

APPENDIX A

MINIMUM PHOTON COMPUTATION

The number of photons that are incident on the nerve [30] can be calculated by using

$$\int_{\lambda_1}^{\lambda_2} N_I hc \frac{d\lambda}{\lambda^2} = E_T \tag{1}$$

where
N_I is incident number of photons
$1-2 = 400 \text{ m} - 700 \text{ m}$, wavelength range of visible spectrum
$h = 6.62 \times 10^{-27}$ erg-sec, Planck's constant
$c = 3 \times 10^{10}$ cm/sec, speed of light
$E_T = kPAT$, total energy in ergs incident on nerve

Here,
$P = 0.06 \text{ lm/mm}^2$, power of light source
$A = 2.6 \times 10^{-2} \text{ mm}^2$, stimulated area of nerve fiber
$T = 0.4$ sec, minimum stimulation time
$k = 1.61 \times 10^4$ (ergs/sec)/lm, conversion factor

Thus the incident number of photons is

$$N_I = 4.5 \times 10^{10} \text{ photons} \tag{2}$$

The methylene blue dye absorbs approximately 20% of the incident illumination; therefore the number of photons absorbed by the dye is

$$N_A = 0.9 \times 10^{10} \text{ photons} \tag{3}$$

In a methylene blue (1 : 10,000) solution the number of dye molecules per cubic centimeter is approximately

$$n = 1.8 \times 10^{17} \text{ molecules/cc} \tag{4}$$

At this concentration and for a staining period of 90 min, the nerve fiber takes up approximately 10% of the dye. Therefore the number of dye molecules in the stimulated segment of nerve can be calculated from N_M = 10% nV, where V, the volume of the stimulated nerve fiber, is 2×10^{-7} cc. Thus $N_M = 3.6 \times 10^9$ dye molecules.

We finally can calculate the minimum number of photons per dye molecule needed to evoke a response

$$No = \frac{N_A}{N_M} = \frac{9 \times 10^9}{3.6 \times 10^9} \simeq 2.5 \text{ photons/mole} \tag{5}$$

Chapter 3

Nerve Impulse Code

THE MODEL

In order to explain the signal transmission and coding and decoding demonstrated experimentally in this chapter, the following model has been proposed (see Fig. 38) [35, 37, 57]. Messages sent from point A to point B are first coded, then transmitted over two parallel channels to a decoder, and finally decoded. The coded message is transmitted along the parallel channels at the rate of five letters per minute, but only one letter of the five per minute is correlated between channels. The dynamic characteristics of the input–output function are satisfied by this correlation of one letter per minute. Either channel alone is adequate for behavioral input–output, and the effect of either channel alone is indistinguishable from the other.

From such a model one can then conclude that: (a) Both coding and decoding take place at the rate of one letter per minute. (b) The other four letters per minute are not signal information but may be used to study the processes of noise generation in the coder.

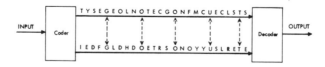

Fig. 38. Model with coder, decoder, and parallel communication channels serving as conceptual framework for experiments defining crayfish nerve impulse code.

MATHEMATICAL ARGUMENT

It has been pointed out [29, 60] that the use of a detailed pulse pattern as the nerve impulse code can provide a much higher information capacity than other coding systems. The maximum information contained in
a unit of time with a code that has n symbols with a probability, P_j, for each
of them and average symbol duration, b, is according to Shannon [49]:

$$C = -\frac{1}{b} \sum_{j=1}^{n} P_j \left(\log_2 P_j\right) \tag{1}$$

Under the simplifying assumption that the n symbols are equally frequent,
we have

$$C = \frac{\log_2 n}{b} \tag{2}$$

We would like to compare the capacity of a pulse pattern code with
that of an average frequency code by using crayfish nerve signal characteristics. The observed intervals usually range from 10 to 100 msec.
Thus, if we assume that the system might distinguish intervals with a precision of approximately 5 msec, we have $n = 20$. It should be noted that
these arbitrary assumptions cannot influence the results strongly, since n
appears only logarithmically in the formula. Taking the average symbol
duration b to be approximately equal to 50 msec, we have

$$C = \frac{\log_2 20}{0.050} \simeq 100 \text{ bits/sec} \tag{3}$$

Also, it may be noted that the assumptions made in calculating the
capacity of the detailed pattern code are very conservative. Wall et al.
[60], and also MacKay and McCulloch [29], with somewhat different assumptions based on faster mammalian preparations, derived rates one order of
magnitude higher.

Let us now examine the capacity of the code based on the average
frequency. Again, the frequency is found to vary from approximately 10
to 100 pulses per second. If we assume that the decoder can distinguish
levels 5 pulses/sec apart, we have approximately 20 symbols and, thus,
$n = 20$.

We must now decide what can be considered as the average duration
of a symbol. It seems reasonable to assume that the system can establish
or measure the average firing frequency for a time approximately equal to

its time constant. Stark and Hermann obtained the transfer function of the system in the first chapter,

$$G(s) = \frac{32e^{-1.0s}}{(1 + 1.2s)^2} \text{ pulses sec}^{-1} \text{ mlm}^{-1} \tag{4}$$

The time constant thus equals 1.2 sec. We finally have

$$C = \frac{\log_2 20}{1.2} \simeq 3 \text{ bits/sec} \tag{5}$$

The capacity of this average frequency code is thus more than one order of magnitude lower than for the case in which detailed pattern is the carrier of information.

If the variance is considered as a possible signal, we see that b, the average symbol duration, must be at least as long as that specified for the mean. Thus, at the cost of at least doubling the complexity of the decoding mechanism, the gain is only doubled. Similarly, various elementary trans-formations such as differentiation to obtain the mean rate of change will not add to the maximum capacity of the code. This latter alternative prob-bly accounts for some of the quantitative results by earlier workers on pulse spacing, but in no way changes the order of magnitude difference be-tween the simple code and the detailed pattern code.

THE MAIN EXPERIMENT

The pulse trains from the two photoreceptor cells were recorded simultaneously under constant light stimulation. The light intensity infor-mation is transmitted at the main firing frequency, which is approximately the same in both fibers, as shown in the interval histograms of Fig. 39 (bot-tom). The highly periodic characteristics of the firing pattern are reflected in the periodic features of the autocorrelation functions in Fig. 39 (middle). Nevertheless, it is clear from Fig. 39 (top) that the cross-correlation func-tions do not show significant deviations from the mean. We feel that this is the crucial experimental result.

THE SYSTEM

Methods

In order to have the fiber activity in a form easily usable by a com-puter, the output of the electronic window, which selected the desired pulses from the light-sensitive cells, was fed to a flip-flop circuit that produced an on-off type of record as presented in Figs. 40 and 41.

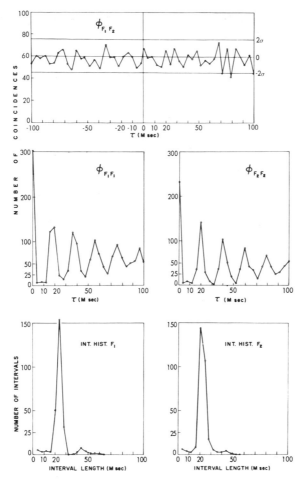

Fig. 39. Absence of correlation, periodic signals during light stimulation.

Each change of level in the output of the flip–flop circuit corresponds to a recorded pulse coming from the nerve. The output of the flip–flop was recorded on tape and then played back at one–eighth the original speed in order to be recorded with a paper recorder.

Directional Sensitivity of Photoreceptor

In previous studies dealing with the detailed timing of the nerve signal from the sixth abdominal ganglion of the crayfish, we assumed, on morphological grounds, that light stimulation of the ganglion produced the same effective stimulation of both ganglion cells. The results presented here substantiate that assumption, since relatively large deviations of the

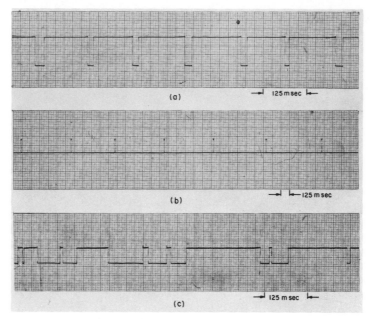

Fig. 40. Pattern of spontaneous activity observed in different isolated fibers. Records (a) and (b) in vitro, (c) in vivo.

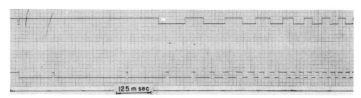

Fig. 41. Pattern of transient response of two fibers after onset of light.

light-stimulating spot did not produce substantial differences in the average frequency response of one of the cells, even when the light was moved toward the contralateral side during recording.

In a first group of experiments, the directional discrimination of the sixth abdominal ganglion was studied by recording the electrical response of the nerve axon of one photosensitive neuron when it was stimulated by a beam of light coming from a bundle of optical fibers used to carry the stimulating light. When the tip of the bundle of optical fibers was 1 mm above the ganglion, a spot of light 2 mm in diameter was produced. This spot was used to explore the photoreceptive response to different positions of the light source. When the light spot was displaced away from the mid-

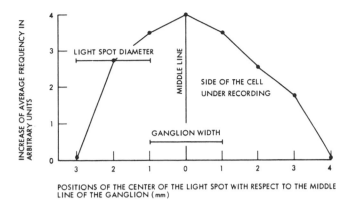

Fig. 42. Increase in average frequency discharge of one of
photosensitive cells in response to different positions of stimulating spot of light. Ordinate: steady-state average frequency response of right photosensitive neuron. Abscissa: central
position of tip of bundle of light fibers carrying light stimulus.
Also represented is approximate width of sixth abdominal ganglion and diameter of light spot. Note slight asymmetry of response to changes of light spot position.

dle line of the photoreceptor, only a small asymmetry in the average frequency steady-state response of the fiber was recorded with respect to the
symmetrically opposite position of the light (Fig. 42).

The first step was to establish that there are only two light-sensitive
fibers in the nerve cord, as has been indicated in earlier chapters. The
pulse trains obtained under light stimulation, before splitting the cord,
looked like two superimposed periodic pulse sequences. This is made clear
by the interval histograms presented in Fig. 43a. This figure contains
histograms of intervals between successive pulses (A), between every second pulse (B), and every third pulse (C). One can observe that the distribution becomes very sharp as we go from the single-interval to the double-
interval histogram, and widens again when triple intervals are taken. In
order to be able to compare the simultaneous activity of the two light-sensitive fibers, the ganglionic chain was split longitudinally in two halves
through the fourth and fifth ganglia, reaching almost as far as the sixth
ganglion. As soon as the surrounding connective tissue was dissheathed,
the chain could be easily separated into two cords, isolated from each other,
even through the fifth and the fourth ganglia. When the cord is dissected,
the single-interval histograms of each separated branch show a very sharp
distribution similar to a double-interval histogram of the complete cord
(Fig. 43b). The apparent randomness of the combined signal is due to the
superposition of two fairly periodic signals [3, 4].

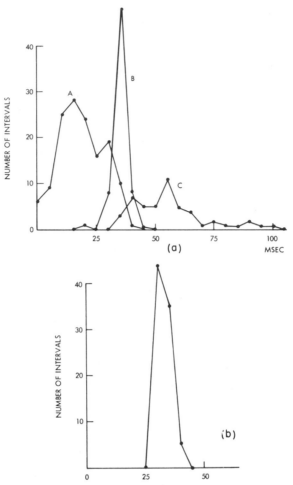

Fig. 43. Interval histograms of the activity under stimulation of two light-sensitive cells. (a) Two cells recorded
by same electrode. Histograms are presented for intervals
between successive pulses (curve A), between every second pulse (curve B), and between every third pulse (curve
C), before dissection. (b) Histogram of intervals between
successive pulses in one of cells recorded separately after
dissection.

An essential experiment was performed to establish that the two
light-sensitive cells transmit the same information and therefore independently produce the same walking movements after one of them is illuminated.
Indeed, as shown in Fig. 44a, the hemisection of the ganglionic chain does
not suppress the walking response and only produces an increase in the
latency of the response. The preserved response is similar in both sides of the
animal.

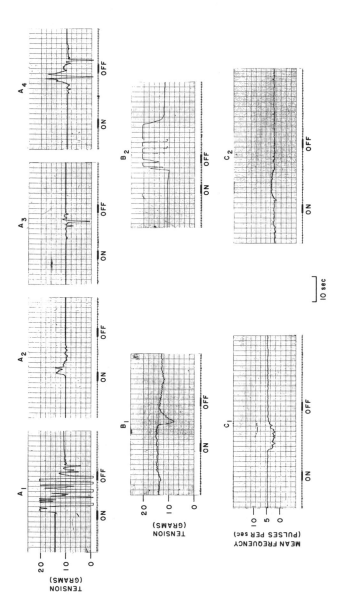

Fig. 44. Walking responses produced by light stimulation.

In Fig. 44, responses can be noted with alternating flexion and extension movements (A), mainly flexion movements (B_2), and purely extension movements (B_1). When the average frequency of the discharge of the motor neurons of the leg muscles was recorded, we found that the response consisted mainly of either an increase (C_2) or a decrease (C_1) in the average frequency.

Our results show that both fibers convey the same information to the decoder, and it is therefore reasonable to assume that the same code must be used by the two signals.

Approached in another way, if the periesophageal ring branches are disconnected from the supraesophageal ganglion, it is possible to pick out from these branches one fiber (B) whose average frequency discharge increases with the light stimulation of the sixth abdominal ganglion. This fiber can be found in either of the two branches of the periesophageal ring, even when one of the photoreceptor fibers has been suppressed by abdominal ventral cord hemisection. The response of these fibers is completely deterministic, and its average frequency is very close to that of the photoreceptor cell fiber (A) (Fig. 45). Cross–correlation studies between (A) and (B) pulse trains during light stimulation show its higher positive correlation at $\tau = -48$ msec, the distance between the electrodes being 5.5 cm (Fig. 46).

Structure of the Neural Chain

In another group of experiments, the structure of the neural chain as shown in Fig. 47 was elucidated. The crayfish walking movements were

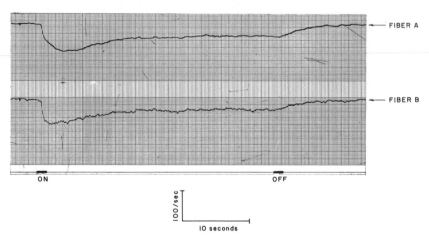

Fig. 45. Similar time course of responses in simultaneous recording of average frequency light response of nerve fiber (A) coming from sixth abdominal ganglion photoreceptor, and a nerve fiber (B) in homolateral side of periesophageal ring, disconnected from supraesophageal ganglion.

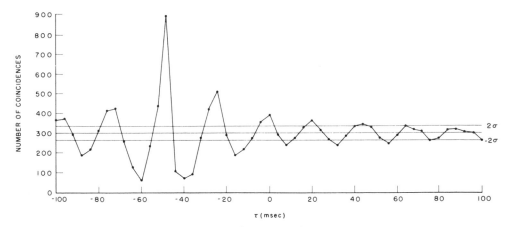

Fig. 46. Cross-correlation function between fiber A and fiber B. Note positive correlation at τ =-48 msec, and oscillatory nature of function corresponding to close values of average frequency discharge of two fibers.

investigated by means of the following procedures: (a) enucleation of the eyes; (b) periesophageal ring transection; and (c) transection immediately below the infraesophageal ganglion. Enucleation of the eyes was accomplished by cutting the stalk of the eyes; periesophageal ring transection by uncovering ventrally the region below the rostrum and cutting the two branches of the ring under the microscope; and infraesophageal section by uncovering the region between the origin of the food–handling legs and transecting the ganglionic chain. Survival of the animal after the surgical procedures was almost 100% if, instead of removing skeleton pieces, one uncovered the region, fractured it, and allowed the fractured lips to regain their original position following the procedure.

The walking movements were recorded by having one of the legs of the animals pull a lever with a mirror attached to it. A beam of light reflected by the mirror stimulated a photocell once during each walking swing of the legs.

The voltage generated by the photocell triggered a wave-form generator, producing a pulse with each swing, which was displayed on a paper recorder. The animals were secured on their backs to a small table and immersed in a tank of water at 20°C. Constant bubbling of air through the water provided adequate ventilation. The animals and the recording device were located in a dark chamber with a hole through which the stimulating light from a filament projector lamp was allowed to pass. When required, the light stimulation was automatically switched on by a clock device for 30 sec every 15 min.

During the first 12 hours after extirpation of the eyes, the animals showed prolonged and frequent walking which in unrestrained animals consisted of exploratory walking along the walls of the tank with the help of their antennae. The abdominal region and the uropods of the telson showed characteristic extension. When the distribution of walking movement trains was studied, no clear periodicity could be observed. Figure 48 shows in shaded bars the probability of getting a walking train in each interval of 21 arbitrary cycles of 35-min duration. Each of the nonshaded bars represents the probability of getting a response to a 30-sec light stimulation produced approximately every 15 min. The animals with periesophageal ring transection remained motionless most of the time with a permanent flexion of the abdominal region. If they were mechanically stimulated in the dorsal abdomen, they would walk for a while, and then remain in one place for a very long period. Without displacing themselves, however, they displayed spontaneous walking-like movements of the legs when lying on one side. These walking movements can be reliably evoked by light stimulation; however, they are not deterministically produced.

A typical experiment showed that the probability of a light-evoked walking movement in a blind animal changed from 0.57 to 0.77 after periesophageal transection. The animals with a transection below the infraesophageal ganglion remained motionless after the operation. They only showed weak, coordinated movements of the legs when mechanically stimulated on the ventral thoracic surface. Light stimulation produced no leg movements.

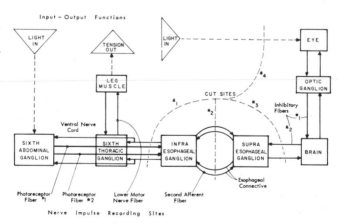

Fig. 47. Experimental arrangement for dissection studies to elucidate and define structure of the neural chain underlying the photoreceptor walking movement system. Note that classical neurophysiological nerve impulse monitoring and ablation techniques are embedded within a multiple input-output systems approach.

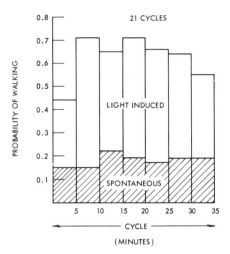

Fig. 48. Effect of light stimulation in
blinded crayfish on probability of walking.
Twelve hours and 15 min of walking recording was divided into 21 cycles of 35 min each.
Each cycle was subdivided in 7 numerated
periods of 5 min each. Shaded bars represent
probability of production of walking each numerated period. Nonshaded bars represent probability of light-induced walking
in each numerated period. Note absence
of cyclic walking activity and the increase of walking probability with light
stimulation.

From the experiments with ganglionic chain transection below the
infraesophageal ganglion, one sees that coordinated movements of the legs
can be elicited without the infraesophageal ganglion. Actual walking movements are only possible, however, with the control of the infraesophageal
ganglion as shown in the periesophageal–ring–transected preparation.

The higher probability of getting a walking movement response to
light stimulation in the periesophageal–ring–transected preparation than in
the blinded preparation shows that there is a tonic inhibitory influence of
the supraesophageal ganglion upon this response. From the experiments
in which B fiber activity was recorded, it is clear the infraesophageal ganglion region is the place at least where the signal crosses the middle line
of the animal. This might explain the walking-movement response bilaterally produced by the activation of a single sixth ganglion photoreceptor cell.

As mentioned earlier, the response of the B fibers is completely deterministic, and its average frequency is very close to that of the photo–

receptor cell fiber A. Illumination of the main eyes has no effect on the discharge of fiber B when that fiber is isolated without sectioning any other fibers in the periesophageal ring.

In those preparations in which one of the branches of the periesophageal ring is disconnected near the infraesophageal ganglion, the branch now attached to the supraesophageal ganglion shows several C fibers with a different response pattern from light stimulation of the photoreceptor. This pattern, in an average frequency recording, can be described as an increase of frequency which fades out before the light stimulation is turned off. A ringing-like phenomenon can be seen superimposed on top of this overall increased frequency (Fig. 49a). For most of these C fibers, stimulation of the main eyes causes them to discharge in a fashion similar to that already described, but with additional "on" and "off" bursts (Fig. 49b). In some experiments it was found that the diffuse stimulation of the lower surface of the eyes inhibits the C discharge produced by the sixth abdominal ganglion light stimulation.

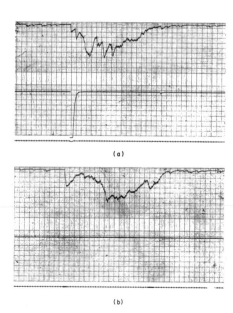

(a)

(b)

Fig. 49. (a) Average frequency response of fiber coming from the supraesophageal ganglion (fiber C) in response to light stimulation of sixth abdominal ganglion. Note overall increase of frequency, its transitory nature, and superimposed ringing-like effect. (b) Response of same fiber to stimulation of eyes. Note "on" and "off" bursts.

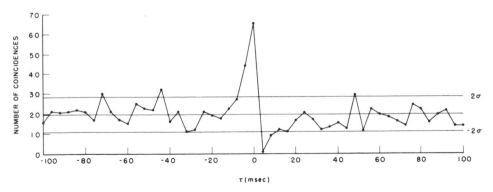

Fig. 50. Cross-correlation function between fibers B and C. Note positive correlation at $\tau = 49$ and 0 msec.

The cross-correlation analysis between B and C trains (Fig. 50) shows that there is positive cross correlation at $\tau = 8$, 4, and 0, the highest correlation being one-third of the pulses at $l = 0$. It is more likely, however, at $\tau = -2$, because of the grid size used (2 msec). One can obtain a C discharge even with hemisection of either side of the abdominal ventral cord. The reliability of the response of these fibers, although not yet numerically evaluated, is higher than any walking response studied thus far, but the probability of the response is clearly sensitive to the interval between two subsequent stimulations of the sixth abdominal ganglion.

In another group of experiments, the electrical signals from single fibers were recorded at different interganglionic regions. The methods of microdissection, electrical recording, and computer processing have been described in Chapter 2. Access to the fibers in the esophageal ring was made by producing a circular window, 6 mm in diameter, in the exoskeleton below the rostrum. The uncovering and isolation of the ring was easily done by pushing the green glands backward with cotton plugs.

Considering that the speed of conduction of the photosensitive fibers is approximately 3.8 m/sec, it is tempting to suggest the presence of at least one synapse between A and B. This would account for the difference between the calculated delay (14.4 msec), with A and B considered as the same fiber, and the actual delay (48 msec) that was found. More experiments, of course, will elucidate the point.

Considering C fibers as inhibitors of the walking movement system in the infraesophageal ganglion would account for the tonic inhibitory effect of the supraesophageal ganglion and for the common observation of walking movement suppression by the illumination of the eyes of the intact animal.

The small time delay in maximum correlation that was found between B and C suggests a very simple neuronal circuit that might be monosynaptic.

From the experimental data we feel that the infraesophageal ganglion acts as a center that gates a true sixth abdominal photoreceptor walking movement reflex. The gating depends on several inputs coming from other receptors, of which the eye would have the highest priority.

ADDITIONAL RESULTS

Different types of spontaneous activity were observed in our experiments. A feature that was present in several cases (although not always) was the appearance of coupled pulses (see Fig. 40) with a time separation of $10-40$ msec. The distance between the couples can be rather regular, ranging from approximately 150 msec (Fig. 40a) to 1 sec (Fig. 40b), or can be quite irregular (Fig. 40c). Usually the *in vitro* preparations presented higher regularity than the *in vivo* ones. Often the *in vitro* preparations, after a few hours, presented practically zero spontaneous activity.

When the activity of the two dissected branches was recorded simultaneously, no correlation between the timing of the pulses of the two fibers was observed (Fig. 51c,d). In Fig. 51e,f,g, several histograms are given of the interval between a pulse in one fiber and the next pulse in the other fiber. We see that the histograms are approximately linear on a semilogarithmic scale, as they should be if the pulses of the two fibers were uncorrelated; therefore, the intervals follow a Poisson distribution.

Further evidence of the anatomical independence of the two fibers is the fact that in some cases the average frequency of the spontaneous activity is very different in each of them.

When the light stimulus was applied in the form of a step with a short rise time, the two fibers responded with different transients, as can be seen in Fig. 41. This may be taken as evidence regarding lack of redundancy. In this particular experiment the fiber of the lower trace, which showed a sharp increase in frequency, had a lower spontaneous activity. At the steady state both fibers reached approximately the same frequency.

A general feature of the transient responses to a high level of light is that they present a peak in the average frequency approximately 1 sec after the start. The peak frequency is twice as big as the steady-state frequency that follows.

The main feature of the steady-state response to light is the increased regularity in the intervals between pulses in both fibers in contrast with the spontaneous activity that presents a high irregularity (Fig. 51h,i,j). The light-response regularity is more accentuated in the *in vitro*

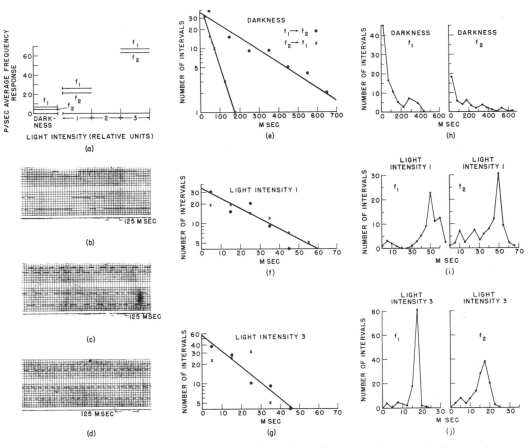

Fig. 51. Comparison of activity of two light fibers (f_1 and f_2) recorded in vivo simultaneously, but separately under various light intensities. (a) Average frequency response of the fibers in darkness and at two different light levels. (b), (c), and (d) Firing pattern of two cells in darkness (b) and at lower (c) and higher (d) light levels. (e), (f), and (g) Histograms of intervals from one pulse in one fiber to next pulse in other fiber. (h), (i), and (j) Interval histogram for activity of two fibers (f_1 and f_2) in darkness and at two different light levels.

preparations, as can be seen by comparing the two histograms given in Fig. 52.

When the level of the stimulating light was very low, we observed a slow increase of the average frequency which, after several seconds, settles to a slightly higher cycle than the spontaneous activity without a considerable change in the irregularity of the intervals. More experiments in this region of low light levels are now under way.

One can, however, find experiments in which in darkness and at different light intensities, the average frequency differs little between the two

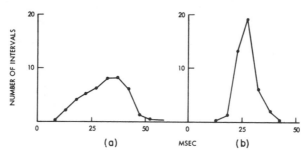

Fig. 52. Interval histograms of activity under light stimulation of single fiber in vivo (a), and in vitro (b).

fibers (Fig. 51a). Direct inspection of the records shows a completely different pulse pattern of activity between them (Fig. 51b−d). The lack of
correlation of the interfiber interval histograms, between a pulse in one
fiber and the next pulse in the other, still can be shown (Fig. 51e,f).

The interval histograms of each fiber show, in the same experiment,
that the mode interval, the mean interval, and the dispersion of the intervals can be correlated with the light levels (Fig. 51h,i,j) and thus with each
other.

FIBERS SHOWING MINOR CORRELATIONS

Occasionally, minor correlations were observed, statistically significant with these averaging techniques, but not strong enough to revive
the idea of the detailed-pattern code with a high information rate. An
example of minor correlation is shown in Fig. 53 which also shows the
highly periodic autocorrelation functions [Fig. 53 (middle)]. Here, however, the cross-correlation function [Fig. 53 (top)] shows periodicities that
extend beyond the ±2 standard-deviation lines on the graphs. The most
significant peak in Fig. 53 (top) at $\tau = 0$ represents a negative correlation
of −3.3 pulses/sec as compared with the 44 pulses/sec average rate and
9.2 ± 0.6 pulses/sec chance coincidence rate, which is determined by the
grid size (2 msec) chosen for coincidences.

The possibility that this periodic fluctuation resulted from a periodic
error in the cross-correlation computation suggested a "control experiment" in which we compared two impulse trains from different crayfish,
but with quite similar average frequencies. The result is shown in Fig. 54
and the cross-correlation function stays well bounded by the ±2 standard-
deviation lines as expected [Fig. 54 (top)]. It is important to note that although the autocorrelation functions have strong periodicities [Fig. 54
(middle)], this alone is not sufficient to produce significant cross correlations.

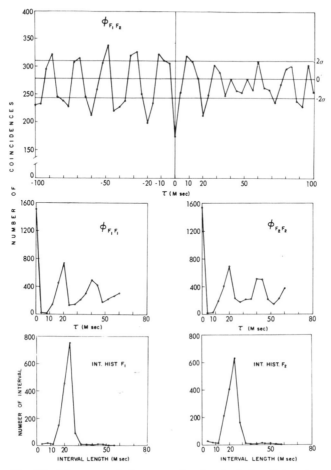

Fig. 53. Minor correlation, periodic signals during light stimulation.

    In Fig. 55 a collection of experimental results show noncorrelations
(Fig. 55b,d), as well as minor correlations (Fig. 55a,c); light and dark
conditions are included, as well as another control (Fig. 55e).

    Figure 56 shows the broad and bimodal interval histograms for one
unstimulated case. Accordingly, the autocorrelation functions [Fig. 56
(middle)] show little periodic structure. The cross correlation [Fig. 56
(top)] shows minor but statistically significant peaks and valleys. The
highest peak at −4 represents a positive correlation of 0.7 pulses/sec,
which may be compared with the 15 pulses/sec average rate and the 0.9
± 0.2 pulses/sec chance coincidence rate (determined by the 2-msec grid
size chosen for coincidences).

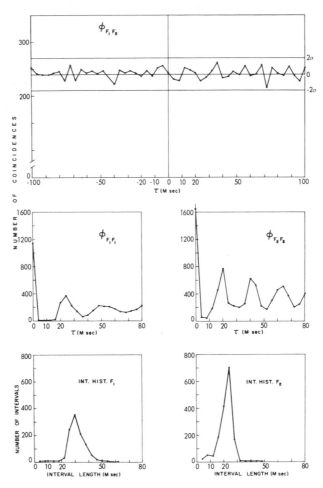

Fig. 54. Control: no correlation between two crayfish; periodic signals during light stimulation.

Figure 57 shows the frequency of occurrence of the significant cross-
correlation points found in our experiments. With the preparation in darkness (Fig. 57a), positive correlations appear at 0, ±4, and ±8 msec τ shifts
frequently enough to be noticed. During light stimulation (Fig. 57b) there
is a clear periodicity at the beginning which fades somewhat at higher values
of τ. This observation agrees with the individual cases discussed above.

DISCUSSION

Analysis of Noise Generation in Coder

The observed pattern of the pulse trains, particularly during spontaneous activity, may contain some information on the nature of the pulse-

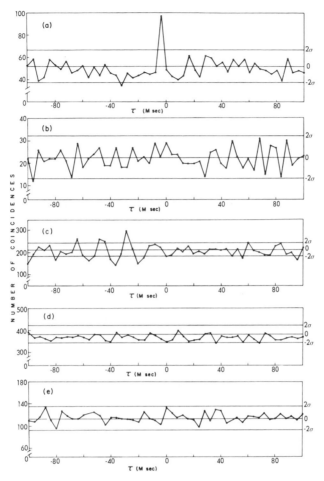

Fig. 55. Representative cross-correlation results.

generating process. The presence of paired discharges was a characteristic feature in several preparations. The interval between the paired pulses ranged from 10 to 40 msec, being rather constant in some cases and fairly varied in others. The intervals between the pairs covered a much wider range (from 50 msec to more than 1 sec), and were also rather regular or more-or-less random, in different cases. Some times, when the light stimulus was applied, the pairs started to appear at increasingly smaller intervals until a regular pulse sequence resulted. The interval then was almost equal to the short interval between the paired pulses of the spontaneous activity (Fig. 41, lower trace). This picture seems to indicate that in the generation of the pulses, two processes may be involved:

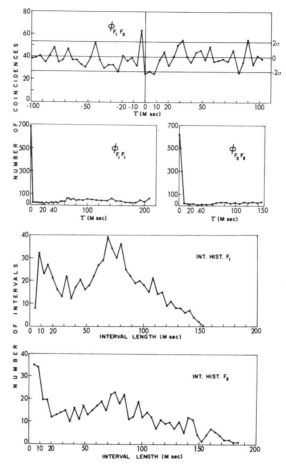

Fig. 56. Minor correlation: aperiodic signals in darkness.

   1. A "fast process" responsible for the short interval between the
paired pulses. The size of the small interval may be determined either by
the refractory period of the neuron or by an excitatory oscillation in the
dendrites.

   2. A "slow process" connected to the light-sensitive mechanism and
responsible for the long intervals between the pulse pairs. Noise in the
light-sensitive mechanism may cause the variations that are sometimes
seen in the size of the long interval during spontaneous activity, and so account for the more irregular pulse patterns observed. Under strong light
stimulation the cells are firing at a maximum rate that corresponds approximately to the small interval set by the fast process.

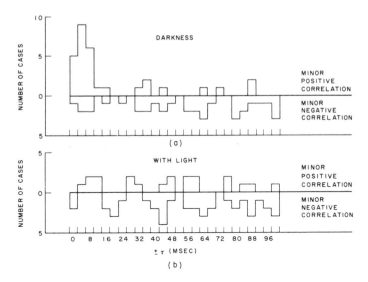

Fig. 57. Frequency of occurrence of significant cross-correlation points. With preparation in darkness positive correlations appear at 0, ±4, and ±8 msec τ shifts: with light stimulation there is a periodicity in the beginning which fades at higher values of τ.

If we wish to investigate the use of detailed-pattern information by a decoding mechanism in a natural neural system, then it should first be attempted in a paucicellular system such as the crayfish [33]. Otherwise, with many input fibers to each complex dendritic tree and soma, and with recurrent feedback elements, the synchronization of the many detailed impulse patterns will require an inordinately large and precise system of measurement and a complicated decoding device or process. Thus, one should look at a favorable invertebrate preparation rather than utilize the mammalian cortex.

The system of the photosensitive cells in the sixth abdominal ganglion of the crayfish also presents many advantages for the experimental approach to this problem, i.e., that there are only two receptor cells that can be reliably submitted to the same stimulus and be shown to carry the same information. Indeed, we have seen that cutting one of the fibers did not affect the walking response evoked by the illumination of the ganglion. Furthermore, the activity of the two cells can be recorded simultaneously and separately and the individual pulses can be compared with each other. If the detailed pattern were the code through which the information is transmitted, a high correlation between the two pulse trains at the level of the detailed pattern should be observed.

Instead, we found no correlation (Figs. 39 and 55d). (Occasionally we have seen that only approximately 7% of the pulses show some cross-correlation dependence as in Fig. 53.) This absence of correlation must indicate that the detailed timing of the pulses is not a carrier of useful information and thus leaves mean pulse frequency to be considered as the code.

Different approaches have previously been used to investigate the detailed pattern in a nerve pulse train as a possible carrier of useful information. One method has been to examine the response of a cell to different patterns of controlled input nerve impulses. It has been reported that in *Aplysia* ganglion cells [34, 47, 48] and in the neuromuscular preparations of crustacea [44, 64] pulse sequences with the same mean frequency evoke different responses that are dependent on the input pulse pattern. In another approach [38] the probability of "postimpulse firing" in a single neuron was examined, and it was argued that since the intervals are neither Poisson distributed nor independently distributed, they convey useful detailed-pattern information.

Signal Information Carried by a Train of Nerve Pulses

The photosensitive sixth abdominal ganglion of the crayfish has been used before in this laboratory to study the relationship between the light signal and the generated nerve pulses. The mean-pulse frequency was found to be proportional to the logarithm of the light intensity and a transfer function has been established for this system [16].

The purpose of this research was to find out if, besides the mean-pulse frequency, there are other carriers of signal information in the pattern of the generated nerve pulses. The crayfish sixth ganglion is particularly suitable for this type of study as it contains only two independent photosensitive units [23] that can be compared with respect to the way in which they transmit the common light signal. Separating the two light-sensitive fibers by microdissection and recording their pulse trains simultaneously, we found them to be uncorrelated as far as the detailed pulse pattern is concerned. If the detailed timing of the individual pulses carried useful information, there would be a very large amount of information transmitted through the generated pulse trains. Our results indicate that this is not the case in this system, and that the light signal information is transmitted through the mean-pulse frequency.

SUMMARY

Nerve-impulse patterns from the two photosensitive neurons of the sixth abdominal ganglion of the crayfish have been recorded and we have presented evidence that the signals carried by each of these two photoreceptors elicit the same walking movement response of the legs. These new results reinforced our view that the code carried by each fiber passes through the same decoding mechanism.

Our first results showed the absence of correlation in the detailed pattern of interval sequences between nerve impulses of the two parallel nerve trains coming from the photoreceptor. Noncorrelation was demonstrated by visual comparison of the two nerve impulses in one fiber and the next pulse in the other fiber. These histograms were linear on a semi-logarithmic scale, as they would be if the two fibers were uncorrelated and produced a Poisson distribution of these interfiber intervals.

We have presented further evidence in the form of both autocorrelation and cross-correlation functions to substantiate the absence of correlation noted previously, and to give a quantitative estimate of some minor correlations that occasionally appear.

If the nerve impulse code depends upon the exact timing of the pulses, then the detailed pattern of the two pulse trains should be very closely correlated. Our results seemed to be crucial in this aspect because no consistent, significant correlation had been observed. The 3 bit/sec average frequency code is put forward as a sufficient information rate to satisfy the dynamic characteristics of the photoreceptor-walking-movement response in this particularly well-defined paucicellular neurological system.

REFERENCES AND FURTHER READING

1. Allen, E. J.: Studies on the nervous system of crustacea. I. Some nerve elements of the embryonic lobster. Quart. J. Mil. Sci. N. S. 36: 461−497 (1894).
2. Barnes, T. C.: Peripheral tonus associated with impulse discharge in crustacean nerve. J. Physiol. 70: 24−25 (1930).
3. Cox, A. R., and Smith, W. L.: Superposition of several strictly periodic sequences of events. Biometrika 40: 1 (1953).
4. Cox, A. R., and Smith, W. L.: On the superposition of renewal processes. Biometrika 41: 91−99 (1954).
5. Cobb, S.: Observation on the comparative anatomy of the avian brain. Perspectives Biol. Med. III: 383−408 (1960).
6. England, T. E., and Hermann, H. T.: Transmission of neural impulses in the ventral cord of the crayfish. Tech. Report, Research Laboratory of Electronics, M.I.T. (in preparation).
7. Fernandex-Moran, H.: Fine structure of the light receptors in the compound eyes of insects. Exptl. Cell Res. Suppl. 5: 586−644 (1958).
8. Goldsmith, T. H., and Philpott, O. E.: Microstructure of the compound eyes of insects. J. Biophys. Biochem. Cytol. 3: 429-440 (1957).

9.   Groen, J. J., Lowenstein, O., and Vendrik, A. J. H.: The mechanical
      analysis of the responses from the end organs of the semicircular
      canal in the isolated elasmobranch labyrinth. J. Physiol. 117: 320−
      346 (1952).

10. Hama, K.: A photoreceptor-like structure (ventral) nerve. Anat.
      Rec. 140: 329−336 (1961).

11. Hartline, H. K.: A quantitative and descriptive study of the electrical
      response to illumination of the arthropod eye. Am. J. Physiol. 83:
      466−483 (1928).

12. Hartline, H. K., and Graham, C. H.: Nerve impulses from single re-
      ceptors in the eye. J. Cellular Comp. Physiol. 1: 277−295 (1932).

13. Hartline, H. K.: Intensity and duration in the excitation of single
      photoreceptor unit. J. Cellular Comp. Physiol. 5: 229−247 (1934).

14. Hermann, H. T., and Stark, L.: Random walk response of the cray-
      fish. Quart. Prog. Rept., Research Laboratory of Electronics, M.I.T.
      61: 230−234 (1961).

15. Hermann, H. T., and Stark, L.: Prerequisites for a photoreceptor
      structure in the crayfish tail ganglion. Anat. Rec. 147: 209−217
      (1962).

16. Hermann, H. T., and Stark, L.: Single unit responses in a primitive
      photoreceptive organ. J. Neurophysiol. 26: 215−228 (1963).

17. Hermann, H. T., Stark, L., and Willis, P. A.: Instrumentation for
      processing neural signals. Electroencephalog. Clin. Neurophysiol.
      14: 557−560 (1962).

18. Hesse, R.: Untersuchungen über die Organe der Lichtempfindung
      bei Lumbriuden. Z. Wiss. Zool. 61: 393−419 (1896).

19. Hesse, R.: Untersuchungen über die Organe der Lichtempfindung
      bei Niederen Tieren, die Sehorgane des Amphioxus. Z. Wiss. Zool.
      63: 456−464 (1890).

20. Johnson, G. E.: Giant nerve fibers in crustaceans, with special ref-
      erence to cambarus and palemonetes. J. Comp. Neurol. 36: 323−375
      (1924).

21. Johnson, G. E.: Studies on the functions of the giant fibers of crus-
      taceans, with special reference to cambarus and palemonetes. J.
      Comp. Neurol. 42: 19−33 (1926).

22. Kennedy, D.: Responses from the crayfish caudal photoreceptor.
      Am. J. Ophthalmol. 11: 19−26 (1958).

23. Kennedy, D.: Physiology of photoreceptor neurons in the abdominal
      nerve cord of the crayfish. J. Gen. Physiol. 46: 551−572 (1963).

24. Kennedy, D., and Bruno, M. S.: Spectral sensitivity of crayfish and
      lobster vision. J. Gen. Physiol. 44: 1089−1102 (1961).

25. Kennedy, D., and Preston, J. B.: Activity patterns of interneurons in
      the caudal ganglion of the crayfish. J. Gen. Physiol. 43: 655−670 (1960).

26.   Krieger, R.: Über das Centralnervensystem des Flüsskrebses. Z.
       Wiss. Zool. 33: 527−594 (1880).
27.   Kropp, B., and Enzmann, E. V.: Photic stimulation and leg move-
       ment in the crayfish. J. Gen. Physiol. 16: 905−910 (1933).
28.   Lamport, H., Mauro, A., and Stark, L.: How is tension transmitted
       from striated muscle fiber to tendon? Abstracts of Communications,
       20th International Physiological Congress, Brussels, Belgium.
29.   MacKay, D. M., and McCulloch, W. S.: The limiting information ca-
       pacity of a neuronal link. Bull. Math. Biophys. 14: 127−135 (1952).
30.   Millecchia, R.: Artificial photosensitization of the crayfish ventral
       nerve cord. Quart. Prog. Rept., Research Laboratory of Electronics,
       M.I.T. 70: 345−346 (1963).
31.   Miller, W. H.: Morphology of the ommatidia of the compound eye of
       limulus. J. Biophys. Biochem. Cytol. 3: 421−428 (1957).
32.   Moody, M. F., and Robertson, J. D.: The fine structure of some ret-
       inal photoreceptors. J. Biophys. Biochem. Cytol. 7: 87−91 (1960).
33.   Moore, G. P.: Personal communication (1964).
34.   Moore, G. P., and Segundo, J. P.: Stability patterns in interneuronal
       pacemaker regulation. Symposium for Biomedical Engineering, San
       Diego, California (1963).
35.   Negrete, J., Yankelevich, G. N., and Stark, L.: Component analysis
       of the abdominal photoreceptor walking movement system in the cray-
       fish. Quart. Prog. Rept., Research Laboratory of Electronics, M.I.T.
       76: 336−343 (1965).
36.   Negrete, J., Yankelevich, G. N., Theodoridis, G., and Stark, L.:
       Light inhibitory effects in the crayfish sixth ganglion. Quart. Prog.
       Rept., Research Laboratory of Electronics, M.I.T. 74: 252−254
       (1964).
37.   Negrete, J., Yankelevich, G. N., Theodoridis, G., and Stark, L.:
       Signal information carried by a train of nerve pulses. Quart. Prog.
       Rept., Research Laboratory of Electronics, M.I.T. 74: 255−261
       (1964).
38.   Poggio, G. F., and Viernstein, S. J.: Time series analysis of im-
       pulse sequences of thalamic somatic sensory neurons. J. Neuro-
       physiol. 27: 517−545 (1964).
39.   Pringle, J. W. S., and Wilson, V. J.: The response of a sense organ
       to a harmonic stimulus. J. Exptl. Biol. 29: 229 (1952).
40.   Prosser, C. L.: Action potentials in the nervous system of the cray-
       fish. I. Spontaneous impulses. J. Cellular Comp. Physiol. 4: 185−
       209 (1934).
41.   Prosser, C. L.: Action potentials in the nervous system of the cray-
       fish. II. Responses to illumination of the eye and caudal ganglion. J.
       Cellular Comp. Physiol. 4: 363−377 (1934).

42.   Prosser, C. L.: Action potentials in the nervous system of the cray-
        fish. III. Central responses to proprioceptive and tactile stimulation. J.
        Cellular Comp. Physiol. 4: 495−505 (1934).
43.   Prosser, C. L.: Effect of salts upon "spontaneous" activity in the
        nervous system of the crayfish. J. Cellular Comp. Physiol. 15: 55−
        65 (1940).
44.   Ripley, S. H., and Wiersma, A. G.: The effect of spaced stimulation
        of excitatory and inhibitory axons of the crayfish. Physiol. Comp.
        Ecol. III: 1−17 (1953).
45.   Roberts, T. M.: Discussion of speculations on servo control of move-
        ment. In: The Spinal Cord. J. L. Malcolm and J. A. B. Gray, eds.,
        Churchill (London) (1953), pp. 255−258.
46.   Robertson, J. D.: The molecular structure and contact relationships
        of cell membranes. Prog. Biophys. 10: 343−418 (1960).
47.   Segundo, J. P., and Moore, G. P.: Functional significance of neuronal
        spike discharge parameters. Bol. Inst. Estud. Med. Biol. (Mex.) 21:
        371−373 (1963).
48.   Segundo, J. P., Moore, G. P., Stensaas, L. J., and Bullock, T. H.:
        Sensitivity of neurons in Aplysia to temporal pattern of arriving im-
        pulses. J. Exptl. Biol. 40: 643−667 (1963).
49.   Shannon, C. E.: Mathematical theory of communication. Bell Sys-
        tem Tech. J. 27: 379−423 (1948).
50.   Sjöstrand, F. S.: Retinal rods and cones: the ultrastructure of the
        retinal receptors of the vertebrate eye. Ergeb. Biol. 21: 128−160
        (1959).
51.   Stark, L.: Stability oscillations, and noise in the human pupil servo-
        mechanism. Proc. Inst. Radio Engrs. 47: 1925 (1959).
52.   Stark, L.: Transfer function of the biological photoreceptor. Wright
        Air Develop. Ctr. Tech. Rept. 59-311: 1−22 (1959).
53.   Stark, L., and Hermann, H. T.: Light transfer function of a biological
        photoreceptor. Nature (London) 191: 1173−1174 (1961).
54.   Stark, L., and Hermann, H. T.: The transfer function of a photore-
        ceptor organ. Kybernetik 1: 124−129 (1961).
55.   Stark, L., and Hermann, H. T.: Review of photoreceptor responses
        to wavelength variables. IRE Trans. on Electronic Computers
        EC-11: 806 (1962).
56.   Stark, L., and Sherman, P. M.: A servoanalytic study of the consen-
        sual pupil reflex to light. J. Neurophysiol. 20: 17−26 (1957).
57.   Theodoridis, G., Negrete, J., Yankelevich, G. N., and Stark, L.:
        Photosensitive neurons of the crayfish sixth ganglion as a dual sys-
        tem, each neuron carrying same signal information. Quart. Prog.
        Rept., Research Laboratory of Electronics, M.I.T. 75: 197−209 (1964).
58.   Utall, W. R., and Kasprzak, H.: The caudal photoreceptor of the
        crayfish: adaptation and the luminosity function. Abstracts of Bio-
        phys. Sec., Washington, D.C., TE 7 (1962).

59.   Van Harreveld, A.: A physiological solution for fresh water crus-
       taceans. Proc. Soc. Exptl. Biol. Med. 34: 428—432 (1936).
60.   Wall, P. D., Lettvin, J. Y., McCulloch, W. S., and Pitts, W. H.: Fac-
       tors limiting the maximum impulse transmitting ability of an afferent
       system of nerve fibers. In: Information Theory. C. Cherry, ed.,
       Academic Press, New York (1956), pp. 329—343.
61.   Welsh, J. H.: The caudal photoreceptor and responses of the cray-
       fish to light. J. Cellular Comp. Physiol. 4: 379—388 (1934).
62.   Wiersma, C. A. G.: Function of the giant fibers of the center nervous
       system of the crayfish. Proc. Soc. Exptl. Biol. Med. 38:661—662 (1941).
63.   Wiersma, C. A. G.: Giant nerve fiber system of the crayfish. A con-
       tribution to comparative physiology of synapses. J. Neurophysiol.
       10: 23—38 (1947).
64.   Wiersma, C. A. G., and Adams, R. T.: The influence of nerve im-
       pulse sequence on the contraction of different muscles of crustacea.
       Physiol. Comp. Ecol. II: 20—33 (1950).
65.   Wiersma, C. A. G., and Norritski, E.: The mechanism of the ner-
       vous regulations of the crayfish heart. J. Exptl. Biol. Med. 19: 255—
       265 (1942).
66.   Wiersma, C. A. G., Ripley, S. H., and Christensen, E.: The central
       representation of sensory stimulation in the crayfish. J. Cellular
       Comp. Physiol. 46: 307 (1955).
67.   Wolken, J. J.: Retinal structure. Mollusc cephalopods: octopus,
       sepia. J. Biophys. Biochem. Cytol. 4: 835—838 (1958).
68.   Wolken, J. J., Capenos, J., and Turano, A.: Photoreceptor structures.
       III. *Drosophila melanogaster*. J. Biophys. Biochem. Cytol. 3: 441—
       447 (1957).

THE PUPIL

INTRODUCTION

As a clinical neurologist I was impressed by the possibility that various disorders of movement might be explicable in terms of control theory. As a neurophysiologist I was looking for a way to embed the functioning of the neural systems in a mathematical framework as sophisticated as those used in physics or biophysics and at least partially adequate to give some real understanding to these elegant systems. Cybernetics, with its analogies between natural and artificial control and communications systems, and its mathematical language shaped by numerous engineering applications, seemed to answer this need. Irrespective of cybernetics, if one wishes to analyze the more subtle of "normal" features of the nervous system, one cannot work with such minimal preparations as decerebrate or heavily anesthetized animals. It seemed intellectually and ethically preferable to go to awake, intact man and develop the necessary instrumental and experimental design procedures to compensate for the inherent difficulties. These problems include the inability to "wet dissect" into the black box, the "nonstationarity" or time-varying characteristics of a human being, the large amounts of disturbing fluctuations in responsiveness, the interaction from other ongoing neural functions, and the difficulties in precise measurement.

The pupil of the eye had already been studied from a clinical point of view by Lowenstein, and using his excellent infrared movie camera and scanning film reader, together with appropriate electronic apparatus, it seemed possible to devise a direct reading pupillometer to permit continuous precise output recording, while at the same time using light as the stimulus input. In discussion in the fall of 1955, with Professor Peter Schultheiss at the Electrical Engineering department at Yale, I learned the value of "open-loop" transfer function analysis and thought of the optical trick, the Maxwellian view for so opening the loop. Shortly thereafter, one of Schultheiss' students, Philip Sherman, came to the medical school and together we started the pupil experiments while I simultaneously attended the Professor's graduate courses in control and communication theory. We were fortunate in being able to circumvent instrumental difficulties and especially in obtaining absolute calibrations and a system frequency response that led to the transfer function described in Chapter 1.

My physiology and clinical colleagues asked me, "Of what use is such elaborate mathematization?" "What does the transfer function tell you that you didn't know before?" "Can it predict a new phenomenon?" Professor Harold Lamport suggested that oscillations were a source of puzzlement to physiologists. This started me on the next stage together with Tom Cornsweet who had helped me greatly with the optical design of the pupillometer. Adjusting gain, again by means of an optical trick, we were able to produce high gain instability oscillations. Another student, Frank Baker, and I changed the pupil system pharmacologically and showed the ability of the transfer function model to predict oscillation frequency quantitatively under these altered conditions.

The optical techniques for opening the loop and producing high gain oscillations, of course, cannot be used in studying other biological servomechanisms, so I spent considerable effort working with Mr. John Atwood in devising the environmental clamping technique which can be used to "dry dissect" in a general fashion. This work on stability and oscillations in the pupil system is described in Chapter 2.

The continual fluctuations in pupil area that occur spontaneously interfere with the precision of one's results, but if they themselves are considered as a second output of the system, they can be exploited to yield more information about the system. Together with Fergus Campbell, visiting me from Cambridge, and John Atwood, we tried to analyze this phenomenon and were directly led into the autocorrelation functions of Norbert Wiener and of the challenging world of digital computers. Professor Morris Davis of the Yale Computing Center provided a gentle and helpful hand and our first experimental analysis, done in the summer of 1957 (with its mistaken argument), is presented in Appendix A of Chapter 3. Further evidence obtained with Frank Kuhl, another student of Professor Schultheiss, is discussed in Appendix B of Chapter 3. Only after many years of work and successive developments of our on-line, real-time computer facility, have we been able to refine our understanding of the experimental evidence in order to resolve the difficulties. Chapter 3 represents research with Dr. Saul Stanten on the stochastic properties of the pupil and shows the essential multiplicative properties of pupil noise, its apparent additive characteristics, ensemble as well as temporal statistics, and physiological analysis leading to a nonlinear system model and a physical and neurophysiological model. *En passant*, we were able to correct the earlier mistaken interpretation by showing a particular normalization implied the assumption of a definite model.

Our present on-line computer system developed out of visits I made in the spring of 1959 to large computer installations situated in various industrial locales. Dr. Walter Bauer and Dr. Montgomery Phister of the Ramo-Wooldridge Co. were interested in our work and arranged for partial donation to Yale University of a TRW-300 computer (drum memory and solid state electronics) with integral analog-to-digital and digital-to-analog converters since this was designed as a chemical process plant on-line controller. Later at M.I.T., we expanded to a GE-225 computer (core memory with special integral analog in-out channels).

The pupil is a class 1 neurological servomechanism and displays all the properties of such a system — essential nonlinearities, noise, continuous operation, output dynamics and range limitations, biological reference adaptation, multiple inputs. In 1959, I modeled its scale compression and asymmetrical nonlinearities and noise characteristics on the Yale computer. Formula evaluation with table look-up or Runga-Kutta method on the quasi-linear differential equations were tried, but it was not until we had a hybrid analog-digital computer simulation operating in our lab at M.I.T. that realistic and useful model comparisons with experimental pupil behavior could be made. The original nonlinear model and equations were described in my 1959 review article [34].

Further definition of the nonlinearities and model studies performed over the last 5 years by Allen Sandberg, Irwin Sobel, and myself are touched upon in Chapter 4, which includes the block diagrams from the 1959 review, first- and second-order Wiener kernels, and a description of an interesting compensatory nonlinearity in the output muscular element. The Wiener kernels represent an attempt to obtain a statistical bound on the behavior of the pupil and our studies are directed, ultimately, toward examining possible physical insights arising from this representation.

We have endeavored to go beyond black-box studies. We have developed "dry dissection" techniques such as environmental clamping, multi-input, multi-output analysis, and pharmacological studies. Although technically difficult to arrange, the "wet dissection" approaches are more direct. Dr. Frank Baker attempted some studies of this type in cats and Mr. John Simpson worked with isolated iris preparations in an effort to understand the mechanical arrangements of sphincter and dilator, and the complex, smooth dynamics.

The pupil remains, after many years of research, an ever-fresh source of phenomena and problems: each new approach is rewarded by the discovery of yet another elegant mechanism in this beautiful organ. I, myself, never dreamed that the pupil, almost 10 years after our open-loop transfer function experiment, would be the subject of studies as diverse as Wiener higher order kernel analysis and multiple-neuron shot noise models for the role of the Edinger-Westphal nucleus in pupillary contraction.

Chapter 1

The Transfer Function

INTRODUCTION

Many biological processes, especially those involving the functioning of the central nervous system, behave as self-regulatory devices or servomechanisms [34, 44]. The pupil reflex to light is an example of such a process. This paper approaches the problem of a quantitative study of applying servoanalytic concepts and techniques [4, 11]. A servomechanism is an automatic regulatory device actuated by the difference of "error" between a desired or reference input and the actual value of output. A controlled quantity is maintained dependent upon reference input despite disturbances within and external to the system. A "loop" is formed as the output quantity is fed back to the input. In Fig. 1 a block diagram of a simple servosystem resembling the pupil reflex is shown. The amount of light

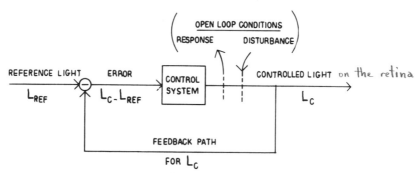

Fig. 1. A simple servosystem. This shows forward and feedback paths in the servo loop and different components therein. Symbols are explained in diagram and text. Dashed lines indicate where loop might be opened. A disturbance could be introduced and response around the loop measured.

flux falling on the retina is the controlled quantity (L_C). A reference light flux quantity (L_{REF}) exists which is compared with actual light (L_C) falling on the retina. Any difference or error (L_C-L_{REF}) which occurs is measured, and it is this quantity which actuates the control system, or pupil neuro-muscular apparatus. The control system then operates, varying pupil size and so changing controlled light (L_C) to reduce error.

It is of interest to note the historical development of these concepts in biology. The operating characteristics of the "milieu intérieur" of Claude Bernard and of "homeostatis" as proposed by Cannon can be restated in such a way as to make the relationship between this development and servotheory clear [3]. A homeostatic mechanism is one which senses the difference between desired and actual states and then puts into force a se-ries of processes which in turn produces opposing effects to minimize er-ror. A more subtle conceptual comparison is made when one transforms the reflex arc into the servo loop by adding a continuous sensory input and thus "feeds back" information concerning final state or "closes the loop" [23]. The title of Charles Bell's 1826 paper, "On the Nervous Circle Which Connects the Voluntary Muscles with the Brain," illuminates the point [2].

More recently, "nervous circles" have been further discussed [7, 8, 13, 22] and several interesting experiments have been performed. In one, Merton applied a quantitative explanation in terms of servotheory to the silent-period phenomenon of the myotatic reflex [14, 15]. In another, Pringle and Wilson determined the transfer function for the response of a sense organ to a harmonic stimulus [21]. These and other experiments [5, 6, 10, 27, 28] suggest the application of servoanalytic concepts to the study of the nervous system.

Experimental techniques with which a servoengineer analyzes an automatic control device are somewhat similar to methods used by the physiologist. First, the engineer draws a block diagram or an anatomical sketch of the system and its functional components, as in Figs. 1 and 7. Then he disturbs the system and traces response through the loop. This procedure requires continuous measurement of responses, as well as quan-titative control over inputs. A further, more sophisticated, technique in-volves "opening the servo loop." We shall now discuss how these methods have been adapted to a study of the pupil nervous system.

The pupil of the eye acts as a regulator of light impinging upon the retina. The transfer function and noise characteristics of this stable-type zero servomechanism (Fig. 1) will be presented in this chapter. The nor-mal behavior of the pupil system can be modified by changing the experi-mental conditions. Then such interesting phenomena as instability and os-cillations can be demonstrated.

The pupil was chosen for study from a host of possible examples of biological servosystems for several reasons [2, 3]. First, its motor mechanism, the iris, lies exposed behind the transparent cornea for possible measurement without prior dissection. This had previously been exploited for scientific and clinical researches by using high-speed motion picture cameras. Further, by employing invisible infrared photographic techniques, measurements can be made without disturbing the system, because its sensitivity is limited (by definition) to the visible spectrum. Second, the system can be disturbed or driven by changes in intensity of visible light, a form of energy fairly easy to control, and painless in its administration to the subject. The first two advantages lead to still a third: the possibility of performing experiments on awake, unanesthetized animals whose nervous system is fully intact and functional. In fact, all of the experiments to be discussed below have been performed on human subjects. Last, the system responds with a movement having only one degree of freedom, a change in pupil area, which simplifies the system equation analysis.

The pupil is so widely observed an organ that most persons are already acquainted with certain basic facts of its anatomy and physiology. The pupil is the hole in the center of the iris muscle which enables light to enter the eye and impinge upon the retina, the sensitive layer of the back of the eye. The retina is comprised of primary sense cells containing photosensitive pigments which trap photons and subsequently stimulate nerve cells. The retina is part of the central nervous system and possesses a complex multineural integrative (i.e., information transforming) apparatus. The optic nerve leads mainly to the visual cortex of the cerebral hemispheres via a relaying station, the lateral geniculate body. However, some fibers, called the pupillomotor fibers, go directly to the brain stem and relay in the pretectal area and thence to the Edinger-Westphal nucleus. This nucleus contains the nerve cells, part of the parasympathetic system, whose fibers (after an external relay in the ciliary ganglion) control the powerful sphincter muscle of the iris. Fiber tracts also go to the sympathetic system in the spinal cord. Here, nerve cells send fibers back to the orbit, after relaying in the superior cervical ganglion. The dilator of the pupil is controlled by these sympathetic fibers and is responsible for the wide dilatation of the pupils after the administration of adrenaline.

Excitation of the Edinger-Westphal nucleus produces constriction of the pupil and it is also probable that inhibition, i.e., decrease in the operating level, of this nucleus is also the most important mechanism for dilating the pupil.

Any further relevant and necessary details of neuroanatomy will be discussed as they occur. The reader is assumed to possess a knowledge of linear servomechanism as presented in a college text such as that of Schultheiss and Bower [3].

EXPERIMENTAL METHODS

In order to obtain careful, quantitative data from the human pupil
system under a variety of experimental conditions, we felt it would be impossible to use the older infrared photographic techniques [5, 6], and developed a simpler modification of this technique. In Fig. 2 the essential
nature of the experimental arrangement of our pupillometer is shown.

The pupil area was measured continuously by reflecting infrared
light from the iris onto a photocell. The pupil is ordinarily black because
most of the light passing through the pupil into the eye is absorbed by pigment layers behind the retina. Thus, when the pupil is large (and the iris
small) less light is reflected from the front of the eye onto the photocell.
When the pupil is small (and the iris large) more infrared light is reflected
onto the photocell. In this manner we obtained a convenient and continuous
measurement of the system response. At first an attempt was made to use
specular reflection from the iris in order to improve linearity of the response, but retinal specular reflection was also obtained and this was not
negligible. Therefore the use of scattered light reflection from the iris
onto the infrared-sensitive photocell was an important part of the experimental instrument design. Another way of increasing the signal-to-noise
ratio was by the use of a relatively small area of infrared illumination of
the iris. This was arranged to be only slightly larger in diameter than the
largest diameter of the pupil. The use of a narrow spectral band of infrared
light shaped to the infrared spectral sensitivity of the photocell also aided
in this experimental approach. The elimination of the long wavelength in-

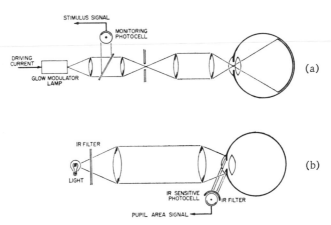

Fig. 2. Experimental apparatus. Two main portions of the
apparatus are (a) servo-controlled light stimulation, and (b)
direct-recording, area measurement by invisible infrared
energy.

frared meant that most of the heat and discomfort to the subject was removed. The output of the photocell, a vacuum-type No. 917, was fed directly
into the high impedance input of the recording amplifier. There was a small
capacitance across it in order to remove high-frequency noise, and the tube
and leads were shielded. The photocell housing could be shifted for studying the right or left eye. The infrared-sensitive photocell was shielded with
an infrared Wratten filter to eliminate the effects of stray visible light.
Whenever beam splitters were placed in the path of the infrared light these
were made dichroic to minimize attenuation of infrared light. The infrared light source was a 35-mm slide projector with a built-in fan for cool-
ing, and it was supplied from a constant voltage transformer to obtain stability of light intensity.

In order to translate the photocell currents into pupil area measurement, it was necessary to calibrate the instrument. Such a calibration is
shown in Fig. 3. This was obtained by taking flash photographs of the pupil
and at the same time noting the amount of photocell current. The flash of

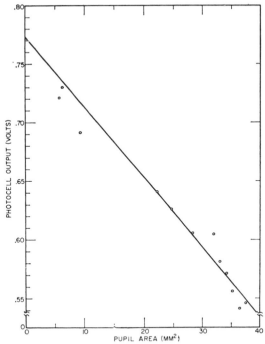

Fig. 3. Calibration. A linear fit was used for simplicity. Calibration is necessary because of variation in iris-infrared reflectance from subject to subject.

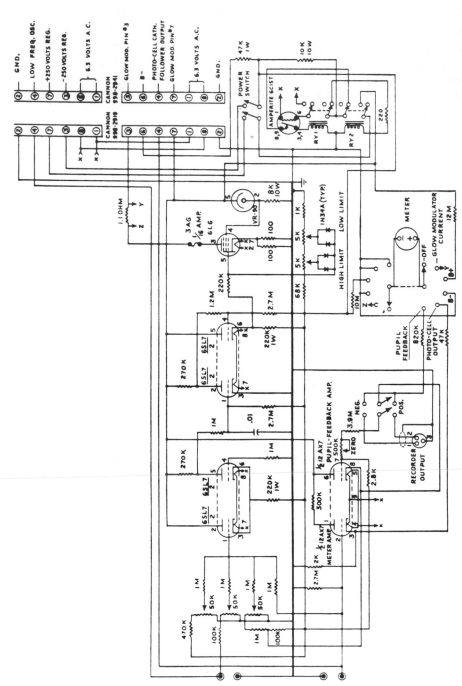

Fig. 4. Circuit diagram of light control unit, a servoamplifier with feedback from output of monitoring photocell which linearizes power amplifier and glow modulator tube.

light naturally produced a pupillary response in the subject but the photo-graphic measurements were over and completed before the pupil had a chance to respond. There had been found a fairly proportional relation-ship between pupil area and photocell current, but in the figure one can see that there is a diversion from linearity. However, this is small and not significant within the range of most of the experiments, and was not cor-rected for. The calibration camera was a permanent part of the apparatus and was a single lens reflex camera with a built-in viewfinder. A dichroic beam splitter was placed so that most of the visible light reflecting from the iris was transmitted to the camera while interfering relatively little with the measuring infrared light. The electronic photoflash light was heavily filtered to permit only blue light to illuminate the iris, a practice which markedly reduced intensity and discomfort for the subject, as well as the effect on the infrared measuring photocell. An event marker auto-matically indicated on the recording graph the instant the photograph was taken. The camera viewfinder had excellent split-image focusing so that the subject could be accurately determined to be in the focal plane of the camera. Further details concerning the positioning of the subject's head and eyes will be given in a later portion of this section. Each individual that was studied in these series of experiments was separately calibrated because of the differences in infrared reflectivity of various subjects' irises. It was noted that subjects with brown eyes generally gave more in-frared reflection at a stated pupil diameter than subjects with blue eyes.

The stimulus light had to be controlled in a quantitative manner. Originally, rotating Polaroid films were used to obtain sinusoidal intensity changes. However, since the original work was reported, an electronically controlled light source has been developed to provide a flexible stimulus system in which light intensity can be varied sinusoidally or in a stepwise fashion, the d-c level may be adjusted, and other experimental arrange-ments which will be discussed later are made possible. The central fea-tures of the stimulating light system are shown in Figs. 4 and 5. Current passing through a glow modulator lamp (Sylvania 1131C) changes the light intensity. In order to linearize the power amplifier and glow modulator lamp, a vacuum photocell (No. 929 with a wide linear range) was inserted into the system to sample a portion of the visible light stimulus. The out-put of this monitoring photocell was fed back into the first stage of the high gain amplifier. Thus, the high gain amplifier and the power amplifier, controlling current through the glow modulator, were made to follow the output of the monitoring photocell. Other aspects of this circuit included a filter to reduce high-frequency response and keep the amplifier stable and the use of limiters to protect the glow modulator tube. Various inputs were used to drive the light source. They included a low-frequency sinu-soidal oscillator, step generator, and d-c level changes.

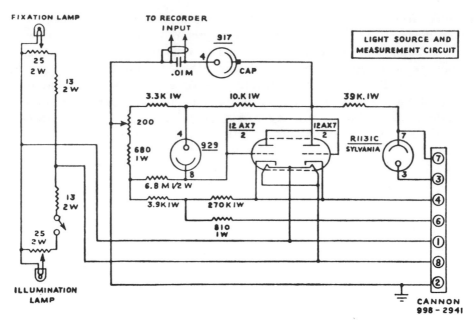

Fig. 5. Circuit diagram of glow modulator monitoring photocell and cathode follower unit.

It was also possible, since the measuring system provided a voltage proportional to the pupil area, to feed back this pupil area voltage to control light intensity. For example, if the pupil became larger we could use the pupil area signal to decrease the light intensity. In this way we could oppose the effect of the pupil system and so open the loop. Again, the polarity of the signal representing the pupil area could be changed and then fed back to the light control amplifier. In this way a higher gain system could be obtained. For example, if the pupil area became smaller, the light intensity would be turned down. In this way the pupil area change would have a much greater effect than it would have from consideration of the optical geometry. It was possible to control quantitatively the amount of pupil area feedback to the light control system. Increasing the gain, as shown later, makes the entire system unstable and oscillations of the pupil system are obtained.

The optical arrangement of the stimulus light was most important. Several arrangements of the stimulating light are shown in Fig. 6. Under normal closed-loop operating conditions light was diffuse so that changes in the pupil area could affect the amount of retinal flux. Provision was made, however, for careful focusing of the stimulus light so that the entire light beam entered the pupil in the form of an image of a small disc whose diameter was smaller than the smallest diameter of the pupil. Changes in

pupil size could then have no effect on flux on the retina. This arrangement of stimulus light, shown in Fig. 6b, produced the open-loop operating
condition by removing the influence of system response over stimulus.

The pupillometer measures continuously the amount of light impinging upon the eye of the subject. In this manner, a continuous pen recording of both stimulus and response is provided. The pen recorder is able
to be modified so that four channels are in use. The extra two channels
exhibit the response of the system in a modified way. For example, if the
gain of the amplifying system is turned up, then small fluctuations of the
pupil can be seen. This high gain measuring system, however, may be
compared with the lower gain continuous record of the regular response
recording channel. An additional channel might display the response after
passing through a narrow band-pass filter, so that the fundamental response
is emphasized.

In order to obtain accurate measurements of a human subject's pupillary area, it is of course necessary that the subject remain absolutely
still. In order to enable a cooperative subject to do this, two varieties of
bite boards were prepared. For most of our careful quantitative experiments on normal subjects, individual tooth impressions made of dental wax
were placed on a metal angle and clamped to the pupillometer. Another
method used was to split and glue a rubberized bite ring onto the metal bite
board. In this way the subject did not need to have a personal bite board
made and this method was widely used for studying neurological patients.
Both of these arrangements were found to be quite satisfactory in preventing head movements as the weight of the subject's head rests on his upper
teeth, and there is a relative fixity in contact between the subject's head
and the apparatus. There is no pressure against the soft parts of the head
such as obtains in a chin rest, and contact of the jaw musculature does not

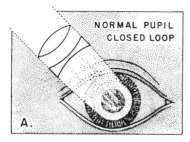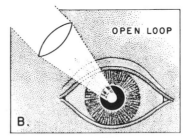

Fig. 6. Two techniques used for stimulation. (a) Normal condition in which
movement of iris changes intensity at retina. (b) Technique generally used for
obtaining amplitude-phase frequency data. Light entering center of pupil is
unaffected by movement of iris. Maxwellian view.

shift the head. In order to keep the subject's eye still, a fixation point was provided with a special light and placed at an infinite distance (optically) from the subject. In this way the subject was able to fixate comfortably on a small spot and relatively little eye movement appeared. The amount of residual eye and head movement could be determined by observing changes in recorded area under experimental arrangements in which no changes in pupil area would be expected. Under these conditions the base line remains extremely stable. Trained subjects were able to remain in the apparatus for several hours, during which time repeatable measurements could be obtained. Untrained subjects such as patients from the Neurology Clinic could also be adapted to the instrument since there was little or no discomfort in the procedure. Removing the heat-radiating long infrared spectrum from the infrared lamp was an important part of the experimental design in this respect. Another detail found to be helpful was the arrangement for quickly changing the configuration used as the fixated point. In this way the subject's attention and his eye could be again brought to focus on the fixation point.

The viewfinder of the calibration camera was used together with a special illumination light which lit up the subject's eye, in order to position the subject carefully and to check on the subject's condition again and again throughout the course of the experiment. Millimeter lines were ruled on the viewfinder and a good calibration could be obtained by estimating the size of the subject's pupil and comparing this with photocell output at that instant. As the apparatus has developed toward more stable operation we have gradually shifted from continuous to stepwise controls. In this way a trained technician can run a careful experiment in a short time on a clinic patient.

TRANSFER FUNCTION ANALYSIS

Initial experiments were performed using open-loop injection of small sinusoidal signal stimuli to obtain a linear transfer function of the pupil system. Studying a system under open-loop operating conditions is an important dissection technique widely used by the servoengineer. This method is indicated in Figs. 1 and 7 by dashed lines representing a break in the servo loop. A disturbance is injected, transmitted around the loop, and measured at the point of break. Thus, system response has no influence over the disturbance, which remains entirely in the control of the experimenter. The input-output relation, that is, the transfer function, is simplified, and for this reason an essential part of this experimental approach to the pupil reflex was the development of the method for open-loop operating conditions discussed in the preceding section describing the pupillometer. There were three reasons for our use of sinusoidal stimuli: experimental

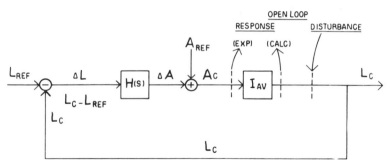

Fig. 7. Linearized approximation to pupil servosystem. Actual pupil system is more complicated than this figure indicates. However, we linearized by using small variations about a fixed operating point. Thus the necessary calculations are simplified and linear servoanalytic methods may be applied. A_{REF} is reference area, A_c is controlled area, ΔA is a change in area generated by control system whose transfer function is $H(s)$, I_{av} is average intensity value used to multiply controlled area to yield controlled light flux. This approximation is explained in text.

techniques for obtaining given level accuracy are simple, the mathematical analysis is well understood and relatively easy to manipulate, and system design and performance are evaluated readily. As an example of the first point, once the retina had adapted to mean light intensity, we were able to vary sinusoidal modulations over the entire frequency range while the pupil system remained in a steady state. The small signal approach enables us to have a linearized approximation to the pupil servosystem. The actual pupil system is much more complicated than Fig. 7 indicates and this will be discussed in a later section. However, in attempting to obtain a value for absolute gain it is necessary to utilize this small signal approach. The pupil system is also very sensitive to operating levels. Thus, by using a small signal, the operating level remains relatively constant, and the response of the pupil system can be made quite reproducible. Furthermore, there are several nonlinear operators in the pupil system; for example, the log transducer which represents the Weber-Fechner law, and the actual geometrical multiplication of pupil area by light intensity to obtain retinal flux. By using small signals we are able to ignore the logarithmic operator, and also to linearize and so substitute a subtraction operator for the multiplication operator.

At this point clarification of the concept of gain and the operational definition given it in our experiments is necessary. Open-loop gain is the ratio of the magnitude of the response of the servomechanism in open-loop condition to a signal injected into it, providing that the signal has completely transversed the loop up to the point of injection. It is of course a dimen-

sionless quantity since the injected signal and the response measured at
the same point must be in dimensionally similar terms. It is irrelevant
where the break in the loop is, as a fundamental theorem in linear servoanalysis proves the identity of all overall loop transfer functions.

The break point in the pupil loop is just before the injection of the
disturbance (see Fig. 7). The disturbance is a change in retinal flux caused
by a change in external intensity. The change of light intensity must be
multiplied by the area of the open-loop disc of light (provided that the intensity is measured at this point).

$$F_e = \overline{A} \cdot \Delta I \tag{1}$$

The response is the change of flux due to the pupil area change which would
have occurred if the pupil area had been allowed to operate on the light entering through the pupil. Thus the change in area is multiplied by the average intensity considered as if it were evenly distributed over the entire
pupil area.

$$F_i = \overline{I} \cdot \Delta A \tag{2}$$

This assumes that the retina cannot distinguish between various light intensity distributions at the place of the pupil, which is reasonably valid for
a Maxwellian view. Reference to Fig. 6, showing the open-loop light distribution, and to Figs. 1 and 7, showing the open-loop break position, will
help to clarify this argument. The small signal approximation is needed
for the gain calculation

$$G(s) = \frac{\Delta F_i}{\Delta F_e} = \frac{\overline{I} \cdot \Delta A}{\overline{A} \cdot \Delta I} = \frac{\Delta A/\overline{A}}{\Delta I/\overline{I}} \tag{3}$$

The numerator (response of the system) and the denominator (injected signal) are both in dimensions of flux; thus a dimensionless gain is obtained.

A sample experiment is shown in Fig. 8. The two oscillatory traces
represent oscillations in light intensity of the stimulus and oscillations in
the pupil area of the response. It will be noted that the response curve
contains noise, and a discussion of this will form a later section of this
paper. There is also some harmonic distortion noticeable and this again
is evidence of the nonlinearity of the pupil system and introduces an error
of up to 10% in our phase shift measurement [46].

The basic data obtained in the analysis of this simple experiment are
the relative amplitude of input and output and the phase shift between input
and output. These two values are obtained for each of a number of fre-

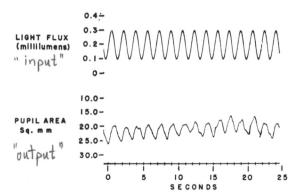

Fig. 8. Pupil response to steady-state, light flux changes. Ordinates and abscissas as marked. This is a typical response showing dominant fundamental response with harmonic distortion, high-frequency noise, nonstationarity of response, and low-frequency drifting. Open loop, 0.6 cycles/sec.

quencies of intensity modulation. In summary, the experimental procedure was designed to adapt servoanalytic methods to study the pupil reflex. Small sinusoidal light stimuli of varying frequencies were applied. The sinusoidal response was measured and its amplitude and phase relationship to the stimulus was determined. Data were obtained in open–loop and closed–loop operating conditions. A number of frequencies could be studied quickly using our pupillometer on a subject in one sitting.

From these amplitude and phase data several displays of system behavior can be obtained. The first is the Bode plot in Fig. 9. From this graph one can readily see certain qualitative features of the pupil system: low gain, steep attenuation of gain at high frequencies, and a large phase shift.

In addition to the above qualitative and semiquantitative discussion of the pupil system, it is also desirable to have a full but concise mathematical description of system behavior. Such a canonical expression is the transfer function, for convenience written as a function of the "complex frequency" operator s. In Fig. 7 loop elements of the transfer function, $G(s)$, are shown. Included are $H(s)$, the transfer function relating area change output to light change input, and the linearized intensity multiplier, I_{av}. $G(s)$ is independent of the actual break point, provided only that the response has traveled completely around the loop to the point of injected disturbance. The data displayed in the Bode diagram will now be used to derive the open–loop transfer function, $G(s)$. Low-frequency gain is 0.16. The attenuation curve appears to have an asymptotic slope of 18 db/octave be-

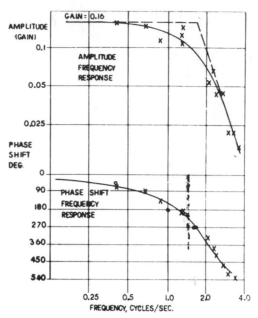

Fig. 9. Open-loop frequency response. Amplitude is plotted on log-log scale while phase shift
(lag) is on log-linear scale. Points are experimental and continuous lines are fitted. Dashed
lines are asymptotes. Vertical dashed line represents frequency at which spontaneous high
gain oscillations were obtained at beginning and
end of experiment with high gain operating condition illustrated in Fig. 11 of Chapter 2 in this
section.

yond the break frequency. This slope can be represented by three time lag
factors. The actual values of the time constants are hard to determine exactly from the experimental data, but as a rough approximation we set each
equal to 0.1 sec. These time lags account for a portion of the phase shift
at higher frequencies. By referring to the phase shift of the Bode diagram,
it can be seen that the actual phase shift is 540° at 4 cycles/sec. The minimum phase shift discussed above accounts for only 270° of this. The remaining 270° can be attributed to a nonminimum phase shift, equivalent to
a time delay of 0.18 sec, expressed as $e^{-0.18s}$.

These calculations now may enable us to write the transfer function

$$G(s) = \frac{0.16e^{-0.18s}}{(1 + 0.1s)^3} \qquad (4)$$

Several parameters in the open-loop transfer function can be compared with data from closed-loop and step experiments. The closed-loop transfer function is related to the open-loop transfer function by

$$F(s) = \frac{G(s)}{1 + G(s)} \tag{5}$$

Low-frequency closed-loop gain is calculated to be 0.14. Closed-loop gain was experimentally determined at several operating points as 0.15. Time delay was measured in step function experiments and found to be 0.18 sec. The agreement shown here is satisfactory, especially considering our difficulty in achieving experimentally a good small signal approximation (10% modulation in open-loop experiments). Further experiments have shown that both the consensual and direct pupil response have a similar transfer function.

CONCLUSION

In order to study biological systems operating in their normal physiological range, it is often necessary to use unanesthetized and intact animals — one very cooperative species is man. Under these conditions the variability of the system, including both noise and adaptive changes, poses problems for input-output analysis that may be met by small steady-state signals and averaging techniques.

An example of this was illustrated in Fig. 9 which displays a steady-state quasi-linear sinusoidal response of the pupil servomechanism, operating in "open-loop" condition, to a small sinusoidal forcing function of illuminance change. The harmonic distortion, the high-frequency irregularity or noise, the time-varying characteristics of the response of the irregularity from cycle to cycle, and the low-frequency drift of the base line may all be noted. All of these can be to some extent reduced by the averaging of the fundamental response over the entire length of the record and extracting only two measurements, gain and phase. The linearized transfer function obtained in this way is a solid basis for understanding the complicated pupil system and a point of departure for studying nonlinearities. Its applicability must be demonstrated, however, when used for operating conditions different from those from which it was derived.

Chapter 2

Stability and Oscillations

INTRODUCTION

The pupil reflex to light falls into the category of control devices called servomechanisms. The feedback pathway which characterizes servosystems necessitates a system analysis in order to elucidate its physiology. By this is meant that there are certain properties such as stability and oscillations which cannot be attributed to any individual component of this system, but are properties of the entire system [34, 36]. In the particular example to which this study is devoted, the nature of "induced pupillary hippus" becomes comprehensible when considered as the sustained oscillations in a pupil servosystem made unstable experimentally by greatly increasing the gain of the loop. Further, the nature of the servo approach permits the design of quantitative experiments to test predictions based on these concepts. The experiments to be described are of this type.

Stability is a fundamental property of a servomechanism which is closely related to other important characteristics such as gain and phase lag. Indeed, servoengineers devote much effort toward eliminating instability in the design of useful servosystems, even paying for stability by reducing the size and speed of response. The characteristic fault or failure of an unstable servomechanism is exhibited as sustained or increasing oscillations.

The pupil servo loop, as shown in work previously presented, is stable because of its low gain despite large phase lags [8]. However, it is possible to increase the gain of the pupil system experimentally. When this is done, the instability of the modified pupil loop shows itself as sustained oscillations with a frequency in accordance with the prediction of linear servoanalytic theory [6]. In this chapter these instability oscillations are studied and their general nature elucidated. Further, the pupil system is changed by means of drugs and it is shown how the oscillations are affected in a manner predicted by alterations in the normal low gain pupil system. The accuracy of these predictions is further evidence showing the validity of the application of linear servoanalytic theory to this neurological servomechanism.

OPTICAL METHOD

The general nature of the experimental arrangement which has been used for the study of the pupil has been described in the previous chapter. In the present chapter, the direct pupil reflex to light has been studied with the use of an only slightly modified instrument which permits both input and output to be measured on the same eye. The experimental procedure was also essentially similar for part of the present experiments.

The difference between open and closed loop has been explained previously and the experimental arrangement whereby open-loop conditions were obtained was described in Chapter 1. This consists in focusing the stimulating visible light on a small disc at the plane of the pupil. The diameter of this small disc is kept smaller than the smallest diameter of the pupil. In this way the pupil cannot modify the stimulus light and the system

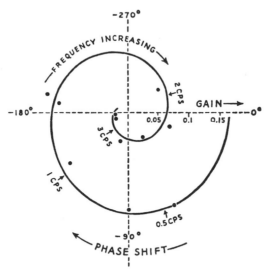

Fig. 10. Nyquist diagram of pupil response, a vector plot of gain and phase shift. Scale of modulus and a few frequencies are indicated. Curve is derived from fitted lines from gain and phase frequency-response graphs, while points are experimental.

94

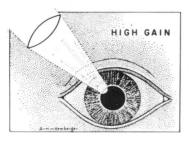

Fig. 11. Technique used for stimulation. Light is here focused on border of iris and pupil. Small movements of iris result in large changes in light intensity at retina.

is operating in open-loop conditions. An additional experimental arrangement has been used in this study. This consists of focusing the visible stimulating light again in a small disc and situating this spot of bright light on the edge of the pupil. Under these conditions, which are illustrated in Fig. 11, very small movements of the iris muscle cause a very large change in retinal illumination. Thus the gain of the pupil system has been greatly increased. The operation of shining light on the edge of the pupil must have been performed many times by ophthalmologists studying the anterior chamber of the eye by means of a slit lamp. However, the first published description of such a procedure was that of Stern in 1942 [47]. Further studies with this type of stimulation were performed by Campbell and Whiteside [6] and by Wybar [53].

A further set of experimental conditions resulted from the application to the eye of topical autonomic drugs, which enabled us to have an altered pupil system on which we could also test the linear servoanalytic approach.

EXPERIMENTAL RESULTS AND ANALYSIS

Another set of experimental data using the direct pupillary response was obtained with the same characteristic long-phase lag and shape of the amplitude curve as was found previously in the consensual pupil reflex to light. In addition, another method of displaying system behavior or the consensual pupillary system is the Nyquist diagram shown in Fig. 10. This vector plot of gain and phase angle is often employed to show clearly the desired characteristics of a servomechanism: stability of operation with adequate speed and accuracy of response. Stability means that the system must not oscillate excessively as it attempts to correct for error. Other characteristics of interest are degree of stability and range of frequency over which the system will respond. Enclosure of the critical point at 180° phase shift and unity amplitude by the Nyquist curve indicates the system instability and predicts sustained or divergent oscillation in response to any disturbance. The fact that the curve in Fig. 10 does not so enclose the critical point indicates that the system is stable. Furthermore, the degree

of stability of the system can be determined by the distance of the curve to
the critical point. Since the gain is 0.12 at 180° phase shift, the pupil system is clearly very stable. The frequency at which 180° phase lag is obtained is called the 180° phase crossover frequency; in this experiment it
is 1.4 cycles/sec.

When the experimental arrangement was shifted to the high gain operating condition, the pupil developed sustained oscillations. A sample of
such sustained oscillations is shown in Fig. 12 where pupil area is plotted
as a function of time. It should be carefully noted that light intensity was
kept constant throughout this experimental operating condition as shown in
the figure. Thus this oscillation is very different from the driven oscillation whereby the frequency-response data are obtained. It was almost possible to produce these oscillations in normal individuals although in some
cases an adjustment of light intensity has to be made. The frequency of
these high gain oscillations was determined by a straightforward method of
averaging several waves and the vertical line in Fig. 13 or is a plot of
this frequency on the phase-frequency diagram. It can be noted that this
is quite close to the frequency at which 180° of phase lag was found in the
normal low gain pupil system. In Table I are shown the results of a
number of experiments of the sort described above with the 180° phase
crossover frequency and the high gain oscillation frequency noted.

Although the association of these two frequencies is quite close, it
was determined that an effort should be made to alter the pupil system and
measure the frequencies in the altered pupil system. It was felt that this
task of being able to shift the two frequencies in parallel manner would be

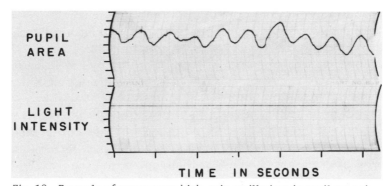

Fig. 12. Example of spontaneous high gain oscillations in pupil area obtained with constant light stimulus using high gain operating condition illustrated in Fig. 11. Frequency in cycles/sec of several such runs determined and averaged to obtain high gain oscillation data given in Tables
I and II.

Table I. Frequency Data in Normal Subjects

Experiment	180° Phase crossover frequency in cycles/sec	High gain oscillation frequency in cycles/sec
JM351	1.9	1.5
JM332	1.6	1.4
JM331	1.4	1.5
PA322	1.4	1.4
CM311	1.5	1.4
SB291	1.6	1.5
JM289	1.4	1.3
BC286	1.4	1.4
BM282	1.4	1.3
DL277	1.5	1.5
JS275	1.5	1.0
FP271	1.4	1.4
SS270	1.5	1.0
Average	1.50	1.35

a more stringent set of conditions which should be met if one were to have
confidence in the linear servoanalytic approach. Therefore several topical
autonomic drugs were applied to the eye in the following way: a 1% solution of hydroxyamphetamine hydrobromide was applied as five successive
1-drop doses spaced by 2-min intervals. In approximately 5 to 10 min the
pupil was widely dilated. Then a 1-drop dose of a 0.1% solution of eserine
was given. This had the result of reducing pupil diameter again to about
normal, that is, 3–4 mm. Thus many parameters of the pupil system such
as d–c operating level (average pupil area) remain constant. The pupil
seemed to respond fairly well to both direct and consensual stimulation but,
as can be shown in Fig. 13, a quite definite change had occurred. These
pharmacological studies have been used mainly as a tool to illustrate clearly
the correlation of the frequency of the high gain oscillation with the 180°
phase crossover frequency. However, it is possible to use this sensitive
system to define drug action. Figure 13, which is the phase curve of the
Bode diagram, plots the experimental results from these drug experiments.
The solid line represents the pupil and its high gain oscillation in an experiment before drugs, and the dashed line represents the pupil of the same
eye 30–60 min later after the application of drugs. It will be noted that
the high gain oscillation frequency and the 180° phase crossover frequency
shifted in a parallel fashion. (The undrugged pupil remained unchanged.)
The results of several experiments of this type are grouped and displayed
in Table II where a consistent correlation can be noted.

Table II. Frequency Data from Drug Experiments

Experiment	180° Phase crossover frequency in cycles/sec		High gain oscillation frequency in cycles/sec	
	Normal	Drug	Normal	Drug
FB16N	1.5	1.0	1.35	1.0
ML14N	1.55	1.15	1.45	0.95
FB220	1.45	0.90	1.45	1.0
Average	1.50	1.02	1.42	0.98

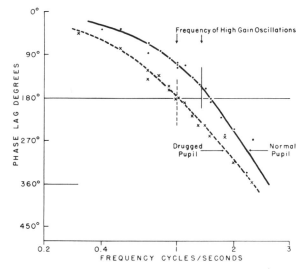

Fig. 13. Data from experiment (FB220) with the use of drugs to shift system parameters. Solid curve represents phase-frequency data from normal pupil in low gain operating condition. Solid vertical line marks frequency of spontaneous oscillation in high gain operating condition which corresponds to 180° phase crossover frequency. Dashed curve represents phase-frequency data from drugged pupil in low gain operating condition. Dashed vertical line marks frequency of spontaneous oscillations of drugged pupil in high gain operating conditions, again showing correspondence with the new 180° phase crossover frequency.

DISCUSSION

It has already been noted that stability is a fundamental property of a servomechanism. The characteristic fault or failure of an unstable servomechanism is exhibited as sustained oscillations or the process of building up to violent oscillations limited in magnitude only by the limits of linearity of the servomechanism. Among the methods that servoengineers have developed for dealing with this problem, the Nyquist stability criterion stands highest because of its graphic representation of the physical state.

The Nyquist criterion has a fundamental mathematical meaning based upon the analysis of the characteristic equation of the system and the distribution of the roots of this equation. If the roots of the characteristic equation do not have a positive real component, then the response of the system to a bounded input will approach zero at some finite time; thus the system is stable. When the roots of the characteristic equation have positive real values, then instability exists and the system will respond to a bounded input by giving an unbounded response with increasing time. Although the Nyquist criterion is based upon this mathematical consideration, it allows one to analyze the system stability without solving the equation by using a graphic display of the open loop response determined experimentally. Furthermore, unlike the analysis of the root distributions themselves, this criterion gives a visual indication of the degree of stability and indeed suggests methods to correct for possible instability in the design of a control system.

The Nyquist plot as shown in Fig. 10 plots the open-loop response of the system as a vector quantity with gain as modulus and phase shift as angle. The critical point representing a gain of 1 at 180° phase lag is not encircled by the curve of a stable system. This is illustrated by the inner, small curve in Fig. 14 and the Nyquist curve in Fig. 10. In an unstable system, in particular in the pupil system which has been modified as in the high gain experiment described above, the Nyquist curve encircles the critical point. This is shown in the outer, large curve of Fig. 14 where the gain of the system is greater than 1 at a frequency where 180° phase lag is obtained. Any error is added to and enlarged each time the signal goes around the loop, because at 180° phase lag the control device is operating in the opposite direction to that required to cancel out the error. This regenerative or positive feedback produces system instability and oscillations. Thus it is possible to determine the degree of stability from the graphic display of the open-loop system data in the Nyquist diagram.

In our experiment, the gain of the system was increased by changing the light distribution at the plane of the pupil as shown in Fig. 11. This changes no other system characteristic but gain. Thus the Nyquist plot of the system response will have only the scale of the modulus changed. We are, of course, assuming that no gross frequency-dependent nonlinearities develop with these changes in the operating condition of the pupil. Figure 6 shows both the Nyquist diagram of a low gain stable system and the Nyquist diagram of a high gain unstable system. Frequency is a monotonically in-

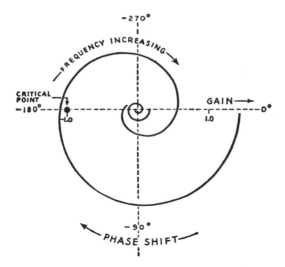

Fig. 14. Illustrative Nyquist diagram comparing plot
of unstable (large, outer curve) with stable (small,
inner curve) servosystem. Curve of stable system lies
between Nyquist critical point and origin, indicating
stability. Conversely, curve of unstable system en-
closes critical point, indicating instability. As noted
in text, critical point is that point in graphical plane
of open-loop transfer function which represents gain
of 1 at phase lag of 180°. Since no frequency-dependent parameters have been changed, 180° phase
crossover frequency of high gain and low gain systems are identical.

creasing function around the curve. Since, to a first-order approximation,
only the scale factor has been altered in this transformation, we would expect the high gain unstable oscillations to develop at the frequency of the
180° phase crossover of the low gain system. Indeed, this has been our
experimental finding. Such is the Nyquist stability criterion and its application in the pupil servosystem.

    The Nyquist stability criterion as well as the ability to use frequency
response experiments depends on the assumed linearity of the servosystem studied. By this is meant that if twice the magnitude of a signal is injected, then twice the response will be observed; in other words, the theory
of superposition holds for this system. It is realized of course that almost
all biological and indeed physical systems must be nonlinear. However, it
is important to attempt to approach these fundamentally nonlinear systems
by means of a linear approximation, because only in the linear field are we
able to transform and manipulate relationships in an easy manner. There-

fore in our experiment we used small signal approximations to aid in this linearization approach. This avoided large signal responses, permitted us to make an approximation for the multiplication function in the control mechanism [8], and enabled us to neglect the harmonic distortion and other frequency-dependent nonlinearities. The accuracy of the prediction obtained from this straightforward linear analysis is a good estimate of the accuracy of the approximation of a linear model to the real pupil servosystem. Thus the evidence presented here strongly justifies our application of linear analysis to the pupil servomechanism.

SUMMARY

Induced pupil hippus is an oscillation in an unstable pupil servomechanism. The Nyquist stability criterion is utilized to demonstrate how this instability is experimentally achieved by increasing loop gain. Quantitatively accurate predictions from servoanalytic theory have been put forward and confirmed experimentally. This validates the use of the linear servoanalytic approach in this analysis of the pupil reflex. The direct pupil reflex to light has been studied and compared with previous studies on the consensual pupil reflex. Certain autonomic drugs have been found useful in modifying system parameters.

ENVIRONMENTAL CLAMPING METHOD

The optical trick for increasing gain and indeed for opening the loop of the pupil system was a fortunate possibility which enabled the experimenter to perform a "dry dissection" of the system. However, a favorable situation like this is not always likely or even possible for most systems and we developed a more generally applicable method which is next described. This method enables the experimenter to clamp the system if (a) the output is measurable; (b) the input can be a controlled physical quantity; and (c) an optical mechanical electrical apparatus can be constructed with more dynamic range and power and faster response time than the biological system itself.

This other experimental approach, different from straightforward input-output analysis of Chapter 1, is applicable when the biological system under study can be controlled by artificial physical apparatus with greater range and power. The method consists of "clamping" the value of a particular variable or parameter in the mechanism to be studied; an example is the squid-axon-membrane voltage clamp of Marmont and Cole [10, 25].

A generalization of the "clamping" method and an experimental application to the induction of instability oscillations of the pupil system will be described. In essence, the system loop is opened and a generalized filter

capable of variable phase lag and gain is inserted, and this new loop re-
closed. The combined system can then be driven to in stability oscillations,
as described in this paper.

Part of the experimental results will be shown to justify the applicability of the linear transfer function to the domain of "clamped" behavior;
another part will show the limitations of such application due to important
nonlinearities. The quantitative argument will be developed so that it is
clear to those who, while not specialists in control-system theory, still are
somewhat conversant with complex function theory.

The experimental apparatus used in studying the pupil light servomechanism as discussed in Chapter 1 has means for forcing the light stimulus to be proportional to a given voltage (V_L), means for producing a
voltage proportional to pupil area (V_A), and means for opening the pupil
loop. If V_A, the pupil-area voltage, is fed to a "clamping box" with adjustable gain and phase lag, and then to the light-control amplifier, a new closed
loop is formed which is composed of the pupil system and the "clamping
box" in tandem (Fig. 15).

The electronic means for accomplishing the gain and phase changes
are relatively straightforward. The gain control is a potentiometer that
adjusts the area signal voltage (V_A) which is brought into the input stage
of the light-control amplifier. The phase control is a symmetrical lattice
circuit with variable capacitances, and impedance-matching amplifiers.
This enables one to vary phase lag over 180° without altering gain. A crossed
bipolar switching system also permits an additional 180° of phase change
to be introduced.

System Equations

The incremental dimensionless gain G of the pupil is defined as the
ratio of the magnitude of retinal-luminous flux change due to alteration in

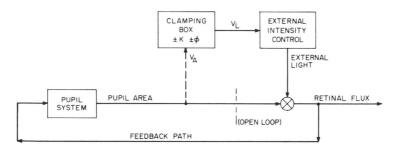

Fig. 15. Block diagram of the "clamping box" arrangement. Note the
new closed loop formed from the pupil system and the clamping box ex-
ternal-intensity control in tandem. The locus of the "cut" for opening
the pupil loop is also shown.

pupil area to the magnitude of retinal–luminous flux change due to external illuminance variation. That is,

$$G = \Delta F_i / \Delta F_e \tag{1}$$

where ΔF_i is change of flux due to pupil–area alteration, and ΔF_e is change of flux due to external light variation.

If we assume small signal conditions and ignore second–order changes, this can be written as

$$G = \Delta A \cdot \overline{I} / \Delta I \cdot \overline{A} \tag{2}$$

where ΔA is change of pupil area; $\overline{A}$, average pupil area; ΔI, change of external illuminance; and $\overline{I}$, average illuminance at the plane of the pupil. In open–loop operating conditions, we set $\overline{I}$ as equal to the average illuminance considered as if the light were distributed over the entire area (see Fig. 6b). This assumes that the retina cannot distinguish between distributions of light at the plane of the pupil.

The open–loop transfer function of the pupil system as a function of the Laplace transform variable "s" has been found to be

$$G(s) = 0.16^{-0.2s} / (1 + 0.1s)^3 \tag{3}$$

with some changes in parameters for different subjects and particularly for patients with neurological abnormalities. The low–frequency range where retinal adaptation dominates will be discussed elsewhere and is not included in this transfer function.

An alternate form of the equation defining gain is

$$G = (\Delta A / \overline{A}) / (\Delta I / \overline{I})$$

or

$$\text{percent area change/percent illuminance change} \tag{4}$$

We have found gain as defined above to be inversely proportional to the logarithm of illuminance change (saturation), but to be relatively little affected by average area or average illuminance. It is clear that the ratio of illuminance values at the plane of the pupil is equal to that ratio at the plane of the retina, even though absolute magnitude of illuminance changes considerably as the stimulus light diverges to include a large solid angle of 30° to 50°. Hereafter, illuminance may not be specified as pupil or retinal because the difference is not crucial.

Stability

From the system equations it can be seen that the pupil servo loop, close to its operating frequency range, contains a good deal of phase lag contributed both by minimum phase elements, i.e., the triple lag, and by non-minimum phase elements, i.e., the transport delays. It is stable, however, because of the less-than-unity gain. From general considerations, such as the Nyquist stability criterion, it can be demonstrated that if system loop gain were increased sufficiently to unity, then oscillations would occur at the 180° phase crossover frequency, about 1.4 cycles/sec.

The gain of the open-loop pupil system G_p multiplied by the gain of the clamping box G_{cb} equals the gain G_{ts} of the combined tandem system and this must equal unity during the condition of sustained oscillation.

$$G_p \times G_{cb} = G_{ts} = 1 \tag{5}$$

The phase lag of the combined tandem system G_{ts} equals the algebraic sum of the phase lags of its component parts,

$$\varphi_p + \varphi_{cb} = \varphi_{ts} = 180° \tag{6}$$

and this sum must equal 180° during the condition of sustained oscillations.

In previous papers we have demonstrated experimentally the production of instability oscillations by the use of an optical method of increasing gain. The frequency of those oscillations agreed with that predicted from the linear-system equations. However, the optical approach is not as powerful as the environmental clamping method of the present paper. It lacks the ease of experimental manipulation as well as the possibility of extension beyond simple change in gain. These features of the present system will be discussed in conjunction with the experiments described below.

EXPERIMENTAL RESULTS

High Gain Instability Oscillations

With the experimental arrangement as discussed above, illustrated in Fig. 15, the gain control of the clamping box was carefully turned up until the occurrence of oscillations that could be easily maintained was noted. These oscillations are illustrated in Fig. 16.

One difficulty present in these experiments is the lack of control over mean pupil area, and thus mean retinal illuminance. If it were possible to control mean light intensity (and allow relatively small changes in mean pupil area to be ignored), gain would be determined entirely by potentiometer setting on the control box. Under present experimental conditions, the

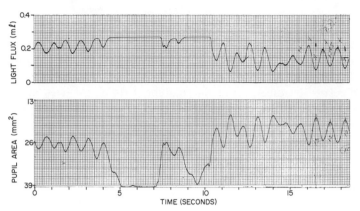

Fig. 16. Instability-oscillation experiment. The development of
maintained sinusoidal oscillations can be noted. Record selected to
show difficulties which can arise, namely, saturation of range of
light-intensity control by slow fluctuations of area. The absence of
phase shift in electronic portions of tandem system is clearly illustrated. Pupil phase lag is 180°; frequency, 1.4 cycles/sec; pupil
gain, −6 decilog.

illuminance variable, since it was multiplied by the external gain of the
clamping box, often exceeded the operating range of the stimulus lamp and
then saturated. An instance of this is shown in Fig. 16. Also, since gain
as defined includes mean area and mean pupil illuminance, the gain of the
entire tandem system fluctuates with these base-line fluctuations. It is almost always possible, however, to cause sustained oscillations to develop,
and indeed Fig. 16 is selected to be representative rather than excep-
tional.

It should be noted that the oscillations are more sinusoidal and have
less high-frequency noise and less harmonic distortion than the optically
and classically driven oscillations of Chapter 1. This is due to the self-
filtering action of the resonance peak, which forms as the system becomes
unstable and acts as a quite narrow band-pass filter.

Because the base line shifts, it is necessary to calibrate during these
experiments and to determine open-loop pupil gain as defined in Eqs. (1),
(2), and (4). Gain of the "clamping box" (area change driving pupil–illuminance change) is clearly the reciprocal of the open-loop pupil gain because
the complete "closed loop" formed by the pupil system and clamping box
in tandem must have a gain of 1 if sustained oscillations exist [Eq. (5)].

The phase lag of 180° between illuminance change and area change is
due entirely to pupil phase lag as the normal electronic phase lags are negligible [Eq. (6)]. This is seen to be so in Fig. 16. The frequency can be

easily measured and is, as predicted from the transfer function, about 1.4 cycles/sec.

Phase Changes

An important further generalization of the method is to introduce phase lags or leads by means of the clamping box. For example, if -20° phase lag (identical to a +20° phase lead) is added, the pupil system must then contribute 200° phase lag to obtain the 180° total phase lag [Eq. (6)]. This means that the pupil–clamping box system must oscillate at that frequency at which the pupil system has the value of phase lag which it is contributing to the total phase lag. Figure 17 shows such an experiment and it is again seen that the oscillation is at a frequency which conforms to the prediction from the transfer function. Because of the use of the lattice network for phase changes, the actual phase shift introduced by the clamping box is a function of the frequency of oscillation. In the experimental analysis procedure, the oscillation frequency is first measured, the phase lag introduced in the clamping box determined from a calibration graph, and then phase lag contributed by the pupil system deduced. However, because both area and light fluctuations are recorded, these computations can be checked against the actual measured-phase relations. As expected, these agree completely.

The gain of the clamping box in the experiment of Fig. 17 is 3.3 so that the pupil system gain must be 0.3, again in reasonable agreement with prediction. It should be recalled that low–frequency gain indicated as 0.16 in the transfer function in Eq. (3) is, in fact, a variable parameter (inversely proportional to the logarithm of input illuminance change).

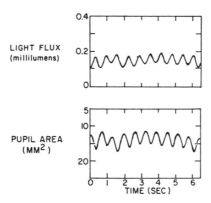

Fig. 17. Instability oscillation with introduced phase shift. Phase shift due to electronic system is 20°. Pupil phase lag is 200°; frequency, 1.6 cycles/sec; pupil gain, − 5 decilog.

Double Oscillations of Pupil Servomechanism

Figure 18 shows another interesting phenomenon. Here, two oscillations develop simultaneously, each with its own gain, phase lag, and frequency. This is possible because the migration of the pair of complex poles over the imaginary axis of the root-locus plane is restrained by nonlinearities (logarithmic saturation). Further gain increase then permits another pair of complex poles to reach the imaginary axis. We have noted earlier the large phase lags in the pupillary system, suggesting a sufficient supply of poles. This is further assured because of the nonminimum phase transport-delay element.

The pupil-area curve of Fig. 18 can be approximated by Eq. (7). The phase-plane plot of Eq. (7) is shown in Fig. 19.

$$p(t) = 3.5 \sin 2\pi (0.183)t + \sin 2\pi (1.28) t \tag{7}$$

The oscillations of Fig. 18 were achieved by the clamping method. The phase lag of the "clamping box," however, as described above, is a nonlinear function of frequency.

When the "clamping box" is composed of an amplifier and a delay line, arbitrary gain and phase as a linear function of frequency can be induced. The laboratory digital computer was programmed and operated to be a delay line of variable delays. By using the known phase response of the pupil servomechanism, the additional phase delay required to produce a total of 180° and 540° can be calculated. These predicted frequencies

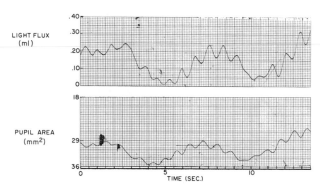

Fig. 18. Double-instability oscillations illustrate certain nonlinear features of pupil system as explained in text. For rapid oscillations pupil phase lag is 203°, frequency, 1.3 cycles/sec; pupil gain, −11.1 decilog. For slow oscillations, pupil phase advance is 65°, frequency, 0.18 cycles/sec; pupil gain, − 9.3 decilog.

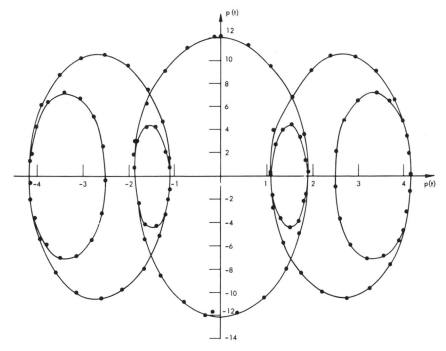

Fig. 19. Phase-plane plot of Eq. (7).

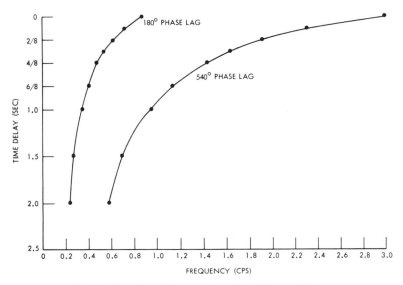

Fig. 20. Predicted frequencies of double oscillations.

of oscillation as a function of time delay are shown in Fig. 20 for 180° and 540°.

COMPARATIVE RESULTS

The Bode plot of Fig. 21 summarizes data from a series of experiments such as those described above. The driven-response results, from which a standard transfer function might be obtained, are also shown for comparison. There is good agreement between phase data obtained from small-signal steady-state experiments and phase data obtained in the clamping experiments. This suggests that the range of applicability of the transfer function (linear equation) can be extended to the phase behavior of the system in the clamped operating condition.

Gain data show less agreement. This is because of the presence of a known nonlinearity, logarithmic saturation, which acts to reduce pupil gain for larger inputs. Conversely, instability oscillations which are small in amplitude permit higher pupil gains to occur under actual experimental conditions. The steady-state gain curve appears to be the lower bound of the instability-oscillation pupil gains. Although the pupil contribution to instability-oscillation gain is not identical in a straightforward way with the values of pupil gain obtained in the classical manner, both show less-than-unity gain, which fact is important evidence concerning the pupil system and supports the definition of the pupil gain.

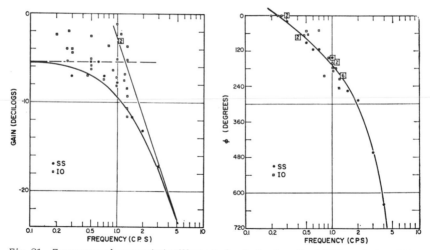

Fig. 21. Frequency characteristics illustrated as Bode plot of both high gain instability-oscillation experiments (squares) and driven-response experiments (filled circles). Heavy solid lines are empirical fits to steady-state experiments; the dashed and thin continuous lines are asymptotes. Numbers indicate number of different experiments whose values fall too closely together to be plotted separately.

Several points should be noted. Because of the limited, controlled illuminance range, and the multiplicative effect of area level on light level, the gain of the clamping box could not always be made sufficiently high to produce instability oscillations. The clarity of the low-frequency oscillations profited by the self-filtering action of the random system at unity gain. Valuable records of very low phase lags, and even phase advances, could thus be obtained with the clamping method, whereas the driven system experiment is hampered by nonlinearities and noise in the low-frequency region.

DISCUSSION

Often it is difficult to dissect a biological system and still maintain its operation in a physiological state. In these cases it may be worthwhile to extract a maximum of information concerning the system by just such manipulative experiments as illustrated by sinusoidal driving input-output and by environmental clamping control. Indeed, by suitably arranging operating conditions and by utilizing the multiple-input pathways that most biological systems possess, a good deal of analytical dissection is possible. By this is meant that a rational-parameter differential equation can be constructed to represent dynamical system behavior in which each element of the equation is related to the physical laws of a particular element in the biological system. Further, because the system equations have been derived from system behavior in physiological operating conditions, the elements are also so defined. Then comparison with studies of isolated physiological preparations (nerve, muscle, synapses) can demonstrate if the physiological behavior represented by the corresponding portion of the system equation can be fully understood from the single cell or isolated tissue function.

The environmental clamping method is complementary to the classical input-output technique which employs sinusoidal, transient, or stochastic driving functions. It has some advantages, which rest mainly on control of and the resultant accessibility of the biological system. For example, the regenerative action of the action potential precluded straightforward analysis and had to be circumvented by the axon clamp. Another advantage will be discussed below in connection with the interaction of linear and nonlinear attributes of a system. Finally, the simplicity of oscillations for routine data-reduction methods suggests the utility of the clamping method in clinical testing.

A still further generalization of this method would be to put a rational-parameter simulation model of the system under study in a feed-forward path parallel to the real system. Any output difference could be used to vary model parameters, and in this way the model would gradually be brought to match the real system identically. Instead of input–output functions, the experimental results would consist of a table of parameters fully describing the real system.

The problem of general analysis of a nonlinear system is an open one. The use of linearizing techniques is a powerful approach, but must be experimentally justified wherever applied. Prediction of phase and frequency relationships in the environmental clamping situation argues for the value of the pupil transfer function, and demonstrates that the linear properties of the physiological and artificial control mechanism determine the frequency of the oscillation. Conversely, other aspects of the oscillation (amplitude, shape) depend upon nonlinear properties of the pupil system. Some nonlinear properties are, in fact, most clearly displayed during such experiments. Thus the linear servoanalytic approach is valuable not because it ignores nonlinear phenomena, but because it provides such a firm foundation from which to attack these phenomena experimentally and conceptually.

Chapter 3

Pupil Noise

INTRODUCTION

The random fluctuations of human pupillary area provide an interesting example of a stochastic process in a biological system. Statistical communication theory and system analysis have guided the authors in performing experiments which reveal important properties involved in the random process. The purpose of this paper is to present the various experiments performed during this investigation and to propose a system model which not only describes the average pupil response to a given stimuli but also the random fluctuations called pupil noise [33]. We classify these fluctuations as noise in the sense that noise usually connotes a signal which does not seem to serve any useful purpose. No attempt will be made here to describe the pupil physiology nor to present the mathematical foundations of the statistical theory. (For this background information, the reader is referred to references 12, 58–61 and Chapters 1 and 2.)

EXPERIMENTAL METHODS

This section will describe briefly some of the experimental techniques and apparatus used to perform the experiments described in this chapter.

The apparatus used in most of the experiments was the pupillometer originally described by Stark [34]. Most of the major features of the pupillometer are shown in Fig. 22. The pupil area is measured continuously by reflecting infrared light from the subject's iris onto a photocell. Since the pupil absorbs most of the light shined upon it, the reflection is essentially from the iris. Therefore, the larger the pupil area, the smaller the photocell current. In addition to the infrared measuring system, a visible light stimulation system built around a Sylvania 1131C glow modulator tube is used.

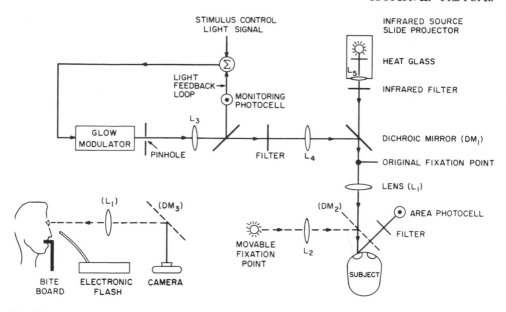

Fig. 22. Pupillometer used to monitor continuously pupil area, including infrared measuring light source, glow modulator with linearizing feedback stimulating source, optical system, fixation points, and calibration arrangement.

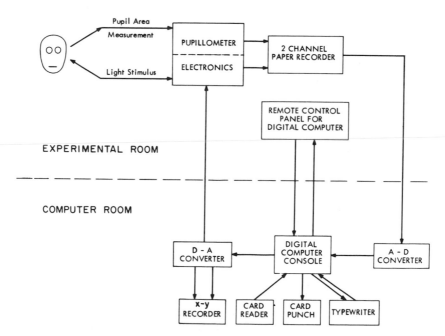

Fig. 23. Block diagram of experimental arrangement including pupillometer, computer and auxiliary equipment, remote control panel, and telephone lines.

An innovation was added to the apparatus in order to provide a controllable fixation point so that the pupil area, which changes with focal point as well as light level, could be changed independently of the illumination level. This addition is shown in dotted lines in Fig. 23 and included a dichroic lens to pass infrared as well as visible light, and a 30-cm double convex lens. The fixation point was constructed from a tiny pinlight bulb, the position of which could be changed in order to obtain different focal points. The total light flux from the pinlight bulb was negligible compared to the flux from the glow modulator tube.

All experiments carried out during this investigation utilized the pupil open-loop system in which the illumination spot, centered in the middle of the pupil, was always kept smaller in area than the smallest pupil area. This was done so that the illumination flux impinging upon the retina would not be affected by the pupil size.

Since the experimental study of a random process requires a large degree of data processing, a general purpose digital computer, the GE-225, was used for both on-line and off-line experimentation. The overall experimental setup, including pupil and computer rooms, is shown in Fig. 23. In order to perform on-line experimentation and to be able to control the computer from the experimental station, a remote control panel which operated in parallel with the computer control switches was installed in the pupil room. It proved to be an extremely useful experimental tool.

All the graphs depicting experimental results were plotted directly from the computer on an X-Y recorder from original data and results. In these computer diagrams the mechanical system of the X-Y recorder tends to smooth out the curves into a continuous curve although the actual data contain a discrete number of points. In situations where curves were fitted to data, the original data points are shown.

The pupillometer was calibrated by taking flash photographs of the pupil and simultaneously recording the photocell current. The recording was performed by the computer by presenting the pupil area signal on one channel of the analog-digital convertor and using the flash as a trigger which caused the computer to sample the pupil area signal and print the results on the typewriter. An example of a typical calibration from pupil area (photograph), to computer number (typewriter output), is shown in Fig. 24. A more detailed description of the calibration technique and the experimental setup is explained in [63]. All the experiments described in this paper were performed on one well-trained subject. The results are typical of ten other subjects and each experiment was repeated a number of times to ensure the integrity of the results.

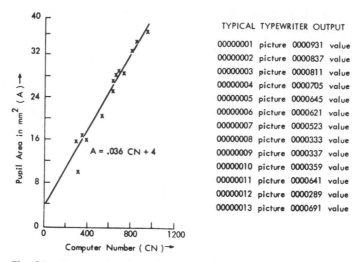

Fig. 24. An example of a calibration relating photographic measurement of pupil area to "computer number" demonstrating linearity
of equipment. Also included is computer typewriter output and calibration equation.

This chapter will present the experimental results obtained during
the course of this investigation. Most of the interpretation concerning an
appropriate model will be presented in a later part of this chapter.

During preliminary work, it was noticed that the pupil noise level
varied with the stimulating light level. Figure 25 shows some calibrated
typical pupil noise records at various light levels, as well as the derivative
of the noise signal. Notice the significant change in the amount of noise at
different light levels. Discontinuities in the signals in this figure are the
result of blinks.

AMPLITUDE HISTOGRAMS

In order to determine the properties of the first-order probability
density function of pupillary noise, an amplitude histogram was calculated,
employing constant illumination conditions. This experiment was performed by first using the computer to digitalize the noise wave form and at a
later time calculating the histograms as the frequency of occurrence of the
amplitude samples.

Figure 26a shows histograms performed directly on the pupil area
signal. Figure 26b represents histograms performed on the derivative of
the area signal used in Fig. 26a. Finally, Fig. 26c shows the histogram
from a different experiment where the pupil area signal was band-pass filtered from 0.3 cycle/sec to 10 cycles/sec in order to eliminate low-fre-

quency drift, which sometimes appears in pupil noise records when the subject tires. The low-frequency drift will tend to create a less confident statistical estimate. The patient will also demonstrate nonstationary characteristics, especially in the experimental mean value. Prefiltering was performed in many cases to eliminate or lessen these effects. In general, it
is very difficult to say how linear filtering changes probability density functions. However, if the original probability density function is Gaussian,
the filtered signal (as demonstrated in Fig. 26) will also be Gaussian.

In all the histograms shown in Fig. 26, 32 quantization levels were
used and data were taken in four independent sections. At low light levels
the gain of the recording device was increased to give the signals a wider
dynamic range in relation to the analog–digital equipment. Notice how in
all cases the experimental curves are close to Gaussian. This really is
not surprising, especially if one can assume that the noise signal is made

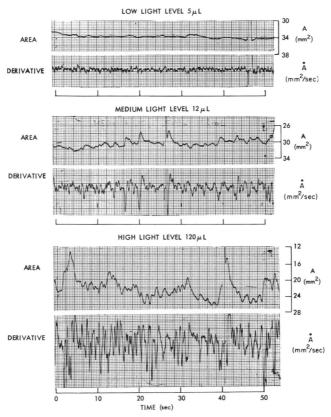

Fig. 25. Pupil area noise and derivative, light levels in microlumens. Eyelid blinks can be seen.

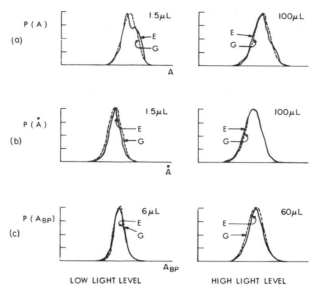

Fig. 26. Normalized amplitude histograms at various light
levels in microlumens, for area signal (A), derivative of
area signal (Å), and area band-pass signal (A_{BP}) demonstrating Gaussian nature of probability density functions. E, solid
lines, represent experimental results plotted by computer.
G, dotted lines, represent Gaussian fitted curve also plotted by computer. Ordinate and abscissa scales in arbitrary units.

up of a large number of small independent events, and thus the central limit
theorem applies. The theoretical Gaussian curves represented by dotted
lines in Fig. 26 were determined by equating the peak values of the experimental and theoretical curves. Because of the closeness of the fits it did
not appear necessary to derive statistics concerning that closeness.

All the curves in Fig. 26 are normalized in order to illustrate the
Gaussian nature of the density function. Figure 27 shows four histograms
at different average pupil areas and at different light levels. Notice in this
representation that noise power, related to the width of the density curve,
increases with decreasing pupil area and increasing light level. These
curves are not normalized and are presented for comparison purposes.
Two histograms at both high and low light levels are shown to illustrate the
fact that the noise level and average area are not completely determined by
the light level. This point is amplified further in the next section in conjunction with the accommodation experiment.

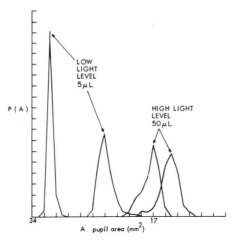

Fig. 27. Calibrated amplitude histograms of
pupil area noise for low and high levels of
light in microlumens demonstrating the
Gaussian nature of the probability density
functions and the change in noise level
(width of density curves) with average area
(peak of density curves). Integrated graphical area under each curve equals 1.

NOISE AS A FUNCTION OF LIGHT LEVEL

The Gaussian nature of the first–order probability density function
indicates that the standard deviation or rms (root mean square) value of
the noise signal may be chosen as the measure of the amount of noise. Of
course, the average value and rms value determine completely the first–
order Gaussian probability density function.

A series of experiments were performed to elucidate the manner in
which noise level varies with stimulating light. The effects of signal drift
were reduced by measuring the noise in many short 10-sec intervals and
averaging the results of the rms calculation on each 10-sec interval. Another technique used was to study the derivative of the area signal. Figure
25 shows typical noise records.

Figure 28 shows the rms value of the pupil noise area signal (σ_A), the
rms value of the derivative of the area signal $(\sigma_{\dot{A}})$, and the average value
of the area signal $(\bar{A})$ as a function of illumination level (L), on a semi-logarithmic scale. Two log units of light level were obtained by using neutral
density filters placed in the path of the glow modulator–stimulating light
source.

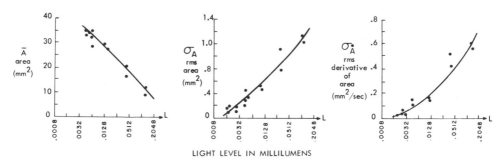

LIGHT LEVEL IN MILLILUMENS

Fig. 28. Average area, $\overline{A}$, noise level of pupil, σ_A, and noise level of derivative $\sigma_{\dot{A}}$ as a function of absolute light level (in millilumens) on a semilogarithmic scale.

The curves depicted in this figure were adjusted by subtracting a constant from the light value. This compensates for the fact that the infrared measuring source contributes some energy to the visible spectrum which is not recorded by the monitoring photocell. Utilizing a calibrated photocell, matched to the spectrum of the eye, the amount of residual light was determined. After compensation for the ratio of average pupil area to the area of the infrared measuring source, the compensated curves were drawn. The logarithmic nature of these curves should be emphasized.

Figure 29 depicts σ_A as a function of $\overline{A}$. Notice the approximately linear relationship. Clearly, physical considerations lead us to the conclusion that as $\overline{A}$ becomes smaller, the noise level must eventually decrease. If $\overline{A} = 0$, then $\sigma_A = 0$, since the pupil area can never become negative. Although this argument seems clear, the effect of decreasing noise with very small $\overline{A}$ has never been noticed, although the σ_A vs. $\overline{A}$ curve has been seen to flatten out on occasion, as $\overline{A}$ became small.

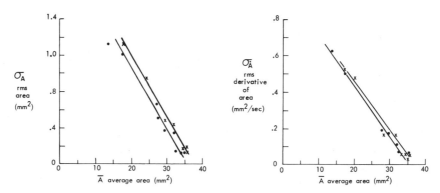

Fig. 29. Noise and noise derivative as a function of average pupil area showing linear relationships. Dots and x's represent separate experiments.

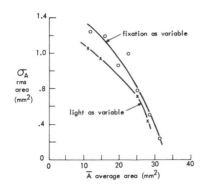

Fig. 30. Noise as a function of average area, both with accommodation (fixation) and with light level as area controlling variables, showing its multiplicative nature.

Figure 29 indicates that during the second experiment there might have been a slight d-c level change in one of the amplifiers, since the average area seems to have changed by a constant value, while the derivative remained relatively the same. In any case, the results of both experiments agree fairly well.

To determine whether the illumination level or the average pupil area was the determining factor in the generation of the noise level, it was necessary to perform an experiment in which the light level was kept at a low value and the pupil size changed by employing the accommodation response. In this experiment we utilize the fact that the average pupil area decreases when the focal plane of the fixation point is brought closer. The results of this experiment and a control experiment in which the fixation point was at optical infinity and the light level changed are presented in Fig. 30. Notice that both curves of noise vs. area, with fixation and light as independent variables, seem to coincide. This particular experiment shows noise with fixation as a variable having a slightly larger rms value. This is quite reasonable when one realizes that it is very difficult for the subject in the pupillometer to maintain close fixation for any length of time. Just the action of varying fixation slightly will add a little more effective pupil noise.

This experiment indicated that the noise is introduced or generated in that portion of the brain common to both the pupil light reflex system and the accommodation response system. The fact that light level is not the only noise-controlling factor has been clearly demonstrated. A comparison between Fig. 27 and Fig. 26c will give the reader a feeling of the variability in experimental results over a period of weeks.

MULTIPLICATIVE NOISE AND THE SINUSOIDAL DRIVE

The approximate linear relationship between σ_A and $\overline{A}$ indicates that the noise is multiplicative, in the sense that the noise is gated into the pu-

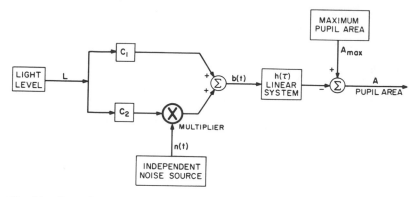

Fig. 31. Simplified model for pupil noise showing multiplicative gating of the noise source as well as a linear system to account for spectral characteristics of noise.

pil control loop in a manner proportional to the stimulus, whether the stimulus be light or fixation. In most experiments light was used as the controlling variable, since it is easier to use and control in a quantitative manner. The model shown in Fig. 31 will yield the linear relationship between σ_A and $\bar{A}$. The two paths in the model of Fig. 31 are necessary because if only the C_2 path existed, the average value of the pupil area, $\bar{A}$, would always equal A_{max} because multiplying a single path by the average of $n(t)$, which is assumed to be zero, would yield an average value of $b(t)$ of zero. The C_1 path allows $\bar{A}$ to be proportional to the average value of L.

To investigate further the multiplicative nature of the noise, the following experiment was performed. The pupil was stimulated by a sinusoidally varying light. The autocorrelation of the response was obtained and the d-c and additive cosinusoidal components subtracted. This was performed by first calculating the autocorrelation of the stimulating sinusoid to obtain a frequency reference and then calculating the average amplitude of the oscillations of the last half of the response autocorrelation. The frequency of the sine wave was chosen to be 2 cycles/sec in order to eliminate the effects of higher harmonics in the response. The amplitude was made large to obtain a large response and the d-c level was made approximately equal to the amplitude.

This experiment will help decide the relative bandwidths between the noise and the transfer function of the linear system. The linear system impulse response is denoted by $h(t)$. Two extreme cases will be considered. The first assumes the noise is white or wide band compared to the linear system. This predicts that the output autocorrelation function will be composed of a continuing sinusoid, a d-c term, and the autocorrelation of $h(t)$, $R_h(\tau)$.

The second case assumes the noise to be very narrow band (colored noise) with respect to the system. This assumption predicts the output autocorrelation function consisting of a d-c term, a continuing cosinusoid, the noise source autocorrelation function, $R_n(\tau)$, and a cosinusoid modulated by the noise autocorrelation.

To anticipate the conclusion, the experiment clearly confirms the first case. The mathematical argument presented below will clarify in detail the effect of the system and the interaction between the signal and the noise upon the output autocorrelation function. This argument will also show how wide-band multiplicative noise passed through a linear system appears similar to additive noise in its contribution to the output autocorrelation function.

If we assume the model in Fig. 31, we would expect the following theoretical results:

Assume

$$L = D + B \cdot \sin(\omega_0 t) \qquad \text{with } D > B \qquad\qquad (1)$$

then

$$b(t) = C_1 \cdot L + C_2 \cdot L \cdot n(t) = D \cdot C_1 + B \cdot C_1 \cdot \sin(\omega_0 t) + D \cdot C_2 \cdot n(t) +$$
$$B \cdot C_2 \cdot n(t) \cdot \sin(\omega_0 t) \qquad\qquad (2)$$

The autocorrelation of $b(t)$ is found to be

$$R_b(\tau) = T_1 + T_2 + T_3 + T_4 \qquad\qquad (3)$$

where

$$T_1 = C_1^2 \cdot D^2 \qquad T_2 = \frac{B^2 \cdot C_1^2}{2} \cdot \cos(\omega_0 \tau)_1$$

$$T_3 = C_2^2 \cdot D^2 \cdot R_n(\tau)$$

$$T_4 = \frac{B^2 \cdot C_2^2}{2} \cdot R_n(\tau) \cdot \cos(\omega_0 \tau)$$

and $R_n(\tau)$ is the autocorrelation function of the independent noise source $n(t)$.

Equation (3) consists of four terms: T_1 a d-c term, T_2 a cosinusoidal term, T_3 a pure noise term, and T_4 which represents a signal times noise term.

The T_1 term when transformed by the linear system and after subtraction from A_{max} yields $(A_{max} - C_1 \cdot D)^2$. The d-c gain of the linear system is assumed to be unity. The cosinusoidal term, T_2, transformed by the linear system yields $\left| H(\omega_0) \right|^2 \cdot B^2 \cdot C_1^2 \cdot \cos(\omega_0 \tau)$. $\left| H(\omega_0) \right|$ is the magnitude of the Fourier transform of $h(t)$. T_3 and T_4 will be considered in limiting cases.

The first case assumes the noise source is white. That is, the noise source spectrum is flat over the frequencies of importance in $h(t)$.

$$R_n(\tau) = N_0 \cdot U_0(\tau) \tag{4}$$

where $U_0(\tau)$ is the unit impulse function and N_0 is the noise power. T_3 then becomes

$$T_3 = C_2^2 \cdot D^2 \cdot N_0 \cdot U_0(\tau) \tag{5}$$

and its contribution to the output autocorrelation is $C_2^2 \cdot D^2 \cdot N_0 \cdot R_h(\tau)$ where $R_h(\tau)$ is the autocorrelation function of $h(t)$. Here the linear system $h(t)$ transforms the autocorrelation of the input, the unit impulse $U_0(\tau)$, into $R_h(\tau)$. T_4 becomes:

$$T_4 = \frac{B^2 \cdot C_2^2}{2} \cdot N_0 \cdot U_0(\tau) \cdot \cos(\omega_0 \tau) \tag{6}$$

$$= \frac{B^2 \cdot C_2^2}{2} N_0 \cdot U_0(\tau)$$

Since $U_0(\tau) \cdot \cos(\omega_0 \tau) = 0$ for $\tau \neq 0$

and

$$= U_0(\tau) \cdot \cos(0)$$

$$= U_0(\tau) \qquad \text{for } \tau = 0$$

Its contribution to the output autocorrelation function is

$$\frac{B^2 \cdot C_2^2}{2} \cdot N_0 \cdot R_h(\tau)$$

The output autocorrelation function is then:

$$R_A(\tau) = [A_{max} - C_1 \cdot D]^2 + \left| H(\omega_0) \right|^2 \cdot \frac{B^2 \cdot C_1^2}{2} \cdot \cos(\omega_0 \tau) +$$

$$D^2 \cdot C_2^2 \cdot N_0 \cdot R_h(\tau) + \frac{B^2 \cdot C_2^2}{2} \cdot N_0 \cdot R_h(\tau) \tag{7}$$

Define $R_{A-S}(\tau)$ as the output autocorrelation minus the d-c and continuing cosinusoidal terms which result from T_1 and T_2.

$$R_{A-S}(\tau) = R_A(\tau) - [A_{max} - C_1 \cdot D]^2 - \left|H(\omega_0)\right|^2 \cdot \frac{B^2 \cdot C_1^2}{2} \cdot \cos(\omega_0 \tau) \qquad (8)$$

Therefore, we obtain for this case:

$$R_{A-S}(\tau) = \left[D^2 \cdot C_2^2 + \frac{B^2 \cdot C_2^2}{2}\right] \cdot N_0 \cdot R_h(\tau) \qquad (9)$$

We find that the shape of $R_{A-S}(\tau)$ should be the same as white noise passed through the linear system.

In the second case, assume the noise is very narrow band and $h(t)$ can be approximated by an impulse or $H(\omega)$ can be considered to be a pure gain compared to the frequencies of interest in $R_n(\tau)$. Call this gain factory unity. We then have for T_4 transformed by the linear system, $\cdot (B^2 \cdot C_2^2)/2 \cdot \left|H(\omega_0)\right|^2 \cdot \cos(\omega_0 \tau) \cdot R_n(\tau)$. T_3 would be unchanged by the linear system since it contains only low frequencies. The effect of T_1 and T_2 are the same here as for the first case.

We then find for this second case:

$$R_{A-S}(\tau) = D^2 \cdot C_2^2 \cdot R_n(\tau) + \frac{B^2 \cdot C_2^2}{2} \cdot \left|H(\omega_0)\right|^2 \cdot \cos(\omega_0 \tau) \cdot R_n(\tau) \qquad (10)$$

If the noise autocorrelation, $R_n(\tau)$, is assumed to be of exponentially decreasing order, we should expect to see in $R_{A-S}(\tau)$ an oscillation of frequency ω_0 that would damp out as τ is increased, since we assume:

$$\lim_{\tau \to \infty} R_n(\tau) = 0$$

The experimental results are shown in Fig. 32. We see from this experiment that the argument of the first case applies quite well. Also shown is the pupil noise autocorrelation under d-c light with the d-c level subtracted. We shall show shortly that the autocorrelation of the noise under d-c light conditions does agree to some extent with that of a pupil impulse response autocorrelation. The signals in this experiment were first band-pass filtered from 0.3 cycle/sec to 10 cycles/sec in order to elimi-

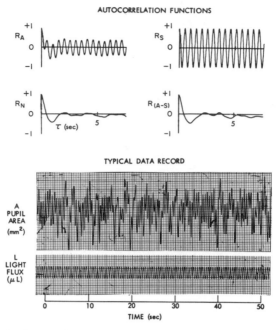

Fig. 32. Additive noise experiment showing typical
data record of sinusoidal drive with 2 cycles/sec light
stimulus and pupil area response. Autocorrelation
functions are of response $R_A(\tau)$, of sine wave alone
$R_S(\tau)$, and of noise under constant light $R_N(\tau)$. $R_{A-S}(\tau)$
is equal to $R_A(\tau) - kR_S(\tau)$, where k is an appropriate
constant calculated to eliminate the additive component in $R_A(\tau)$.

nate low-frequency drift and then sampled at 20 cycles/sec. This filtering
action accounts for the undershoot in the autocorrelation function.*

Figure 33 shows the spectra of both the pupil under d-c light conditions and the spectrum of R_{A-S} (τ). Notice how the spectra agree. Different degrees of smoothing are shown in order to illustrate this technique.

SPECTRAL COMPARISON

In this series of experiments a comparison was made between the
spectrum of pupillary noise, pulse responses, and the magnitude squared
of the Bode plot obtained from a sinusoidal drive.

---

*See pages 146−148 for a discussion of autocorrelation and the multiple
smoothing procedure used on the power spectrum data.

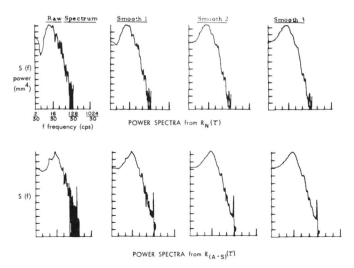

Fig. 33. Power density spectra from $R_N(\tau)$ and $R_{A-S}(\tau)$ show close agreement. Different degrees of smoothing illustrate how general spectral shape is preserved while random fluctuations are reduced.

An average response program was used in order to obtain an averaged pulse response upon which an autocorrelation was performed and a power spectrum obtained. This was done for both small, medium, and large input pulses. See Figs. 34 and 35. The autocorrelations of the pulses all seem to agree quite well.

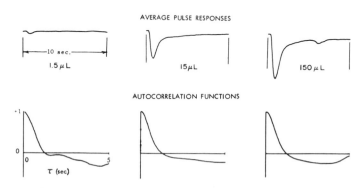

Fig. 34. Averaged pupil pulse responses (seven each) and associated autocorrelation functions; light pulse amplitudes in microlumens above a constant background. All three responses and autocorrelation functions agree in functional form.

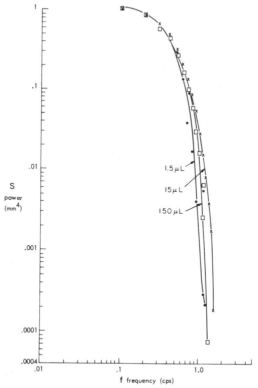

Fig. 35. Power spectra calculated from autocorrelation functions of pulse indicate almost identical shape, independent of the stimulating light pulse amplitudes.

During the same experiment, noise was recorded at a d-c light level for comparison of the noise autocorrelation and spectrum with that of a pulse response. Many short runs were recorded and the normalized autocorrelations (normalized so that the value of the autocorrelation at the origin was equal to 1) averaged. The individual correlations as well as the average are presented in Fig. 36 in order to illustrate the technique of short time correlations. Data for each run were obtained at a sampling rate of 20 samples per second for a 60-sec period.

To complete the spectral comparison, the amplitude response of the pupil to a sinusoidal drive was obtained utilizing a special program which generates the sinusoidal stimulus, records the response, and calculates the magnitude of the gain [39].

Two sine wave experiments were performed, one with a large swing in amplitude (approximately 7.5 mlm), and one with a smaller amplitude

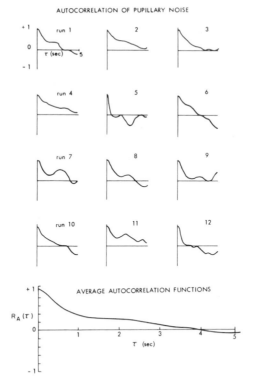

Fig. 36. Autocorrelations of pupil noise. Each
individual run shows large statistical variation
even though each run represents approximately
60 sec of data. Average autocorrelation function
of all twelve runs is smooth and a more stable
estimate of the "true" autocorrelation function.

(approximately 1.9 mlm). Both experiments were performed with a d–c
level of about 7.5 mlm. Figure 37a shows the square of the magnitude of
the gain of the sine wave response for the two experiments. Notice how the
slopes of the spectra are almost identical for the high–frequency portion but
differ in the lower frequencies. This clearly indicates the presence of non–
linearities in the pupil control system.

Before actually comparing the spectra, a word must be said about
scaling and multiplicative constants. All the spectra presented in this
paper are plotted on a log–log scale. Multiplication of the spectral am–
plitude by a constant does not change the basic shape but merely raises or
lowers the curve by a d–c value.

A question arises as to the manner of normalizing the curves in order
to present a comparison in graphical form. Whereas the noise and pulse

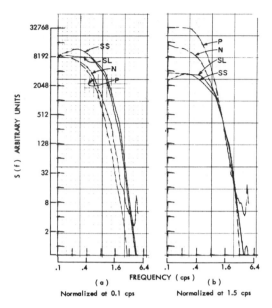

Fig. 37. Spectral comparison of response to large pulse (P), pupil noise (N), large sinusoidal stimulus (SL), and small sinusoidal stimulus (SS) are shown on a log-log scale. Normalization at 0.1 cycle/sec (a) and at 1.5 cycles/sec (b) are shown. Three smoothings used for all spectra.

spectra are in physical units of mm^4, the sinusoidal gain has a dimension of mm^2/mlm. Stark, Campbell, and Atwood [39] in an earlier work on pupil noise chose to normalize the noise spectrum and sinusoidal spectrum by making the curves agree for low frequency. From this comparison it was concluded that the sinusoidal spectrum possessed more high frequencies than the noise spectrum, and therefore the noise was prefiltered before entering the pupil loop. Figure 37a shows our results normalized in this manner and it seems that the pulse spectrum has the smallest amount of high-frequency power, the sine wave experiment the highest, and the noise an amount in between.

Notice, however, that the high-frequency slopes seem to be the same in all cases. Recall that the normalization making all curves equal at low frequency is arbitrary and has no real justification. We could make the curves equal at some other point in order to bring the high-frequency data into closer coincidence. This is done in Fig. 37b. This representation shows that the high frequencies can be made to line up by a vertical shift showing relatively more low frequencies in the pulse spectrum than in the noise spectrum. Similarly, the noise spectrum possesses more low frequencies than the sinusoidal spectrum. A physiological interpretation of these results might be that there is no light adaptation with narrow light pulse and consequently a greater low-frequency response.

Since data do not agree at low frequencies, and the sine wave spectra are either very flat or even decrease for low frequencies, the high-frequency alignment would probably be preferred. This is consistent with the model of Fig. 31, especially if we assume some nonlinearity and filtering of the light signal before the noise is introduced.

Whereas the high-frequency data for noise, pulse, and sinusoids agree fairly well, we ascertained that the low frequencies show definite differences in characteristics. Contradicting the earlier idea of prefiltering of the noise before introduction into the pupil loop, we now conclude that the noise enters as white noise, but that the light stimulus signal is preemphasized for high frequencies and correspondingly deemphasized for low frequencies. This would physically correspond to a form of quick light adaptation.

Since, as has been previously noted, the noise level varies with light level, an experiment was performed in order to determine whether or not other statistical properties such as the autocorrelation function or power spectrum changed their character with illumination level. Figure 38 shows

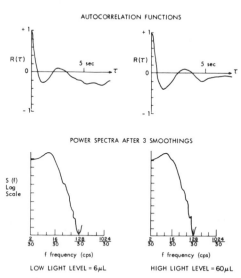

Fig. 38. Noise spectra and autocorrelation functions at high and low light levels in microlumens showing no significant differences in functional form. Raw data signals were bandpassed from 0.3 cycle/sec to 10 cycles/sec.

the autocorrelations and power spectra of pupillary noise at high and low
light levels. The signal was first band–pass filtered from 0.3 cycle/sec
to 10 cycles/sec. Notice that there does not seem to be any significant difference in the spectrum at high or low light levels. This result also agrees
with the model of Fig. 31.

     An experiment was also performed in order to compare the power
spectrum at high and low light levels with near and far accommodation.
The results are presented in Fig. 39. Within experimental error, all four
spectra agree fairly well. This figure presents the data normalized in
three manners. It should be realized that due to limitations in the data
length, the low-frequency portion of these spectra will show an inherent
inconsistency. It is concluded from this experiment that light level and
fixation do not affect the shape of the noise power density spectrum.

INTERACTION BETWEEN TRANSIENT SIGNALS AND NOISE

     In order to investigate the interaction between transient signals and
noise, a special average response program was written which generates

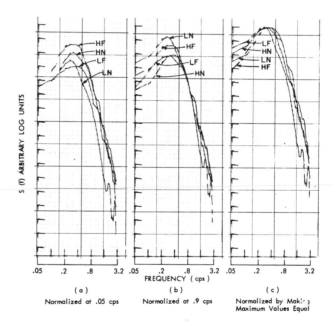

Fig. 39. Noise spectra band-passed (0.3 cycle/sec to 10
cycles/sec), for high (H) and low (L) light levels with near
(N) and far (F) accommodation. Three different normalization frequencies were used. The results agree fairly well,
especially at higher frequencies. Low-frequency results are
somewhat distorted due to filtering and four smoothing operations.

the light-controlling stimuli, introduces the pupil response, calculates the
average of all the responses as well as the standard deviation, punches the
results on cards, and plots the results on an X-Y recorder attached to the
digital—analog converter. The stimuli generated by this program include
pulses, steps, and ramps of specified pulse height, width, initial and final
values. In cases where many stimuli are used in one experiment, the program generates the stimuli in a random order to minimize the effects of
light adaptation, fatigue, and other nonstationary effects.

The rms noise, $\sigma(t)$, which is in general a function of time, will be
called the noise response of the pupil system. Figures 40, 41, 42, 43, 44,
and 45 show the light stimulus, $L(t)$, average area, $\overline{A}(t)$, and the noise response, $\sigma(t)$, for different stimuli. These results are also presented in an
alternative form where the noise is plotted as a function of average area.
The name "noise plane" will be used for this representation. The tics on
these parametric noise plane curves represent half-second time intervals.

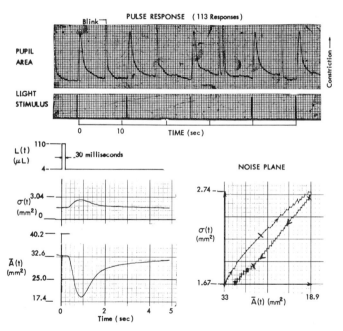

Fig. 40. Noise plane display. Raw data showing light pulse stim-
uli, pupil pulse responses. $\overline{A}(t)$ and $\sigma(t)$ plotted by computer from
average of 113 responses shown below stimulus L(t) in microlu-
mens. Noise plane shows $\sigma(t)$ as function of $\overline{A}(t)$. Tics are placed
at 0.5-sec intervals and arrow marks indicate direction of increasing time.

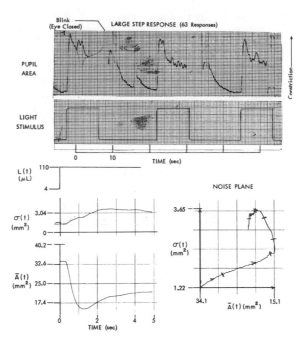

Fig. 41. Noise plane display of large step response.

Figure 40, the response to a narrow pulse, shows an almost linear relation between $\bar{A}(t)$ and $\sigma(t)$, although some hysteresis seems to be present. $\bar{A}(t)$ and $\sigma(t)$ are proportional during the initial decrease in pupil area, but the noise decreases more quickly than the area increases as time progresses. Except for this effect, the crude model in Fig. 31 still remains valid.

Figure 41 (response to a large step) shows significantly different responses as compared to small steps, indicating some form of saturating nonlinearity. Comparison of Figs. 43 and 44 indicates similar area responses and similar noise responses, except for magnitude.

Figure 42 (ramp response) shows the average area response containing a sharp constriction followed by a slower constriction. This effect may be due to the interaction between light adaptation and the increasing light stimuli. The noise response shows a gradual increase from one level to the next. The slight droop in the middle of the $\sigma(t)$ response is probably due to experimental error. The effect is within statistically permissible error.

Reference to the raw data in Fig. 45 indicates that there does not seem to be much noise during an off step of light. Initial experiments with

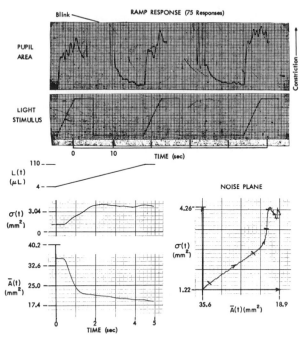

Fig. 42. Noise plane display of ramp response.

off steps reveal a noise response proportional to the average area response. However, the following counter argument was formulated:

Assume the response to an off step is $A_{max} - G \cdot p(t)$, where A_{max} is the maximum pupil area, G is some constant, and $p(t)$ represents the functional form of the response. Clearly, $\lim_{t \to \infty} p(t) = 0$. If the light is shut off each time, when the area is at a different value, G should be considered a random variable. We can then write:

$$\bar{A}(t) = A_{max} - \bar{G} \cdot p(t) \tag{11}$$

$$\underline{\sigma}(t) = \sigma_g \cdot p(t) \tag{12}$$

where $\bar{G}$ is the average value of G and σ_g is the standard deviation of G. Notice that $\bar{A}(t)$ and $\sigma(t)$ are linearly proportional. However, if an effort were made to shut off the light when the area was the same value each time, G should be considered a constant and

$$\bar{A}(t) = A_{max} - G \cdot p(t) \tag{13}$$

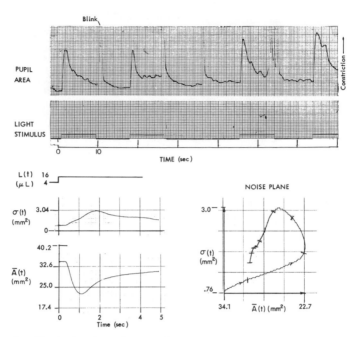

Fig. 43. Noise plane display of small step response, with low constant (d-c) light level.

and

$$\sigma(t) = 0, \text{ independent of time} \tag{14}$$

The above argument assumes that when the light is shut off, no residual light exists and we find that the noise in Eq. (14) equals zero and in Eq. (12) the noise approaches zero. In practice there is always some residual light present and the pupil never reaches A_{max}. Therefore, the noise should not go to zero but to some level dependent upon the residual light level, i.e., some constant value should be added to Eqs. (12) and (14).

An experiment was performed in which the light was shut off at a predetermined time with random initial area. A control experiment was performed simultaneously in which the light was shut off when the area was the same value. Figure 45 shows the results of these experiments. The random initial area experiment shows $\sigma(t)$ proportional to $\overline{A}(t)$ and the control experiment shows an almost constant $\sigma(t)$. Therefore, the above argument indicating no real random noise during the off step transient seems to be validated. The irregularities in the noise plane in the control experiment are present due to the scaling used in the representation.

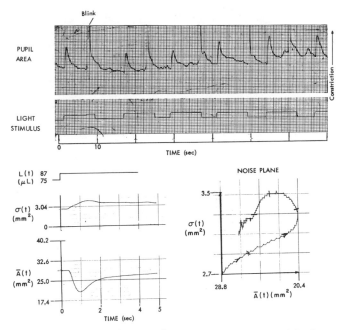

Fig. 44. Noise plane display of small step response, with high constant (d-c) light level.

CROSS CORRELATION BETWEEN PUPIL NOISE IN BOTH PUPILS

In order to pinpoint more accurately the location of the noise source, an experiment was performed in which both eyes were recorded, and a cross correlation between the two noise signals performed.

Both pupils were recorded simultaneously utilizing the dual eye SKF pupillograph designed by Lowenstein, Loewenfeld, and King located at the Massachusetts Eye and Ear Infirmary. Analog data recorded with this pupillograph were sent over telephone lines to the Massachusetts Institute of Technology and were digitalized by the computer. Figure 46 shows the raw data and the result of a normalized cross correlation between signals. Notice that the signals show almost 100% correlation. Any discrepancy is possibly due to a little random noise introduced on the telephone lines. Both raw data and cross-correlation functions indicate that the noise is introduced or generated at some point in the brain in common to both eyes as first shown by the Stark, Campbell, and Atwood study [39].

PUPIL MODEL

In this section of our chapter, we describe a model representing the system whose input is one light level and whose output is pupil area, which could simulate the various experimental results of the preceding studies.

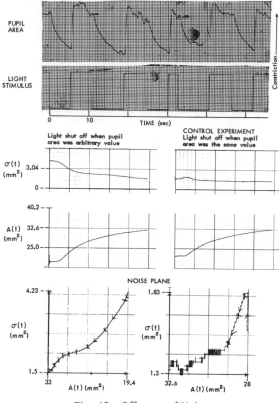

Fig. 45. Off steps of light.

The general experimental properties which this model attempts to describe are:

1. The variation of rms noise as a function of average pupil area and light level, as well as the variation of average area with light level.

2. The spectral character of the noise at high and low light levels.

3. The interaction of pupillary noise with a sine wave stimulus.

4. The spectral comparison between the sine wave spectrum, pulse spectrum, and the noise spectrum.

5. The Gaussian shape of the amplitude histogram.

6. The interaction of noise with transient inputs such as pulses, small and large steps, ramps, and off steps of light.

The model shown in Fig. 47 is an extension of Fig. 31. It has not been simulated and is presented as a possible description of our experi-

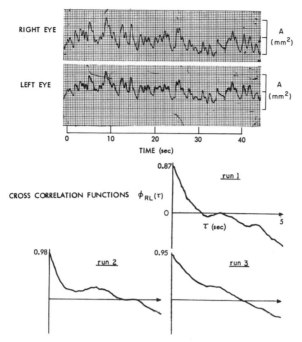

Fig. 46. Correlation between pupil noise in right and left
eyes. Both raw data and three cross-correlation functions
indicate very high degree of correlation. Electronic amplifier gain for left eye raw data record is slightly attenuated.

mental results. Its configuration is based upon the following experimental facts.

1. The $e^{-st_0}/[(1 + \tau_3 s)^3]$ box is used to account for the spectral character of the noise. The factor e^{-st_0} accounts for the time delay of t_0 seconds present in all responses. This box will be called the triple lag.

2. The diode and R–C combination is used to explain relatively slow,
long–term dilation noticed during off steps of light. If $P > 0$ and $(dP/dt) >$
$(-V_c/RC)$, then the ideal diode is shorted $(i_{diode} > 0, V_{diode} = 0)$; therefore,
$P = V_c$ and the gain from P to Q is 1 since we have two paths of a gain of
1/2. When $(dP/dt) = (-V_c/RC)$ the diode is at breakpoint $(i_{diode} = V_{diode} = 0)$
or when $(dP/dt) < (-V_c/RC)$ or when $P < 0$ the diode is open $(i_{diode} = 0,$
$V_{diode} < 0)$, then the signal at Q will be 1/2 the signal generated by the
R–C decay plus the signal from the nonlinear box immediately above the
R–C circuit (see Fig. 47).

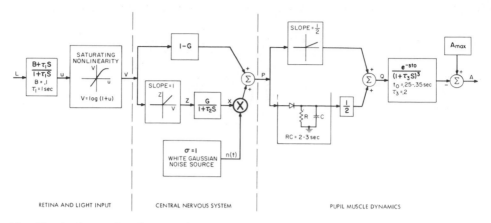

Fig. 47. Analog model of pupil reflex system accounting for pupil noise phenomena. Accommodation input could be injected at V.

3. The $(\beta = \tau_1 s)/(1 + \tau_1 s)$ box is used to supply the overshoot present
in step responses.

4. The logarithmic nonlinearity serves many purposes. It explains
the logarithmic nature of the $\overline{A}$ vs. L curve and the σ_A vs. L curve. Its
saturating effect causes the overshoot to a large step to be less pronounced
than the overshoot to a small step.

5. Noise is gated into the system by means of a multiplier and is assumed to be stationary white Gaussian noise with a zero mean value.

6. The $G/(1 + \tau_2 s)$ box accounts for the slower changes in noise than
in area during transient responses; that is, the noise response usually lags
the area response. This is an additional reason for having a double path
in the central nervous system portion of the pupil model. The first reason
was mentioned earlier: the need for a nonzero average area difference
from A_{max} at different light levels. This would be impossible with only a
single path multiplied by, on the average, $\overline{n}(t)$, which is assumed to be
equal to zero.

The following discussion indicates how the model predicts experimental results.

1. For a constant input light L

$$u = B \cdot L, \quad V = \log (1 + B \cdot L) \tag{15}$$

$$P = (1 - G) \cdot V + G \cdot V \cdot n(t) \tag{16}$$

and the diode is shorted since G is small. The output area will then be

$$A = A_{max} - [\log (1 + B \cdot L)] \cdot [(1 - G) + G \int_0^\infty h(\tau) n(t-\tau) d\tau] \qquad (17)$$

where $h(\tau)$ is the impulse response of the $1/(1 + \tau_3 s)^3$ box.

The average and rms values of the pupil area signal are derived as follows:

$$\bar{A} = A_{max} - [\log (1 + B \cdot L)] \cdot [(1 - G) + G \int_0^\infty h(\tau) \overline{n(t - \tau)} d\tau] \qquad (18)$$

Since the mean value of the noise source is zero,

$$\overline{n(t - \tau)} = 0 \qquad (19)$$

and we have

$$\bar{A} = A_{max} - (1 - G) \cdot \log (1 + B \cdot L) \qquad (20)$$

For the variance we obtain

$$\sigma_A^2 = \log^2 (1 + B \cdot L) \cdot G^2 \int_0^\infty \int_0^\infty i(\tau_1) i(\tau_2) \overline{n(t - \tau_1) n(t - \tau_2)} d\tau_1 d\tau_2 \qquad (21)$$

Since the noise is assumed white and stationary

$$\overline{n(t - \tau_1) n (t - \tau_2)} = N_0 \cdot U_0(\tau_1 - \tau_2) \qquad (22)$$

where U_0 is the unit impulse function and N_0 is the noise power.

Therefore,

$$\sigma_A = G \cdot \log (1 + B \cdot L) \cdot \sqrt{N_0 \cdot \int_0^\infty i^2 (\tau) d\tau} \qquad (23)$$

Solving Eq. (20) for L and substituting into Eq. (23) yields:

$$\sigma_A = (A_{max} - \bar{A}) \cdot \frac{G}{1 - G} \cdot \sqrt{N_0 \cdot \int_0^\infty i^2 (\tau) d\tau} \qquad (24)$$

Equations (20) and (23) predict the logarithmic relationship between σ_A, $\bar{A}$ and L, and Eq. (24) demonstrates the linear relation between σ_A and $\bar{A}$. Figures 28 and 29 depict the experimental results.

2. Under constant illumination conditions the diode will be shorted. The signal P as shown in Eq. (16) is the sum of a constant term and a white

noise term. The output area autocorrelation is found to be

$$R_A(\tau) = G^2 \cdot \log^2 (1 + B \cdot L) \cdot N_0 \cdot R_h(\tau) + \text{d-c terms} \tag{25}$$

where $R_h(\tau) = \int_0^\infty h(t) h(t + \tau) dt$ = autocorrelation of the pulse response of the $1/[(1 + \tau_3 s)^3]$ box.

Subtracting the d-c terms and taking the Fourier transform of Eq. (25) yields the output area power density spectrum.

$$S_A(f) = \frac{G^2 \cdot \log^2 (1 + B \cdot L) \cdot N_0}{(1 + \tau_3^2 \omega^2)^3}, \quad \omega = 2\pi f \tag{26}$$

Notice that the functional shape of the spectrum is independent of the light level and only the total noise power depends upon L. This fact was, of course, demonstrated in Figs. 38 and 39.

3. Even though the input stimulus to the pupil system be a pure sine wave of frequency ω_0, the presence of the log operator will tend to introduce higher harmonics. To preserve the generality of the analysis, the signal at Q due to a sinusoidal input will be expressed as

$$Q = \sum_{K=1}^\infty E_K \sin(\omega_0 \cdot K \cdot t + \theta_K) + n(t) -$$
$$\sum_{j=1}^\infty F_j \sin(\omega_0 \cdot j \cdot t + \phi_j) + \text{d-c terms} + C n(t) \tag{27}$$

where $E_K, F_j, \theta_K, \phi_j$, and C are constants determined by the linear and non-linear system before Q as well as the input.

The autocorrelation of the Q signal assuming $n(t)$ independent of any signal in the system is:

$$R_Q(\tau) = \sum_{K=1}^\infty \frac{E K^2}{2} \cdot \cos(\omega_0 \cdot K \cdot \tau) + R_n(\tau) \cdot$$

$$\sum_{j=1}^\infty \frac{F_j^2}{2} \cos(\omega_0 \cdot j \cdot \tau) + \text{d-c terms} + C^2 \cdot R(\tau)$$

where

$$R_n(\tau) = N_0 \cdot U_0(\tau) = \text{noise source autocorrelation} \tag{28}$$

We, therefore, have for output autocorrelation:

$$R_A(\tau) = \sum_{K=1}^\infty \frac{G_K^2}{2} \cdot \cos(\omega_0 \cdot K \cdot \tau) + \left[\sum_{j=1}^\infty \frac{F_j^2}{2} + C^2 \right] \cdot N_0 \cdot R_i(\tau) + \text{d-c terms} \tag{29}$$

where the G_K's are different constants from the E_K's because of the action of the linear filter. If ω_0 is chosen large enough (2 cycles/sec in the experiment of Fig. 32) so that the G_K's for K greater than or equal to 2 can be neglected, and if we subtract the d–c and additive cosinusoidal components from $R_A(\tau)$, we obtain:

$$R_{A-S}(\tau) = R_A(\tau) - \frac{G_1^2}{2} \cos(\omega_0\tau) - \text{d–c terms} = \left(\frac{F_1^2}{2} + C^2\right) \cdot N_0 \cdot R_i(\tau) \quad (30)$$

Equation (30) verifies the experimental results since we are left with an autocorrelation which has the same shape as the noise autocorrelation under constant illumination conditions. See Eq. (25). Figures 32 and 33 show the experimental results.

4a. The power density spectrum of the noise was derived in Eq. (26). It is simply the spectrum of the triple lag. Figure 48 traces a narrow pulse through the model of Fig. 47 in a qualitative manner. Notice that the signal at Q consists of two components. One is a narrow pulse which will give rise to the same shape spectrum as noise. The second component is a slow exponential which will add some extra low-frequency energy to the pulse spectrum without affecting the high-frequency components. This result is consistent with the experimental results of Fig. 37b.

4b. The experimental sine wave spectral results can also be predicted by the model in a qualitative manner. If we disregard the nonlinearities in the system, we notice the $(B + \tau_1 s)/(1 + \tau_1 s)$ box will emphasize high frequencies and tend to suppress low frequencies. The $1/[(1 + \tau_3 s)^3]$ box will in turn give rise to high-frequency attenuation.

For a sinusoidal stimulus with a small amplitude relative to its d–c value, the log nonlinearity can be incrementally linearized to a first-order

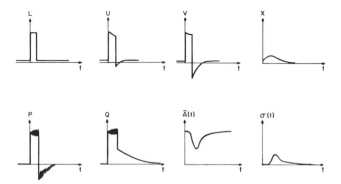

Fig. 48. Sequence of theoretical signal wave forms from analog model for light plus stimulus.

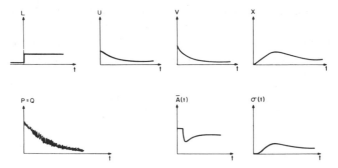

Fig. 49. Sequence of theoretical signal wave forms from analog model for small step stimulus.

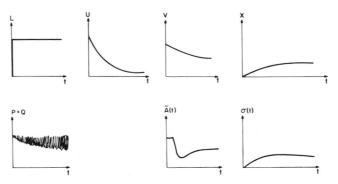

Fig. 50. Sequence of theoretical signal wave forms from analog model for large light step stimulus.

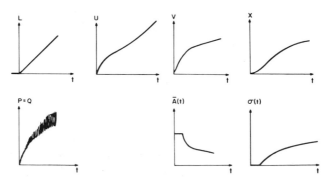

Fig. 51. Sequence of theoretical signal wave forms from analog model for light ramp stimulus.

approximation. Noise is gated into the system in the normal manner, but at the moment is of secondary consequence. Finally, P is always positive so the signal just passes through the diode R–C nonlinearity with very little effect. The triple lag creates the high–frequency character.

A sinusoidal stimulus with a large signal amplitude is greatly distorted by the log nonlinearity. The saturation effect for large input signals will tend to make the signals at V of more equal amplitude. That is, the effect of the $(B + \tau_1 s)/(1 + \tau_1 s)$ box in attenuating low frequencies is greatly reduced. As before, a little noise is picked up and the triple lag creates the high–frequency response.

This model, therefore, predicts the same high–frequency response for small and large amplitude sinusoidal inputs, but a flat large amplitude low–frequency response and an attenuated small amplitude low–frequency response. These conclusions are verified in Fig. 37.

5. Under constant illumination conditions, the signal entering the triple lag is the sum of a d–c component and a white Gaussian noise component. Since Gaussian noise through a linear system is again Gaussian, we would expect the amplitude histogram of the pupillary noise signal to be Gaussian. This is verified by the experimental evidence shown in Figs. 26 and 27.

6. To derive the interaction between noise and transient–type signals from this model, let us consider the following cases separately: (a) positive pulse of light, (b) small positive step of light, (c) large positive step of light, (d) a positive ramp of light, and (e) off step of light.

a. The signals for a positive narrow pulse of light are traced through the model in Fig. 48. As explained previously, the signal at Q is the sum of a pulse with a little noise on top and a noiseless exponential decay. The only variation in output responses will be due to the noise on top of the pulse which will tend to make the effective area under the pulse slightly different each time. The noise response should therefore look like the impulse response of the triple lag, while the average area response should initially look like the impulse response to the triple lag except that the dilation should take longer. This is noticed in Fig. 40 where the noise response and average area initially are the same but the noise seems to decrease faster than the area increases.

b. The signals for a small positive step are shown in Fig. 49. For a first–order approximation the log operator can be linearized and therefore only acts as a pure gain. The noise gating signal, X, is seen to first rise then settle down to its d–c value. The predicted results for $\bar{A}(t)$ and $\sigma(t)$ are in fairly good agreement with the experimental results of Figs. 43 and 44.

c. The signals for a large positive step are shown in Fig. 50. In this case the saturating nonlinearity tends to make the overshoot in the response much less pronounced, which in turn does not give rise to an overshoot in the X signal. Both large and small step responses show similar constriction properties. However, the quick dilation with the small step and the slow dilation with a large step both show a long time constant exponential-type rise due to the $G/(1 + \tau_2 s)$ box. These facts seem to be verified, within experimental error, in Fig. 41.

d. The ramp response signals shown in Fig. 51 show that the combination of the $(B + \tau_1 s)/(1 + \tau_1 s)$ box and the saturating log nonlinearity make V first rise quickly and then at a slower rate. The noise gating signal, X, should be a bit smoother. $A(t)$ shows a quick constriction followed by a slower constriction while the noise signal predicts a fairly rapid increase in noise followed by a slower steady increase. Figure 42 depicts experimental results.

e. Finally, Fig. 52 shows the signals for an off step of light. In this situation the diode R-C plays an important role since the P signal goes negative. The negative signal slipping before the $G/(1 + \tau_2 s)$ box prevents noise from gating the diode on and off when V goes negative. Figure 52 would predict no noise during dilation except an effective rms signal similar to the dilation response if the light were shut off at a different pupil area each time. The experimental results in Fig. 45 seem to agree fairly well.

PHYSIOLOGICAL DEDUCTIONS

On the basis of the various experiments described in the text part of this chapter, certain biological and physiological conclusions may be drawn.

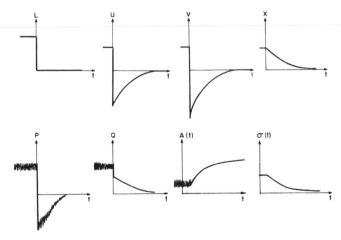

Fig. 52. Sequence of theoretical signal wave forms from analog model for light off step stimulus.

1. The experiment which showed almost 100% correlation between the noise in both pupils indicates that the noise must be introduced at a place in the brain in common to both eyes. Such a region could be either the pretectal area or the Edinger–Westphal nucleus. This experiment eliminates the pupil muscle and the third nerve as sources of pupil noise.

2. Another experiment provides evidence that the Edinger–Westphal nuclei play an important role in the noise generation. The experiment indicates that accommodation increased pupil noise in the same manner as light; the rms noise vs. pupil area graphs for accommodation and light were almost the same. One would, therefore, suspect that the noise–controlling signal was common to the light and accommodation pupil control systems. The junction in the parasympathetic nervous system controlling both these response systems is most probably the Edinger–Westphal nucleus.

3. The accommodation experiment by itself might suggest that the pupil noise is a function of pupil area. However, other experiments such as the noise and area response to transient pulses steps, ramps, and off steps of light indicate that the noise response and area response do not agree in shape in most cases. Therefore, one would conclude that the noise is controlled by some internal signal. This result also agrees with the analog model in Fig. 47.

The number of associations between the structure of the analog model in Fig. 47 and the actual anatomical structures in the pupil control system are suggested: The first section represents the retina of the eye. The $(B + \tau_1 s)/(1 + \tau_1 s)$ box is associated with the retinal light adaptation dynamics. The saturating nonlinearity represents the logarithmic nature of the sensitivity of the retina to light. The middle section represents the central nervous system and contains the white noise source controlled by an internal variable. Exact anatomical associations are very difficult to make in this section. Finally, the last section represents the pupil and associated muscles. The R–C combination is associated with the dilator muscle dynamics while $1/[(1 + \tau_3 s)^3]$ is associated with the constrictor muscle dynamics. The term e^{-st_0} represents the pupil response time delay.

During initial investigations, the multiplicative noise source was thought to be the stochastic variation in nerve impulse firing frequency. A shot noise statistical model was assumed, with the instantaneous firing rate a function of light level. This model was satisfactory in predicting the general trend of all the experimental results but was not completely satisfactory when a detailed comparison with experimental results was made.

a. The shot noise model predicts an rms noise vs. area graph under constant illumination conditions to be a square root curve. However, experimental results indicate an almost linear relationship.

b. The interaction between noise and a sinusoidal drive was predicted very satisfactorily.

c. The prediction for noise and area dynamics with transient input pulses agree fairly well but show that the shot noise model predicts a faster rise time for the noise responses. The amplitude of the predicted responses is sometimes a factor of 2 or 3 different from the experimental results.

These few remarks indicate that the random process involved is more complicated than a pure shot noise process. Possibly the assumption of some correlation between nerve impulse firing times might be appropriate.

CORRELATION AND POWER SPECTRUM

The correlation program actually performs a cross correlation. An autocorrelation is just a special case of a cross correlation $(x = y)$ and the program has facility to perform both. The function calculated is:

$$R_{xy}(K) = \frac{1}{N-K} \sum_{w-1}^{N-K} X_i Y_{i+K} - (AVX)(AVY)$$

$$K = 0, 1, 2, \dots, K_{max} \qquad (31)$$

where N is the total number of points, X_i is the ith data point, AVX and AVY are the average values of the x and y signals, respectively, and K_{max} is the maximum desired shift in points for the correlation function.

Limited memory space in the computer forced the program to be written in a manner in which raw data on binary cards remained in the card reader until it was needed. When data in memory was no longer necessary, it was destroyed so that room was made for the new data. The scheme worked as follows:

First, $K_{max} + 1$ points were read into memory. The partial sums, $R(0) = X(0) \cdot Y(0)$, $R(1) = X(0) \cdot Y(1)$, $\dots$, $R(K_{max}) = X(0) \cdot Y(K_{max})$ were calculated. After these calculations the points $X(0)$ and $Y(0)$ are no longer needed and the next points $X(K_{max} + 1)$ and $Y(K_{max} + 1)$ are read into the $X(0)$ and $Y(0)$ slots and the accumulating partial sums $R(0) = R(0) + X(1) \cdot Y(1)$, $R(1) = R(1) + X(1) \cdot Y(2)$, $\dots$, $R(K_{max}) = R(K_{max}) + X(1) \cdot Y(0)$ are formed. Next the points $X(K_{max} + 2)$ and $Y(K_{max} + 2)$ are read into the

$X(2)$ and $Y(2)$ slots. In general the $X(M)$ and $Y(M)$ points are read into slot number $X(K_{max} + 1 + M)$ and $Y(K_{max} + 1M)$ modulo K_{max} and the accumulating partial sums calculated. This process continues until all points have been read in and used. At any one time only $K_{max} + 1$ data points are in memory.

This program is essentially limited to K_{max} less than or equal to 400 and the total number of points less than 250,000. In actual practice K_{max} is 200 or 300 and the number of points is 2000 or 3000. The calculation time for the GE–225 computer (18 μsec memory cycle time) is approximately 1.2 msec per point per shift or about 8 min for a 2000-point 200-shift calculation.

After the functions are calculated, the results can be punched out on binary cards for future reference, listed on the typewriter, or plotted on an X-Y recorder via the digital–analog converter.

In many experimental situations only short pieces of data can be obtained. The correlation function of any one section of data would indeed show a large statistical variance. In order to alleviate this situation, an auxiliary program was written to accept the binary card outputs of many correlation functions and average the results to give a more statistically stable estimate of the true correlation function.

After an autocorrelation function is obtained, the binary output can be used as the input to the power spectrum program. This program calculates the power spectrum $S_r(f)$, where:

$$S_r(f) = \Delta T \cdot \left[R(0) + 2 \sum_{K=1}^{K_{max}} R(K) \cdot \cos \frac{K \cdot r \cdot \pi}{K_{max}} + R(K_{max}) \cos (\pi - n) \right] \quad (32)$$

where ΔT is 1/sampling rate in seconds; K_{max} is total number of shifts on the correlation function; and $NINT$ is the number of frequency intervals desired. The frequency increment is $1/(2 \cdot \Delta T \cdot K_{max})$. The maximum frequency calculated is $F = NINT/(2 \cdot \Delta T \cdot K_{max})$. The frequency corresponding to r is $r/(2 \cdot \Delta T \cdot K_{max})$.

This program also can output the results on binary cards as well as plot the results. The graph in this case is on a log scale since it is this display that is usually used by system engineers. On these graphs the computer placed tics on the axis every doubling of magnitude. Care was taken to make the scale factors equal so that slopes were preserved.

A unique feature of this program is the multiple smoothing subroutine. Once the raw spectrum $S_r(f)$ is obtained, it usually contains much statistical irregularity due to the finiteness of the data and the discrete data processing methods. A routine employing the Hamming smoothing procedure is used on the results. Smoothing results in a smaller variance of the spectral estimate at the expense of frequency resolution. However, in our work where only the general shape of the spectrum is desired, such as slopes and breakpoints, for system-type models, frequency resolution becomes unimportant. The Hamming smoothing procedure forms a new spectral function in terms of the old by:

$$U_r = 0.23\ S_{r-1} + 0.54\ S_r + 0.23\ S_{r+1} \tag{33}$$

The smoothing subprogram enables the experimenter to perform the smoothing as many times as desired to obtain smoother and smoother curves. Sample spectra with multiple smoothings are shown above in Fig. 33.

SUMMARY

This experimental study of noise in the human pupil is an example of the use of statistical communication theory and control system analysis to dissect a biological system.

On-line digital computer and infrared pupillometer experimental and computation techniques are described.

A detailed system or black box model is presented and shown to account quantitatively for all of the experimental results.

The distribution function of the noise has Gaussian characteristics.

The functional dependence of rms noise upon pupil area, whether the area is controlled by light or accommodation, necessitates a multiplicative noise model.

Certain apparently additive characteristics of noise are shown to depend upon the wide bandwidth of the noise (white noise) at its input to the system.

The spectral characteristics of the pupil area noise (at output) are shown to be independent of average pupil output level and controlling variables such as average light and accommodation input level.

The relationship between the frequency characteristics of the noise and of the response to sinusoidal and transient inputs was shown to be most simply described as identical at high frequencies and differing at low frequencies in a manner predicted by retinal adaptation and scale compression operations at the system and model input.

The relationships between transient pupil responses and transient noise characteristics were displayed in the "noise plane." They indicated the necessity for a parallel path model with a lag before the noise multiplier.

A switching operation which removed noise quickly from the off response was necessary in the model.

Highly correlated noise in both eyes was demonstrated. This, together with the accommodation experiment, suggests the Edinger-Westphal nuclei as the point of injection of the noise.

A modified shot noise model, with the statistical characteristics of the nerve impulse intervals as the flicker source, might be satisfactory as a more physical representation of the system. In this case, gain noise in the path would be equivalent to the multiplicative noise.

APPENDIX A

PUPIL UNREST: AN EXAMPLE OF NOISE IN A
BIOLOGICAL SERVOMECHANISM

The pupil of the human eye continuously undergoes small fluctuations in area, even in steady illumination [39]. These movements may be readily seen on close inspection. We considered it worthwhile to investigate this pupil unrest as an example of noise in a biological system. We have therefore measured and described the phenomenon, performed experiments to decide between possible mechanisms for its production, and finally considered its possible function.

Description of the Pupil Unrest

The preceding material in this chapter contains a more extensive description of pupillary noise.

One-minute samples of pupil area fluctuations were recorded. A part of one such typical time function record is shown in Fig. 25, together with the autocorrelation function (Fig. 36), and power spectrum (Figs. 33, 38, and 39). The autocorrelation function and the power spectra both show that the major component of the pupil unrest is random noise in the frequency range from 0.05 to 0.3 cycle/sec. In the region from 0.75 to 2.5 cycles per sec, all three power spectra we have obtained show a general slope of 36 db/octave with a corner frequency near 0.5 cycle/sec. One power spectrum from a time function of pupil diameter fluctuations showed the same corner frequency and an 18 db/octave slope. Power spectra of diameter fluctuations should have half the slope of area fluctuations.

The Origin of Pupil Unrest

It is possible to make some deductions about the origin of the pupil unrest from information already known about the pupil light reflex, from the new data presented, and from the results of certain critical experiments described below.

The unrest cannot be a manifestation of instability in the pupil light reflex considered as an error-actuated servomechanism. Such systems may oscillate, if the feedback path which completes the loop carries a response delayed by time lags and is sufficiently amplified to produce regenerative action. Considerable lags exist in the pupil system but the necessary amplification is not present because the gain is equal to 0.16. Equation (4) is the transfer function $G(s)$. Thus the pupil will not continue to respond to a stimulus that does not itself continue, and the pupil light reflex servo loop is therefore stable. The absence of a large coherent frequency component at the frequency at which such instability oscillations would be expected is additional evidence against this possibility. Furthermore, as no regenerative instability is possible when the loop is open, pupil unrest should continue and appear unchanged under open-loop conditions. This is found to be true experimentally. Thus, quite independent of the quantitative arguments above, we can disregard the hypothesis that an unstable pupil loop produces the pupil unrest.

Other linear properties of servosystems such as "backlash" or "dead space" also cause oscillations. However, the pupil loop does not show the change in gain which would be found if substantial amounts of backlash or dead space were present. Again, these nonlinearities require closed-loop operation in order to produce oscillations. Thus the open-loop experimental results also exclude these nonlinearities as a possible explanation for pupil unrest.

The above discussion eliminates causes of unrest associated with the closed-loop properties of the light reflex. Therefore, the unrest must originate in or be injected into the loop. Another observation suggests that this point of origin or injection must be into that portion of the loop which is common to both iris muscles. Only central nervous system elements, including the retina, comprise this portion of the loop. Figure 46 shows the unrest of one iris to be highly correlated with the simultaneously recorded unrest of the contralateral iris. This excludes as an otherwise likely source of pupil unrest, spontaneous movements of the iris muscle, of the type seen in a wide variety of smooth muscles when isolated and free from nervous influence.

The distinction between origin within and injection into the pupil reflex system can be made by comparing the frequency response of the open-

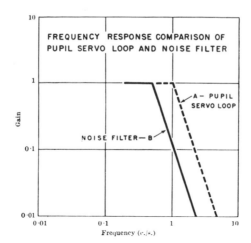

Fig. 53. Approximate asymptotes for (A) pupil
servosystem and (B) lumped parameter network
which would attenuate white noise to produce
a power spectrum similar to Fig. 1. Both are
normalized to unity gain in midfrequency
range.

loop system with the power spectrum of the unrest. Approximate asymptotes are plotted in Fig. 53 for the pupil servosystem (A) and for a lumped
parameter network (B) which would attenuate white noise to produce a
power spectrum similar to Figs. 33, 38, and 39. It is evident that the frequency distribution of the power spectrum of the pupil unrest cannot have
been caused by a filtering action of either the open-loop or closed-loop
pupil system on noise with uniform distribution of power with frequency.
Conversely, the filter which has shaped the power spectrum of this noise
cannot be in series with the flow of information in the pupil system loop,
because the measured frequency response of the pupil system does not
have the 18 db/octave asymptote from 0.5 to 1.0 cycle/sec. It must be
clearly understood that this bandwidth argument implies nothing about the
actual anatomical location of the noise generator. For example, the noise
may be leakage from signals external to the pupil system (cross talk) or
it may arise from part of the neurological apparatus which comprises the
pupil system, as, for example, the retina. In such cases, although the noise
generator and the pupil loop are anatomically united, it must still be true
that they are functionally and physiologically separate; transmission of
light information relevant to the operation of the pupil system cannot pass
in series through the noise generator filter. The erroneous assumption in
this argument is illustrated in Fig. 37 and discussed in the related text.

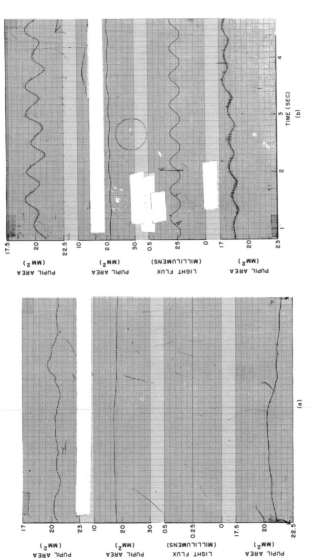

Fig. 54. (a) Pupil noise at 2.0 cycles/sec. First and second traces: same as (b). Third trace: light flux onto retina held constant. Fourth trace: pupil area at high amplification, showing random fluctuations that are not coherent with the d-c input signal and are considered as noise. (b) Pupil response to signal plus pupil noise at 2.0 cycles/sec. First trace: pupil area at even higher amplification in trace 4(a). The signal has been passed through a band-pass filter with 24 db/octave sloped and break frequencies as shown in Table III. Second trace: pupil area at low amplification, which is used to keep a careful record of the average pupil area. Third trace: fluctuations of light flux input in open-loop operating conditions. Fourth trace: pupil area at high amplifications, showing response of the pupil to fluctuations of light intensity superimposed on noise or random fluctuations that are not coherent with the input signal.

Other experiments could localize the noise generator. For example, careful quantitative comparison under open-loop and closed-loop operating conditions could determine whether the noise is injected at the loop input or elsewhere. Because of the low loop gain and the low magnitude of the fluctuations (10% change in area), this experiment is technically difficult and has not been satisfactorily performed. To identify the input which injects noise into the pupil system, it is necessary to establish some measure of cross correlation between the presumptive input and the pupil unrest.

The Role of Pupil Unrest

Pupil unrest might play a specific role in the visual process, or it might represent the tolerance of the pupil servomechanism to uncorrected error. It is unlikely that pupil unrest prevents the disappearance of a stabilized image or produces information of value to the accommodation mechanism since the use of a fixed artificial pupil does not affect the characteristics of these functions. On the other hand, since visual acuity is not appreciably influenced by a 15% change in pupil area, it seems reasonable that pupil area need not be controlled within more precise limits. Perhaps this is an example of economy in construction in the sense that unnecessary requirements are not placed on the noise level in the pupil apparatus.

APPENDIX B

ANALYSIS OF PUPIL RESPONSE AND NOISE

In an attempt to quantify the pupillary frequency response characteristics [44] we were led to a study of the interaction of the signal and the large amount of biological noise present [39, 42]. Several decades of frequencies of interest were studied, with allowances being made for sampling epoch, sampling rate, and band-pass filtering as shown in Table III. Figures 54, 55, and 56 show experimental runs of noise alone. Autocorrelation functions were computed as shown in Figs. 57 and 58. These figures contrast unfiltered and filtered data, showing the strong influence of frequencies outside of the immediate range of interest and, in fact, outside of reasonable sampling measurements. It is of interest to restudy the filtered and unfiltered data in Figs. 54, 55, and 56 with these autocorrelation functions in mind.

Table III

Cycles	0.02 cycle/sec	0.2 cycle/sec	2.0 cycles/sec
Paper speed	0.5 mm/sec	5 mm/sec	50 mm/sec
Filter settings	0.003–0.12 cycle/sec	0.03–1.2 cycles/sec	0.3–12 cycles/sec
Time of run	33.33 min	3.33 min	20 sec

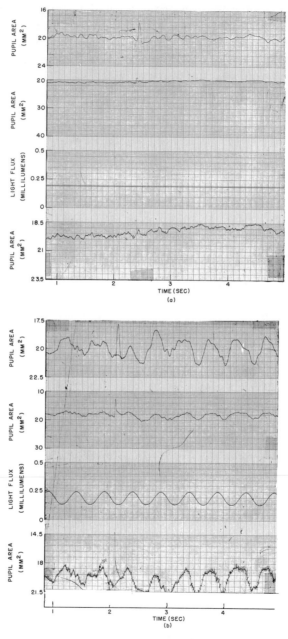

Fig. 55. (a) Pupil noise at 0.2 cycle/sec, showing same outputs and inputs as in Fig. 54a. (b) Pupil response to signal plus noise at 0.2 cycle/sec, showing same outputs and inputs as in Fig. 54b.

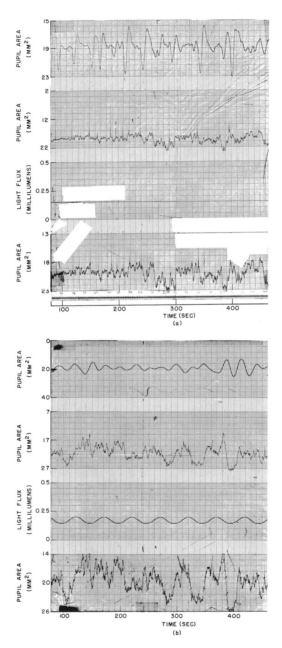

Fig. 56. (a) Pupil noise at 0.02 cycle/sec, showing
same outputs and inputs as in Fig. 54a. (b) Pupil
response to signal plus noise at 0.02 cycle/sec,
showing same outputs and inputs as in Fig. 54b.

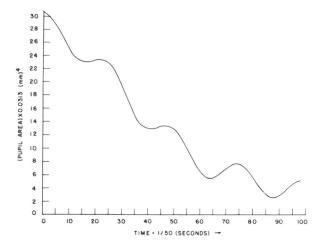

Fig. 57. Autocorrelation function (pupil area)^2 vs. delay
time for 2.0 cycles/sec sine input unfiltered response. The
pupil area is calculated in mm^2; then it is squared and multiplied by the constant 0.0313 to give quantities in mm^4 that are
plotted on the graph. Time, shown on the abscissa, is marked
in seconds and multiplied by the constant 1/50.

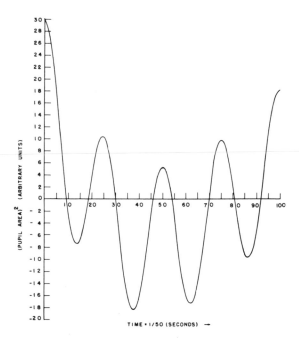

Fig. 58. Autocorrelation function (pupil area)^2 vs. delay time for 2.0
cycles/sec sine input filtered response. The (pupil area)^2 is graphed
in arbitrary units. Time, as shown on the abscissa, is marked in seconds and multiplied by the constant 1/50.

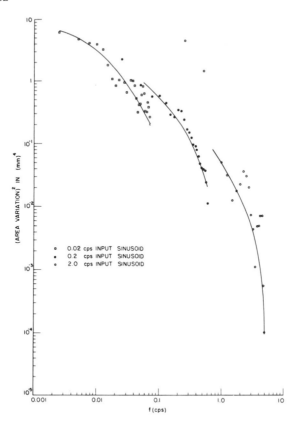

Fig. 59. Power spectrum of noise over frequency range
0 − 4.5 cycles/sec, constructed from data runs at speeds
corresponding to response frequency speeds for 0.02, 0.2,
and 2.0 cycles/sec. Variation in (pupil area)^2 in (mm^4)
vs. frequency (cycles/sec).

Power spectra were computed from the autocorrelation functions and
the additive effect of signal on the noise can be noted by comparing Fig. 59
(noise alone) with Fig. 60 (noise plus signal). These studies extend and
confirm earlier published work on pupil noise [39].

The Fourier coefficients of the signal response were computed cycle
by cycle for 40 cycles, and their statistical distribution was studied. Figure 61 shows this distribution of the alpha and beta coefficients for a noise
signal run and for a noise run in which, in fact, no signal existed. It can
be seen that there is an approximate Gaussian distribution considering the
small population sample of only 40, and that the signal clearly shifts the
alpha coefficient mean, as would be expected. Figure 62 shows the distribution of the amplitude and phase characteristics of the same data. The
amplitude coefficients are still quasi-Gaussian, and shifted upward by the

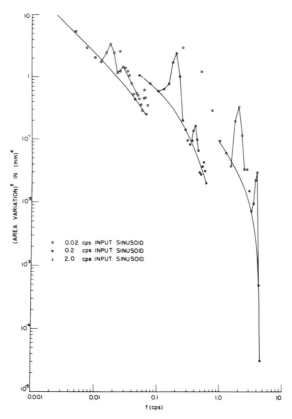

Fig. 60. Power spectrum of input sinusoidal responses of
0.02, 0.2, and 2.0 cycles/sec taken from three different
data runs. Variation in (pupil area)^2 in (mm^4) vs. frequency (cycles/sec).

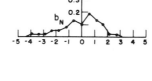

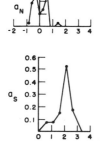

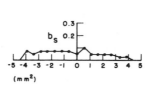

Fig. 61. Normalized probabilities of
Fourier coefficients. These curves show the
distribution of Fourier coefficients for the two
experimental conditions shown in Fig. 54a
and 54b (2 cycles/sec). Curves a_n and b_n
show alpha and beta Fourier coefficients
for the noise case; curves a_s and b_s show
alpha and beta Fourier coefficients for the
signal plus noise case. The ordinates are
normalized and the abscissa represents the
area in mm^2.

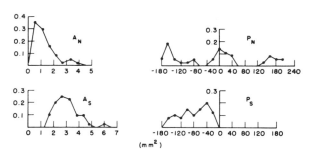

Fig. 62. Normalized probabilities of amplitude and phase coefficients. These curves show the distribution of amplitude and phase coefficients for the two experimental conditions shown in Fig. 54a and 54b (2 cycles/sec). Curves A_N and P_N show amplitude and phase coefficients for the noise case; A_S and P_S show coefficients for the signal plus noise case. The ordinates are normalized and the abscissas represent area in mm^2 for the amplitudes, and in degrees for the phase.

signal, as expected. The phase coefficients for the noise alone show a peculiar and interesting distribution which was predicted from a mathematical study by Mr. Frank Kuhl under the direction of Professor P. M. Schultheiss, of Yale University. With signal added, the phase coefficients show an agglomeration about a phase lag of 300°, as expected from earlier studies [44].

The experimental results, in general, confirm the validity of treating the pupillary response as a signal additive to Gaussian pupillary noise.

Chapter 4

Functional Analysis of Pupil Nonlinearities

INTRODUCTION

In attempting to describe mathematically or to characterize the black box input–output behavior of the human pupil light reflex, we investigated three different basic formulations of the system. The Wiener G-functionals, a recently developed mathematical characterization of systems, will be discussed in detail but we will first briefly present two other, more standard approaches to the problem, describing functions and heuristic modeling [30]. Neglecting the experimental details — some of which will be presented later — let us assume that the experimenter has available transducers capable of generating input light stimuli as wave forms of arbitrary time dependence and monitoring an analog voltage proportional to the output, which is pupil area of a normal awake human subject.

The linear transfer function description attempts to represent the input–output behavior by a linear system. Stark and Sherman [44] first used this method and were able to predict the frequency of high loop gain, instability oscillations [40]. These studies have been extended in a quantitative manner and are in part presented here. Figure 63 shows typical input–output records of the human pupil response to sinusoidally varying light; gain and phase vs. frequency curves for this class of experiments are seen in Fig. 64. The asymptotic slope of the gain vs. frequency curves indicates a system of approximately third order. One is immediately suspicious of any linear model since the gain curves change appreciably with stimulus input amplitude. Note, however, that the phase vs. frequency curves are almost identical for different stimulus amplitudes. It is this very strong behavioral constraint in the pupil light reflex system which enabled Stark, Cornsweet, and Baker [38, 40] to use a linear model to predict correctly the closed-loop oscillation frequency in their studies of the pupil.

160

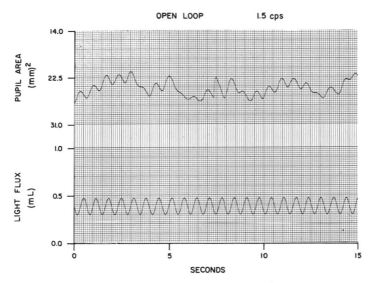

Fig. 63. Pupil response to steady-state light flux changes.

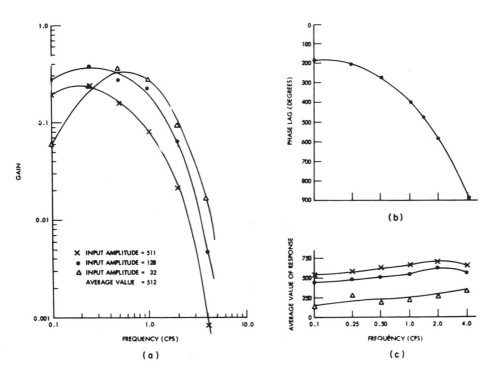

Fig. 64. Sinusoidal steady-state response of pupil system to light as a function of frequency with stimulus amplitude as a parameter.

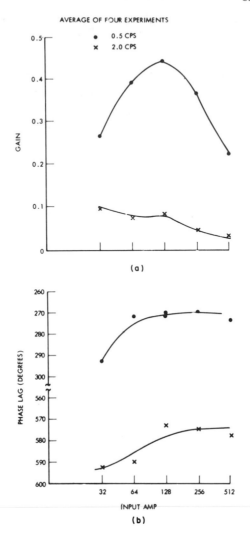

Fig. 65. Sinusoidal steady-state response of
pupil system to light as a function of stimulus amplitude with frequency as a parameter.

In order to present more clearly the pupil nonlinearities, we have
plotted in Fig. 65 the gain and phase vs. input amplitude with frequency of
stimulation as a parameter. Since the gain vs. amplitude curves at the two
stimulation frequencies are appreciably different shapes, the pupil system
cannot be represented simply as a no-memory nonlinearity followed by a
linear system. Therefore it is impossible to describe the system in simple
describing function theory. It can, however, be shown that the previous

sinusoidal response data are consistent with a model comprised of some combination of cascaded linear systems separated by no–memory nonlinearities [34].

An extremely popular and often useful modeling technique is to perform a series of experiments, usually with transient stimuli. One acquires experience with the system which — coupled with other information one might think pertinent, including preconceived notions — thereby enables formulation of a time–invariant, nonlinear heuristic model. The model's primary purposes are to describe the input–output behavior and approximate, at least in some sense, the topology of the signal flow in the real physical system.

In the case of the pupil light reflex, there are several easily observed nonlinear effects. Typical pulse responses are seen in Fig. 66. In addition to the basic wave shape, one immediately observes that a substantial amplitude compression exists fairly early in the signal processing system since the amplitude of response varies over about a factor of 3 while the input signal changed over a factor of 32 and the response shapes are extremely close when all the responses are normalized to the same peak–to–peak height.

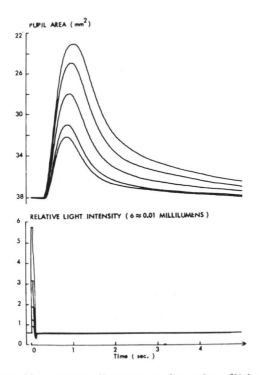

Fig. 66. Human pupil responses to short pulses of light.

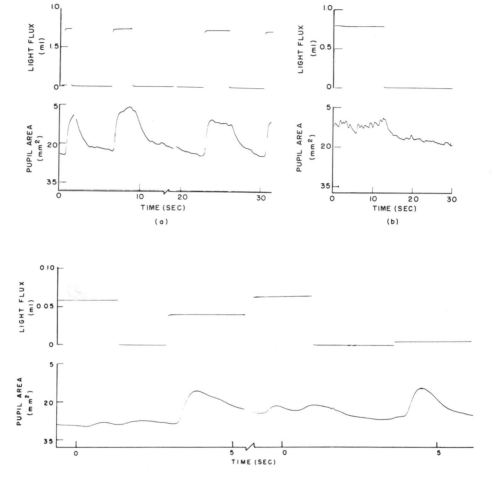

Fig. 67. Pupil light reflex showing asymmetrical nonlinearity effect.

Figure 67 illustrates another important nonlinearity — asymmetry. This nonlinearity seems to show its effect especially when the stimuli approach the small signal region [37]. Because of this, some investigators have raised serious objections to the use of methods which attempt to linearize the pupil system in order to carry out a quantitative analysis [9,34]. The model illustrated in Fig. 68 represents one attempt to form a heuristic model of the pupil light reflex. In such model making one must choose between making the model more complex to incorporate some experimental data or settling for a simpler model, incomplete but more understandable. We will not stress this heuristic method of characterization, but instead proceed to the Wiener method, which lacks close connection with possible physiological processes, but is a general mathematical description capable of representing an extremely wide class of systems.

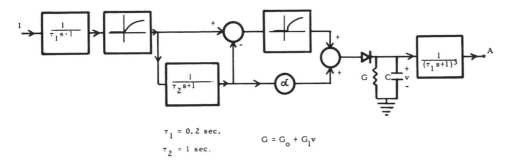

τ_1 = 0.2 sec.

τ_2 = 1 sec.

$G = G_0 + G_1 v$

Fig. 68. Heuristic model of human pupil light reflex.

This characterization, the main subject of this chapter, is essentially a Volterra functional expansion of the pupil system with kernels $h_0(\ ), h_1(\ )$, $h_2(\ )$, etc.* Output and input are related as follows:

$$y(t) = \int h_1(\tau) x(t-\tau) d\tau + \int\int h_2(\tau_1, \tau_2) x(t-\tau_1) x(t-\tau_2) d\tau_1 d\tau_2 + \cdots \tag{1}$$

The computationally difficult task in this characterization is the determination of the kernels $h_i(\tau_1, \tau_2, \ldots, \tau_i)$, $i = 1, \ldots, n$. An even more profound question which arises in this method is whether any physical significance or interpretation can be given the kernels — even if they could be found — or are we to pay the price for mathematical elegance by an inability to interpret our results? Do the kernels have any other identifiable significance in addition to their being the magic kernels that simply crank out the correct output when introduced into the Volterra expansion? First, let's describe two analytic procedures for obtaining the kernels and then present the experimental results pertinent to each theoretical method.

METHODS

Measurements by Use of Random Signals

A linear system as shown in Fig. 69 is represented in the time domain by its impulse response function $h(\tau)$. The output $y(t)$ is related to the input $x(t)$ by the superposition or convolution integral.

$$y(t) = \int h(\tau) x(t-\tau) d\tau \tag{2}$$

One method of experimentally computing $h(\tau)$ is to excite the linear

---

*The term $h_0(\ )$ is the zeroth-order kernel and represents the constant d-c level with zero input; it is ignored in the remainder of our paper.

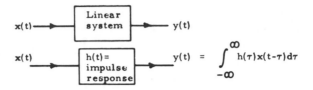

Fig. 69. Representation of linear system by its impulse response h(t).

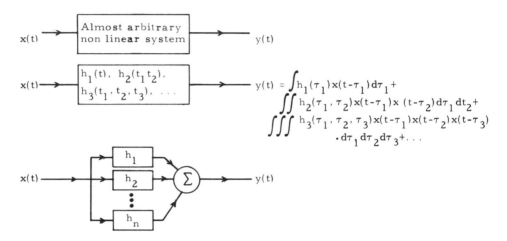

Fig. 70. Volterra representation of nonlinear system. $h_1(\tau_1)$, $h_2(\tau_1, \tau_2)$, etc., are known as Volterra kernels of system.

system with white noise, that is

$$\varphi_{xx}(\tau) = Au_0(\tau)$$

where u_0 is the Dirac delta function, and to cross correlate the output and input [19]. It can be seen that

$$h(\tau) = (1/A)\varphi_{yx}(\tau) \tag{3}$$

The representation of a general nonlinear system is shown in Fig. 70. If the system under study can be suitably approximated by only one of the Volterra terms of the infinite expansion — in general — then a simple technique exists for determining the kernel of that term [18]. Figure 71 illustrates the computation of h_n by multidimensional cross correlation between output and input when the input is white Gaussian noise. The compu-

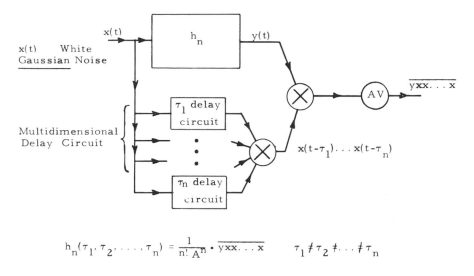

$$h_n(\tau_1, \tau_2, \ldots, \tau_n) = \frac{1}{n! \, A^n} \cdot \overline{y \mathbf{xx} \ldots \mathbf{x}} \qquad \tau_1 \neq \tau_2 \neq \ldots \neq \tau_n$$

Fig. 71. Measurement of isolated high order kernel.

tation of the isolated high order Volterra kernel is a direct extension of the case of a linear system. It should be noted that although we cannot determine the values of $h_n(\tau_1, \ldots, \tau_n)$ along any curve where any two τ's are equal with this method, one can practically assume continuity of h_n to find such values if needed.

An extremely serious drawback to the Volterra series is that it is not an orthogonal functional expansion. Thus if the first n terms have already been computed, then the entire computation must be repeated if $n + 1$ terms are desired. Another important defect of the Volterra method is that computation of the nonisolated kernels is no longer a simple task but involves the solution of a set of simultaneous integral equations [17].

In much the same way that Legendre polynomials are formed to make an orthogonal function set useful for curve fitting, so can a set of orthogonal functionals for nonlinear system characterization be formed. This was first done by Norbert Wiener, whose work was further simplified by Lee and his co-workers [19]. Omitting details of this work, we can say that the general measuring scheme in the determination of Wiener G-functionals is identical to that of determining isolated Volterra kernels. The Wiener representation is illustrated in Fig. 72 where the experimentally determined g_i are used for forming the Wiener G-functional expansion.

After using a Legendre series for minimal square curve fitting, one may go back and, merely collecting appropriate terms, form the

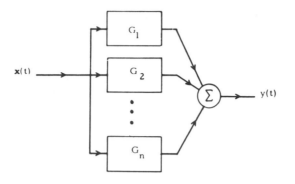

$$G_1\left[g_1 \cdot x(t)\right] = \int g_1(\tau) \, x(t-\tau) \, d\tau$$

$$G_2\left[g_2 \cdot x(t)\right] = \iint g_2(\tau_1, \tau_2) \, x(t-\tau_1) \, x(t-\tau_2) \, d\tau_1 \, d\tau_2 - A \iint g_2(\tau, \tau) \, d\tau$$

$$G_3\left[g_3 \cdot x(t)\right] = \iiint g_3(\tau_1, \tau_2, \tau_3) \, x(t-\tau_1) \, x(t-\tau_2) \, x(t-\tau_3) \, d\tau_1 \, d\tau_2 \, d\tau_3$$
$$- 3A \int \left[\int g_3(\tau_2, \tau_2, \tau_1) \, d\tau_2 \right] x(t-\tau_1) \, d\tau_1$$

Fig. 72. Orthogonal Wiener G-functionals.

algebraic polynomial of the form:

$$y = a_0 + a_1 x + a_2 x^2 + \cdots \tag{4}$$

In a similar manner, after one has first determined the G-functionals —
because of their computational convenience — it is a relatively simple mat-
ter to perform some simple integration and collection of terms in order to
form the corresponding Volterra expansion. Since a finite G-functional
expansion represents the best minimal squares approximation to the system, the derived Volterra expansion must, for the constrained maximum
order of the expansion, also represent the best system characterization in
the mean square sense.

Continuing to relate ordinary power series expansions to functional
expansions, we see that just as the polynomial series is more amenable to
interpretation than the corresponding orthogonal expansions, so the individual terms of a Volterra expansion are more easily interpreted than the
G-functionals themselves. We will examine some of the properties of individual Volterra kernels by applying special transient inputs to the system.

It should be noted that the condition for use of white noise is incom-
patible with actually performing the necessary measurements experimen-

tally since such an input has infinite variance and it can be shown that as a consequence all of the needed averages (high order correlation functions) will also have undefined variances, not to mention the infinite power content of such a signal. The effect of band-limited noise on the measurement will be seen later.

A Volterra or Wiener G-functional characterization can be used to represent a wide class of nonlinear time invariant finite memory systems. The more theoretical aspects such as the exact, necessary and/or sufficient conditions for this type of characterization to converge will not, and in some cases, cannot be derived.

Measurements in the Time Domain

Whereas the previous analysis with random inputs gave rise to a nice mathematical framework, the work now to be presented lacks that foundation but obtains for us some badly needed insight into high order kernels. If a linear system is excited by an impulse, the response is by definition the kernel or impulse response function of the linear system. The generalization of this type of input in order to determine higher order kernels than the first consists of presenting multipulse stimuli where time t and the relative spacing of the pulses are the independent variables tracing out the high order kernels. The method is not orthogonal in the sense that the entire procedure must be repeated if we wish to extend the maximum order of the approximation. A more serious objection to the method is that we will form the Volterra characterization to agree exactly with the real system for one very specialized input and if the system contains more kernels than we assume in the analysis, the derived characterization might prove very poor for other classes of input. The advantage of a random input to characterize the system is that one essentially obtains the best characterization over a wide class of inputs although the representation might not be very good for a particular input if the Volterra series has been truncated. We will treat in detail the case of a double pulse stimuli.

In this case the system is to be approximated by the first two terms of a Volterra series. The word "approximate" could be misleading since we are not using any criterion of goodness *de facto* because the method is going to make $h_1(\tau)$ and $h_2(\tau_1, \tau_2)$ be such that they completely describe the double pulse experiment. The approximation is that the system is being represented for all inputs as the first two terms of a Volterra expansion. The three pertinent experiments necessary for this method are tabulated below:

Experiment 1: $x(t) = Au_0(t)$

therefore: $y_1(t) = Ah_1(t) + A^2h_2(t,t)$

Experiment 2: $x(t) = Au_0(t - T)$

therefore: $y_2(t) = Ah_1(t - T) + A^2 h_2(t - T, t - T)$

Experiment 3: $x(t) = Au_0(t) + Au_0(t - T)$

therefore: $y_3(t) = Ah_1(t) + Ah_1(t - T) + A^2 h_2(t, t) +$

$$A^2 h_2(t - T, t - T) + 2A^2 h_2(t, t - T)$$

where $h_2(\tau_1, \tau_2) = h_2(\tau_2, \tau_1)$ has been used. It is easily seen that

$$y_3 - y_1 - y_2 = 2A^2 h_2(t, t - T)$$

and that

$$-y_3 \big|_{T = 0} + 4y_1 = 2A h_1(t)$$

The term in the output of experiment 3, $h_2(t, t - T)$, represents a nonlinear interaction between the two stimuli pulses. Note that if $h_2(t, t - T) = 0$ for $T \neq 0$, then for $T \neq 0$ the system would obey what we might call time superposition in the sense that we could add the responses due to each stimulus pulse presented separately and correctly predict the output when both were presented together. Thus $h_2(\tau_1 = \tau_2)$ being nonzero off its main diagonal $(\tau_1 = \tau_2)$ line indicates the degree of time nonlinear interaction between different portions of the input signal. A no-memory squarer followed by an arbitrary linear system and possibly shunted by another linear system has this exact property. Any no-memory nonlinearity followed by a linear system obeys the principle of time superposition. In this general case the order of the required Volterra expansion is determined by the order of the nonlinearity and every kernel is zero everywhere except where $\tau_i = \tau_j$; along these lines the kernel is a wall of impulses. Thus the magnitude of the kernels off their diagonals indicates the extent of the nonlinear interaction in time.

Further Topics in Quantitative Measurement:
Cascading and Bandwidth Effects

If an isolated Volterra kernel $h_2(\tau_1, \tau_2)$ is followed by a linear system $h(\tau)$, the following relation between input $x(t)$ and output $y(t)$ is readily derived:

$$y(t) = \iint [h_2(\lambda_1 - \xi, \lambda_2 - \xi) h(\xi) d\xi] x(t - \lambda_1) x(t - \lambda_2) d\lambda_1 d\lambda_2 \qquad (5)$$

The new kernel is therefore

$$h_2(\lambda_1, \lambda_2) = \int h_2(\lambda_1 - \xi, \lambda_2 - \xi) h(\xi) d\xi \qquad (6)$$

For any given (λ_1, λ_2) this can be seen to be just a weighted average of the original h_2 along lines parallel to the main diagonal $\lambda_1 = \lambda_2$; thus we are in effect smearing out h_2 along lines parallel to $\lambda_1 = \lambda_2$.

If an isolated Volterra kernel h_2 is preceded by a linear system h, the following relation between input and output is true:

$$y(t) = \iint [\iint h_2(\lambda_1, \lambda_2)h(\tau_1 - \lambda_1)h(\tau_2 - \lambda_2)d\lambda_1 d\lambda_2]x(t - \tau_1)x(t - \tau_2)d\tau_1 d\tau_2 \qquad (7)$$

The new second-order kernel is therefore

$$\iint h_2(\lambda_1, \lambda_2)h(\tau_1 - \lambda_1)h(\tau_2 - \lambda_2)d\lambda_1 d\lambda_2 \qquad (8)$$

which is simply a two-dimensional convolution integral. This will clearly smear the original h_2 both parallel and perpendicular to the $\lambda_1 = \lambda_2$ diagonal line.

In conclusion, we see that a linear system preceding a nonlinear one will generally increase the nonlinear interaction time whereas the linear system following the nonlinear one will only increase the overall memory of the system.

Using some of the above results, it is interesting to ask if it is possible to find two linear systems, one of which will precede an h_2 and the other which will follow the same h_2 so that the overall behavior of the two resulting systems is identical. The easiest route to a solution is first to define the multidimensional Laplace transform as

$$F(s_1, \ldots, s_n) = \int \ldots \int f(t_1, \ldots, t_n)e^{-s_1 t_1} \ldots e^{-s_n t_n} dt_1 \ldots dt_n \qquad (9)$$

Let $h_2(\tau_1, \tau_2) \leftrightarrow H_2(s_1, s_2)$ and $h(\tau) \leftrightarrow H(s)$, $h'(\tau) \leftrightarrow H'(s)$ define the nonlinear system and the two linear systems, respectively, with h' denoting the one preceding h_2. That is, is it possible for us to find an $h(\tau)$ and $h'(\tau)$ such that the two configurations of Fig. 73 are equivalent. Using the previously derived material, it is not difficult to show that the basic condition for the system of Fig. 73 to be equivalent is that

$$H_2(s_1, s_2)H(s_1)H(s_2) = H_2(s_1, s_2)H'(s_1 + s_2) \text{ or}$$

$$H(s_1)H(s_2) = H'(s_1 + s_2) \qquad (10)$$

for all values of s_1 and s_2. The only nontrivial $H(s)$ and $H'(s)$ for which this is true is $H(s) = H'(s) = e^{-Ts}$ which represents a time delay of T seconds. This result should not be surprising given the previously developed interpretations concerning the difference between linear systems preceding and following nonlinear ones.

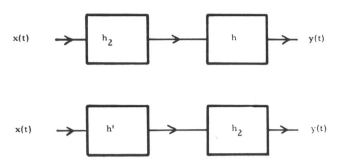

Fig. 73. Permutations of a second-order kernel and a linear
system.

We now turn our attention to a seemingly more pressing problem of
measurement. Functional analysis theory requires white noise whereas
statistical theory tells us that we have an impossible task in obtaining reliable correlation estimates, not to mention the impossibility of generating
white noise from energy considerations. It will turn out, as expected, that
we will only have to use noise of a bandwidth of an order of magnitude
greater than the system's bandwidth in order to make the previously stated
equations for the determination of Wiener G-functionals valid for all practical purposes. We call noise of this bandwidth adequate bandwidth noise.
Our basic problem is illustrated in Fig. 74 in which the system we are
trying to characterize is denoted by S and only signals $y(t)$ and $x(t)$ are
available [31]. $h(t)$ is the impulse response of a low-pass filter which filters the imaginary white Gaussian noise $x_w(t)$ to $x(t)$, the real input into
the system. Each kernel of S' and S is related by

$$h_n'(\tau_1, \ldots, \tau_n) = \int \ldots \int h_n(\lambda_1, \ldots, \lambda_n) h(\tau_1 - \lambda_1) \ldots h(\tau_n - \lambda_n) d\lambda_1 \ldots d\lambda_n \qquad (11)$$

which is merely the generalization of Eq. (8). By taking the multidimensional Laplace transform of both sides and dividing through $H(s_1) \ldots H(s_n)$
and then taking the inverse transform, it is, at least in principle, possible
to get h_n, thus characterizing S if we knew h_n'. Characterization of S' by
h_n', however, involves the computation of

$$\varphi_{y x_w x_w \ldots x_w}(\tau_1, \ldots, \tau_n) \qquad (12)$$

which cannot be done since $s_w(t)$ is unavailable in any physical sense.
However $\varphi_{y x_w \ldots x_w}$ is related to $\varphi_{yx \ldots x}$ by

$$\varphi_{xy \ldots x}(\tau_1, \ldots, \tau_n) = \int \ldots \int \varphi_{y x_w \ldots x_w}(\xi_1 + \tau_1, \ldots, \xi_n + \tau_n) \cdot$$
$$h(\xi_1) \ldots h(\xi_n) d\xi_1 \ldots d\xi_n \qquad (13)$$

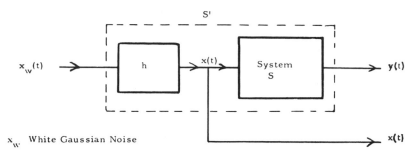

Fig. 74. Measurement of high order kernels with nonwhite Gaussian noise.

The above becomes obvious after we consider the case for a second-order kernel as follows:

by definition

$$\varphi_{yxx}\,(\tau_1,\,\tau_2) = \overline{y(t)x(t-\tau_1)x(t-\tau_2)}^{\,t}$$

and

$$x(t-\tau) = \int h(\xi)x_w(t-\tau-\xi)d\xi$$

therefore

$$\varphi_{yxx}\,(\tau_1,\,\tau_2) = \int\int h(\xi_1)h(\xi_2)\overline{y(t)x_w(t-\tau_1-\xi_1)x_w(t-\tau_2-\xi_2)}^{\,t}\cdot d\xi_1 d\xi_2 \qquad (14)$$

and hence, the result previously stated is proved.

We can again use multidimensional Laplace transform techniques to find $\varphi_{yx_w...x_w}(\tau_1,\,...,\,\tau_n)$ if desired.

In practical measurements, we use adequate bandwidth noise so that multidimensional Laplace transforms and their inversions prove to be unnecessary. The effect of lesser bandwidth of the input can, however, be quantitatively estimated by reference to Eqs. (11) and (13), i.e., Eq. (11) shows that the obtained correlation functions are really the result of convolving the desired correlation functions with the low-pass filter's impulse response. Equation (13) shows the corresponding property for the kernel of interest.

Experimental Methods

An infrared reflecting pupillometer was used to record continuously the pupil area. A glow modulator, voltage-to-light transducer was used as stimulus generator in an optical arrangement which prohibited the pupil from affecting the light reaching the retina as it normally does.

An on-line GE-225 digital computer with integral analog-to-digital and digital-to-analog conversion equipment was used in the experimental phase of the study. The on-line computer was used to generate the Gaussian noise stimulus and to input the corresponding response. Stimulus and response were punched on cards in a highly compressed binary format for further data processing. The necessary high order correlation and averaging programs were written in Fortran and run on an IBM 7094 at the Massachusetts Institute of Technology Computation Center. Computation time for evaluation of the first and second kernels seen in Fig. 77 was approximately 10 min.

The on-line computer was also used in the double pulse experiment to generate the required stimuli in a random order to minimize long-term trends and to average the corresponding responses. At the termination of the experiment, the average response data was punched on cards. Another relatively simple program performed the necessary algebraic manipulations on the data and plotted out the second-order kernel.

EXPERIMENTAL RESULTS

Noise Excitation

Figure 75 illustrates typical input-output data for the pupil system's response to random light excitation. From sinusoidal experiments we have seen that the pupil system has negligible response beyond 4 cycles/sec. As a consequence, the noise bandwidth was made about 10 cycles/sec. The rectificationlike nonlinear behavior is clearly evident by noting the sharp difference between the steady-state pupil area before and during excitation. The computer was also programmed in such a way as to be able to output the same pseudo-random stimulus many times in order to observe

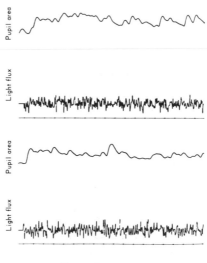

Fig. 75. Effect of asymmetry on pupil
response to random excitation.

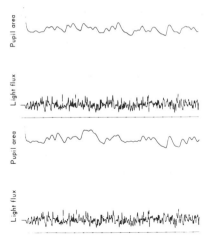

Fig. 76. Pupil response for two identical pseudo-random stimuli functions.

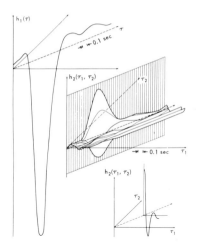

Fig. 77. Wiener kernels $h_1(\tau)$, $h_2(\tau_1, \tau_2)$, and cross section of h_2.

the effect of noise unrelated to the excitation. Input–output data for two identical pseudo-random stimuli functions are seen in Fig. 76. Finite data length and random pupil variation unrelated to the input were the main sources of error. In the taking of experimental data, the subject is instructed not to blink for periods up to 1 min inasmuch as such a strain on the subject often introduces unwanted drift as the run proceeds. It was found very useful to subtract from the data the best – in the least square sense – second-order equation as an attempt to cope with this problem. If this is not done, one may obtain experimental auto- and cross correlations which have large variation from run to run.

Figure 77 shows the first- and second-order kernels which are basically of opposite sign. The shape of the first-order kernel and that of the main diagonal slice of the second closely resemble a pupil pulse response. The width of the second-order kernel off the main diagonal (see right bottom of Fig. 77 for cross section of h_2) is only about $1/2$ sec, indicating that there is no second-order nonlinear interaction longer than this time.

Double Pulse Excitation

Stimulus response records for the double pulse experiment are seen in Fig. 78 [1]. Basically, the relatively noise-free nature of the data is due to two factors: the pulse experiments were performed at relatively low light with the subject fixating at optical infinity, thereby eliminating the large amount of noise related to high levels of illumination and near fixation, and extensive computational averaging of the data.

If one examines the data closely, it would be observed that the latent period is the same for the second as for the first pulse response. This implies that the higher order kernels all have the same inherent time delay, which would seem to indicate that the time delay mechanism is not intimately connected with the nonlinear interaction process.

Figure 79 shows the computed second-order kernel, under the assumption that the system can be completely described by just the first two terms in a Volterra expansion. That the shapes of the slices shown in Fig. 79 closely resemble the elementary pupil pulse responses again emphasizes the importance of nonlinear interaction occurring before the basic time-

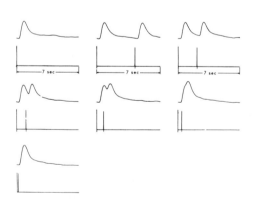

Fig. 78. Double pulse light stimulation of the human pupil system.

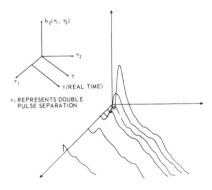

Fig. 79. Second-order Volterra kernel
as derived from double pulse experiment.

shaping mechanism occurs in the system. Note also that the spread
off the main diagonal is appreciably greater in the double pulse experiment
than in the noise experiments.

DISCUSSION

The difference in spread of the two kernels found is surprising. One
can hypothesize that the relatively narrow width of the random kernel is
due to the pupil being at a small average area because other experiments
indicate a large bandwidth of the pupil system when the pupil is small. Another possibility is that the continued excitation put the neurological portion
of the system into a state where nonlinear interactions are greatly reduced.

The basic sign difference between the first- and second-order kernels represents the scale compression nonlinearity mentioned earlier.
This is analogous to the first two nonconstant terms of a Taylor series expansion for a logarithmic-type curve having opposite signs.

For some values of (τ_1, τ_2), h_2 obtained by random stimulation is of
the same sign as the first-order kernel. One might therefore conclude
that there exists some nonlinear interaction whose effect is to enhance or
facilitate the response at a particular time due to the stimulus some time
before. This is in contradiction to the double pulse experiment described
here. It would indeed be interesting to see the results of experiments suitably designed to emphasize this phenomenon if it exists. A more probable,
though less provoking, interpretation is that the negative going portion of
h_2 will have no effect on the output if we were to carry out the function expansion further.

In attempting to characterize a nonlinear system such as the pupil light reflex, it is good to have as many alternative methods of investigation as possible.

Unfortunately, it is often difficult to relate quantitatively the results of the different methods described here. The first two methods of characterization have been used with considerable success in describing biological systems: sinusoidal analysis is a very familiar approach in linear system analysis and people have developed a good feeling for it and its application; heuristic modeling guides one, hopefully, to the threshold of understanding or — which is even more important — into asking questions one would never have thought of without the model.

The functional analysis approach to the human pupil light reflex was attempted for two reasons. First, we wished to develop a canonical representation of the system which would make evident in a mathematical manner certain characteristic nonlinearities of the system. Second, we wanted to explore the usefulness of the Wiener-Lee theory of nonlinear systems on a real problem of characterization where a comparison with other methods could be made. The main application of the theory to date has been to solve problems which would be less elegantly solved by other techniques.

SUMMARY

Three general approaches to nonlinear input-output systems are contrasted: (a) describing function based mainly on sinusoidal excitation functions; (b) heuristic experimental analysis coupled with model building based on intuitive guesses as to component connections, component characteristics, and efficient particular transient inputs; (c) Wiener G-functional expansions based upon stochastic driving functions.

The mathematical methods for experimental measurements of the kernels of a system are described for random signals and multipulse inputs.

The effect of topological arrangement of component boxes and of input noise bandwidth on the methods above are discussed. Apparatus and experimental design using an on-line digital computer is briefly described.

The experimental results are separately displayed for random noise and multipulse excitation functions.

The first-(h_1) and second-(h_2) order kernels are of opposite sign and thus represent the first terms of a saturation nonlinearity. (h_1) and the main diagonal slide of (h_2) resemble the pupil pulse response and thus suggest that the nonlinear interaction occurs before the basic quasi-linear time-shaping mechanism. The width of (h_2) off the main diagonal is less than 1/2 sec, indicating that there is no second-order nonlinear interaction longer than this time. Similarly, the time delay mechanism is shown to be independent of the nonlinear interaction.

The differences noted between the kernels obtained from the random and multipulse experiments include greater (h_2) width off the main diagonal in pulse experiments and (h_2) valleys only in the noise experiments.

REFERENCES AND FURTHER READING

1.   Baker, F. H.: Pupillary response to double pulse stimulation: A study of nonlinearity in the human pupil system. J. Opt. Soc. Am. 53: 1430 (1963).
2.   Baker, F., Emmerich, G. W., Tyler, V., and Stark, L.: Unpublished experiments (1957).
3.   Bower, J. L., and Schultheiss, P. M.: Introduction to the Design of Servomechanisms, Wiley, New York (1958).
4.   Boynton, R. M.: Some temporal factors in vision. In: Sensory Communication, W. A. Rosenblith, ed., Massachusetts Institute of Technology, Cambridge, Mass. (1961), pp. 739−756.
5.   Campbell, F. W., and Robson, J. G.: Unpublished experiments (1958).
6.   Campbell, F. W., and Whiteside, T. C. D.: Induced pupillary oscillations. Brit. J. Ophthalmol. 34: 180−189 (1950).
7.   Clynes, M.: Computer dynamic analysis of the pupil light reflex: A unidirectional rate sensitive sensor. Third International Conference on Medical Electronics, London (1960).
8.   Clynes, M.: Unidirectional rate sensitivity: A biocybernetic law of reflex and humoral systems as physiologic channels of control and communication. Ann. N. Y. Acad. Sci., Pavlovian Conference Monograph on Higher Nervous Activity (1961).
9.   Clynes, M.: The non-linear biological dynamics of unidirectional rate sensitivity illustrated by analog computer analysis, pupillary reflex to light and sound, and heart rate behavior. Ann. N. Y. Acad. Sci. 98: 806−845 (1962), Article 4.
10.   Cole, K. S.: Dynamic electrical characteristics of the squid axon membrane. Arch. Sci. Physiol. 3: 253−258 (1949).

11.   Cornsweet, T. N.: Determination of the stimuli for involuntary drifts and saccadic eye movements. J. Opt. Soc. Am. 46: 987–994 (1956).

12.   Davenport, W. B., Jr., and Root, W. L.: An Introduction to the Theory of Random Signals and Noise, McGraw-Hill, New York (1958).

13.   Elkind, J. I.: Characteristics of simple manual control systems. Lincoln Laboratory Tech. Rept., M.I.T., Number III: 1–145 (1956).

14.   Evans, W. R.: Control System Dynamics, McGraw-Hill, New York (1954).

15.   Hartline, H. K.: A quantitative and descriptive study of the electrical response to illumination of the arthropod eye. Am. J. Physiol. 83: 466–483 (1928).

16.   James, H. M., Nichols, V. B., and Phillips, R. S.: Theory of Servo-mechanisms, McGraw-Hill, New York (1947).

17.   Katznelson, J.: Synthesis of nonlinear filters. Sc. D. Thesis, Elect. Eng. Dept., Massachusetts Institute of Technology, Cambridge, Mass. (1963).

18.   Lee, Y. W.: Statistical theory of nonlinear systems. Class Notes for Massachusetts Institute of Technology Elect. Eng. Course 6.572.

19.   Lee, Y. W.: Statistical Theory of Communication, Wiley, New York (1960).

20.   Lee, Y. W., and Schetzen, M.: Measurement of the kernels of a nonlinear system by cross correlation. Quart. Prog. Rept., Research Laboratory of Electronics, M.I.T. 60: 118–130 (1961).

21.   Lowenstein, O. L.: Pupillary reflex shapes and topical clinical diagnosis. Neurology 5: 631–644 (1955).

22.   Lowenstein, O., and Givner, I.: Pupillary reflex to darkness. Arch. Ophthalmol. 30: 603–609 (1943).

23.   Lowenstein, O., and Loewenfeld, I. E.: Role of sympathetic and parasympathetic systems in reflex dilatation of the pupil. Arch. Neurol. Psychiat. 64: 313–340 (1950).

24.   MacColl, L.: Fundamental Theory of Servomechanisms, Van Nostrand, Princeton, N. J. (1945), p. 107.

25.   Marmont, G.: Electrode clamp for squid axon. J. Cellular Comp. Physiol. 34: 351 (1949).

26.   Mason, S. J., and Zimmermann, H.: Electronic Circuits, Signals, and Systems, Wiley, New York (1960).

27.   McCulloch, W. S.: Why the mind is in the head. In: Cerebral Mechanisms in Behavior, The Hixon Symposium, Wiley, New York (1951), pp. 42–57.

28.   McRuer, D. T., and Krendel, E. S.: Dynamic response of human operators. WADC Tech. Report 56: 524 (1957).

29. Newton, G. C., Gould, L. A., and Kaiser, J. F.: Analytical Design of Linear Feedback Controls, Wiley, New York (1957).

30. Sandberg, A. A., and Stark, L.: Wiener G-function analysis as an approach to nonlinear characteristics of human pupil light reflex. IEEE Trans. Biomedical Engineering (1966). ·

31. Schetzen, M.: Measurement of the kernels of a nonlinear system by cross correlation with Gaussian non-white inputs. Quart. Prog. Rept., Research Laboratory of Electronics, M.I.T. 63: 113−117 (1961).

32. Schultheiss, P. M., and Bower, J. L.: An Introduction to the Design of Servomechanisms, Wiley, New York (1958), 500 pp.

33. Stanten, S. F., and Stark, L.: A biological stochastic process (pupil noise). Transactions of the 4th Colloquium on the Pupil, May 11−12, 1966.

34. Stark, L.: Stability, oscillations, and noise in the human pupil servomechanism. Proc. Inst. Radio Eng. 47: 1925−1939 (1959).

35. Stark, L.: Oscillations of a neurological servomechanism predicted by the Nyquist stability criterion. In: Selected Papers in Biophysics, Yale University Press, New Haven, Conn. (1960).

36. Stark, L.: Environmental clamping of biological systems. Quart. Prog. Rept., Research Laboratory of Electronics, M.I.T. 66: 229−230 (1961).

37. Stark, L.: Biological rhythms, noise, and asymmetry in the pupil-retinal control system. Ann. N.Y. Acad. Sci. 98: 1096−1108 (1962), Article 4.

38. Stark, L., and Baker, F.: Stability and oscillations in a neurological servomechanism. J. Neurophysiol. 22: 156−164 (1959).

39. Stark, L., Campbell, F. W., and Atwood, J.: Pupil unrest: An example of noise in a biological servomechanism. Nature (London) 182: 857−858 (1958).

40. Stark, L., and Cornsweet, T. N.: Testing a servoanalytic hypothesis for pupil oscillations. Science 127: 588 (1958).

41. Stark, L., Iida, M., and Willis, P. A.: Dynamic characteristics of the motor coordination system in man. Quart. Prog. Rept., Research Laboratory of Electronics, M.I.T. 60: 229 (1961).

42. Stark, L., and Kuhl, F.: Analysis of pupil response and noise. Quart. Prog. Rept., Research Laboratory of Electronics, M.I.T. 62: 255−262 (1961).

43. Stark, L., Redhead, J., and Payne, R. C.: Asymmetrical behavior in the pupil system. Quart. Prog. Rept., Research Laboratory of Electronics, M.I.T. 61: 223−230 (1961).

44. Stark, L., and Sherman, P. M.: A servoanalytic study of the consensual pupil reflex to light. J. Neurophysiol. 20: 17−26 (1957).

45.   Stark, L., Vossius, G., and Young, L. R.: Predictive control of eye
        tracking movements. IRE Trans. Human Factors Electron. HFE-3(2):
        52−57 (1962).

46.   Stegemann, J.: On the influence of sinusoidal variations in brightness
        upon pupillary size. Pflügers Arch. Ges. Physiol. 264: 113−122
        (1957).

47.   Stern, H. J.: Simple method for early diagnosis of abnormalities of
        pupillary reaction. Brit. J. Ophthalmol. 28: 276−278 (1944).

48.   Talbot, S. A.: Pupillography and pupillary transient. Doctoral Dis-
        sertation, Dept. of Physics, Harvard University, Cambridge, Mass.
        (1938).

49.   Tustin, A.: The nature of the operator's response in manual control
        and its implications for controller design. J. IEE 94(11A), No. 2
        (1947).

50.   Van Der Tweel, L. H., and Denior Van Der Gon, J. J.: The light re-
        flex of the normal pupil of man. Acta Physiol. Pharmacol. Neerl.
        8: 69 (1959), Fig. 15.

51.   Wiener, N.: Cybernetics, Wiley, New York (1948).

52.   Wiener, N.: Nonlinear Problems in Random Theory, The Technology
        Press, Massachusetts Institute of Technology, and Wiley, New York
        (1958).

53.   Wybar, K. C.: Ocular manifestations of disseminated sclerosis.
        Proc. Roy. Soc. Med. 45: 315−320 (1952).

54.   Young, F. A., and Biersdorf, W. R.: Pupillary contraction and dilata-
        tion in light and darkness. J. Comp. Physiol. Psychol. 47: 264−268
        (1954).

55.   Young, L. R.: A sampled data model for eye tracking movements.
        Doctoral Dissertation, Dept. of Aeronautics and Astronautics, Mas-
        sachusetts Institute of Technology, Cambridge, Mass. (1962).

56.   Young, L. R., Green, D. M., Elkind, J. L., and Kelly, J. A.: Adaptive
        characteristics of manual tracking. Fourth National Symposium on
        Human Factors in Electronics, Washington, D.C. (1963).

57.   Young, L. R., and Stark, L.: Variable feedback experiments support-
        ing a discrete model for eye tracking movements. IEEE Trans.
        Human Factors Electron. HFE-4(1): 38−51 (1963).

58.   Blackman, R. B., and Tukey, J. W.: The Measurement of Power
        Spectra, Dover, New York (1959).

59.   Edwards, L. F.: Concise Anatomy, McGraw-Hill, New York (1956),
        Chaps. 53 and 54.

60.   Laning, J. H., Jr., and Battin, R. H.: Random Processes in Automat-
        ic Control, McGraw-Hill, New York (1956).

61.   Parzen, E.: Stochastic Processes, Holden Day, San Francisco (1962).

62.   Sandberg, A. A.: Sinusoidal light stimulation of the human pupil ser-
      vomechanism. Quart. Prog. Rept., Research Laboratory of Electron-
      ics, M.I.T. 74: 261−265 (1964).
63.   Stark, L., Sandberg, A. A., Stanten, S. F., Willis, P. A., and Dickson,
      J. F.: On-line digital computers used in biological experiments and
      modeling. Ann. N.Y. Acad. Sci. 115: 738−762 (1964).

THE LENS

INTRODUCTION

In Section II we have seen that the pupillary servomechanism has a number of features which contribute to its design characteristics. Among these are the saturation scale–compression nonlinearity which provides for high gain in response to small inputs. This adapts the system to its rather limited output range. Another feature is the multi–input nature of the system; the pupil reacts both to light and to target distance. In this reaction we consider the human lens accommodation system where, as we shall see, these two features, saturation nonlinearity and multiple inputs, are also present, and require careful experiments to define their influence on the lens design characteristics.

While the accommodative response to large inputs is stable and well damped, the small–input gain of the lens system is so great that even under normal conditions the system is in an unstable oscillation. Furthermore, the multiplicity of inputs or clues available to the brain in driving the lens system is great and does not permit easy identification of a "basic" clue. Error distance, or the difference between target distance and the clear vision position, is converted by the physical optics of the lens into blur of the image on the retina. This blur is an "even–error" signal. Does the lens system operate without error–direction information? Do the instability oscillations act as a phase–sensitive demodulator to provide this information? The three chapters in this section attempt to answer these and related questions.

The human lens accommodation mechanism, an example of a neurological feedback control system, plays a vital biological role in processing visual information regarding an organism's spatial relationship to its environment. The gross anatomical structures involved in lens accommodation are shown in Fig. 1, a cross section of the eyeball.

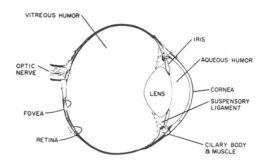

Fig. 1. Cross section of the eyeball.

The crystalline lens is composed of the elastic lens capsule and the single-layered epithelium which together contain the lens substance. The capsule of the lens is held in a state of tension and reduced curvature by the pull of the suspensory ligaments. These ligaments are attached also to the ciliary body and thence to the scleral exoskeleton which is held rigid by intraocular pressure. In this state, the ciliary muscles are at rest, and the eye is focused at infinity. For near vision, accommodation is implemented by contraction of the ciliary muscle which in turn relaxes the suspensory ligaments, permitting the lens to become more spherical and increasing its dioptric power.

The author was introduced to the physiology of this beautiful system by Dr. Fergus Campbell of Cambridge, England. At first we thought that pupillary noise [34] was somehow connected with lens oscillation, or that pupillary noise played a functional role in accommodation. It became apparent that these leads were not fruitful, and that a careful nonlinear open-loop analysis of lens behavior would be necessary. Unfortunately I was able to design and run only a few experiments on Dr. Campbell's optometer, with Dr. John Robson's aid, before returning to the United States. However, certain tentative conclusions seemed apparent early. When I moved to Massachusetts Institute of Technology, I interested Prof. George Zames and Dr. Yoshizo Takahashi in helping me with a describing function analysis based on these experiments.

Our work is presented in the first chapter of this section, Nonlinear Servoanalysis of Human Lens Accommodation, which demonstrates the saturation characteristics of the lens by means of a describing function analysis. This is a nonlinear description of accommodative response as a function of input frequency and amplitude. The analytical model reconciles large and small signal responses from experimental data, and shows the system to be unstable for very small inputs. This nonlinear characteristic permits continual oscillations to occur, even though larger responses are stable. An important question raised in this chapter is whether these os-cillations have a role in converting the even-error signal of the lens sys-tem to an odd-error signal.

Dr. Takahashi and I were interested in exploring the nature of sec-ondary aberrations that the literature [13, 19] suggested were normally used odd-error clues for accommodative tracking, i.e., clues with both sign and magnitude information. However, upon careful examination of our results, we found that only magnitude information was acted upon.

Thus in the second chapter, Absence of an Odd-Error Signal Mech-anism in Human Accommodation, we answer negatively the question raised in the first chapter regarding the role of lens oscillations. Under re-stricted viewing conditions the lens acts as an even-error system, with ret-inal blur apparently providing the even-error signal. The lens oscillations described in Chapter 1 are shown to be nonfunctional in providing error signals. Rather, they serve as another example of the tolerance of these class I nonlinear biological control systems to noise.

The results of the second chapter are confirmed in the third, Accom-modative Tracking: A Trial-and-Error Function, through the use of a more direct measure than the hand tracking used in the second chapter — namely, the accommodative-convergence response to target distance.

Chapter 1

Nonlinear Servoanalysis of Human Lens Accommodation

INTRODUCTION

The automatic focusing system of the human eye is an important biological and neurological servomechanism and has attracted the interest of many physiologists, experimental psychologists, and optometrists. When a man looks at an object his eyes focus on the target and the result is a clear image on the retina. When the eye does not focus on the object the retinal image blurs. The information concerning the state of the retinal image is sent to the central nervous system which processes it and relays to the ciliary muscle orders either to contract or to relax and thus to change the refractive power of the lens. This feedback loop is represented by the block diagram of Fig. 2. The way that the central nervous system processes the information sent from the retina is still unknown, although several hypotheses have been put forward, relating to the effects of aberrations and noise [13, 19, 38, 39].

In order to understand better the mechanism of accommodation, we investigated the dynamic properties of the lens system using a nonlinear frequency-response approach [40]. Our experimental data on the frequency response of the human lens system and the method of analysis will be fully described.

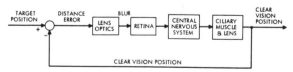

Fig. 2. Automatic focusing system of human eye. Note geometric unity gain negative feedback path.

188

EXPERIMENTAL DATA

The dynamic properties of the lens system may be defined by studying output, changes C of refractive power of the lens, as a function of input,
changes R in object distance. Direct recording of lens power is made possible by using the infrared high-speed optometer of Campbell and Robson
[8], and our data were obtained by means of this optometer. The experimental situation was identical to that of Campbell, Westheimer, and Robson
[9, 10, 13] and has been fully described by them [13].

Transient responses C of refractive power of the lens to stepwise
change R of target position are shown in Fig. 3a. The response to a step
input is approximated by a negative exponential rise of time constant T (assuming a linear first-order system), following a dead time. The responses
are somewhat dependent on direction of the movement of the target. The
average dead time and time constant obtained from these data are 0.36 sec
and 0.4 sec, respectively, for far-to-near accommodation in good agreement with Campbell and Westheimer [13]. This large dead time is the first
difficulty encountered in understanding the dynamics of the lens control
system because a dead time of 0.38 sec would cause a 180° phase lag at
only 1.3 cycles/sec, whereas there is no experimental evidence that instability is taking place at this frequency.

It was felt that careful study of the response of the lens system to
oscillatory movement of the target would be of value. One of the wave
forms of such a response C is shown in Fig. 3b together with a record of
the input. By keeping amplitude R of target motion constant and changing
frequency, the magnitudes C and phase shifts of the refractive power of the
lens were measured and are plotted as the closed-loop frequency response
of the lens system.

Figure 4 shows two gain curves, fitted by eye to each of two separate
experiments on the same subject, D.D. The first, performed on July 1,
1959, utilized moderate input amplitudes, $R = \pm1.0$ diopter, and shows little structure; the lens system appears to be of low-pass type. This, our
first frequency-response curve, confirmed previous fragmentary observations on sinusoidal responses [9, 13] and was in part published by Campbell and Westheimer as Fig. 6 of their paper [13]. The second experiment,
performed on July 9, 1959, used small-amplitude input signals R of ±0.3
diopter. It shows a peak developing at 2 cycles/sec, which follows a considerable dip. This was an exciting new finding which promised to elucidate and bring together much of the scattered observations on the lens
system. The phase curve fits both sets of data points equally well, and
this fit therefore suggests a no-memory nonlinearity. In order to study
these phenomena further, a series of sinusoidal inputs with three ampli-

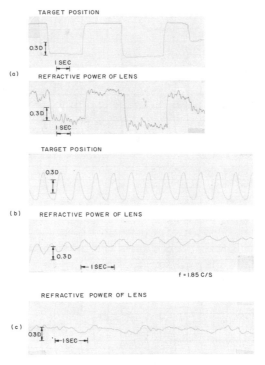

Fig. 3. Actual responses C of lens to (a) step and
(b) sinusoidal changes in target position R. Fluctuations C in lens power in (c) with constant target
position and also in (a) are consistent with very
small signal instability. Contrast with stable responses in (a) and (b). Quasi-sinusoidal responses
in (b) suggest that linear dynamics follow nonlinear element.

tudes and four frequencies were run on another subject, J.R., on July 10,
1962, in a Latin-square design to eliminate trend effects. These results
are shown in Fig. 5 where the peak at 2 cycles/sec with small input signals
is confirmed. Phase data from these and other runs on the same subject
on the same day were pooled to obtain the phase portion of this closed-loop
frequency-response curve.

The refractive power of the human lens is a constantly changing magnitude, even when the subject looks at a fixed target [3, 12, 14]. This fluctuation could be recorded, and an example is shown in Fig. 3c. In Fig. 6,
modified from Figs. 5 and 6 from the paper [10] by Campbell, Robson, and
Westheimer, are displayed power spectral density curves from two noise
records from subject J.R. The power spectral density is concerned with
power not in its physical sense but rather in the mathematical sense of the

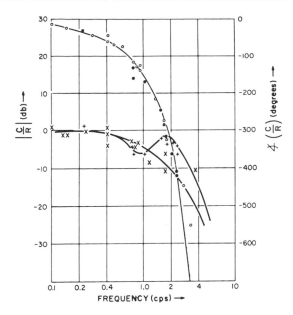

Fig. 4. Experimental closed-loop frequency response
|C/R| in decibels for two input amplitudes are shown.
Large input-amplitude gain is thick line fitted to
crosses. Note lack of structure and only low band-
pass characteristic. Small input-amplitude gain is
thick line fitted to pluses. Note definite peak following a pronounced dip. Phase curve, thin line, fits
equally both sets of experimental points, open circles
for large-input and filled circles for small-input amplitude, indicating presence of a no-memory nonlinearity.

integral square of some function, averaged over time, per unit time (usually normalized by multiplying by the constant $1/\pi 2$). The rigorous definition of spectral density is in terms of the autocorrelation function. However, its approximate intuitive significance is that of the amount of power
(in the mathematical sense) in a given small band of frequencies per unit
of frequency bandwidth. The power spectrum represents the signature of
a random process and enables one to deduce filter structures that may play
a role in shaping the noise. Earlier studies of pupil noise [34], which led
to these elegant experiments by Campbell, Robson, and Westheimer [10],
contain further descriptions of computation methods and their limitations.
The two experimental conditions compared are: a large dilated 7-mm natural pupil, and a small artificial 1-mm pupil. The pupil aperture controls
the depth of focus of the eye so that while the large pupil permits the lens
system to operate in its normal closed-loop operating condition, the small

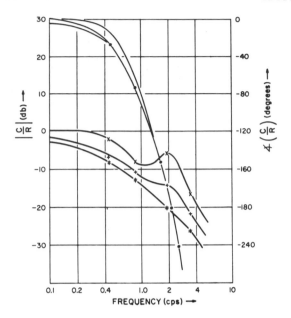

Fig. 5. Experimental closed-loop frequency response,
|C/R| in decibels. Three input amplitudes: ±0.2 diop-
ters (crosses), ±0.4 diopters (pluses), ±0.6 diopters
(double pluses). Note disappearance of peaking with
increase in input amplitude. Phase points from same
subject J.R. at ±0.4 diopters. Dispersion lines about
upper portion of phase curve indicate expected limits
of phase as a function of input amplitude if the phase
curve of the open-loop frequency response (Fig. 9) to
the three input amplitudes is assumed to be a single line.

pupil so increases the depth of focus that the lens system is essentially
open loop [10, 11]. The two curves differ at both low and high frequencies.
At low frequencies the closed-loop system attenuates a good deal of the
noise falling within the effective bandwidth of the lens feedback control
system. At the 2 cycles/sec frequency, a high peak is seen only in closed-
loop conditions, thus clearly suggesting that it is related to loop properties
of the system, and might also be related to the peak appearing in the
closed-loop frequency-response curve of Figs. 4 and 5.

COMPUTATION OF TRANSFER FUNCTION AND DESCRIBING FUNCTION

We began our analysis by assuming that Fig. 7 represents a reasonable model of the lens system; a unity gain feedback path, a frequency-dependent linear part, and a nonmemory nonlinear element [40]. An attempt

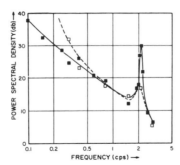

Fig. 6. Power spectral density showing
(solid line and filled squares) peaking of
noise at 2 cycles/sec consistent with reduced small signal stability margin; pupil
at 7 mm diameter. When loop is opened
by using 1-mm-diameter artificial pupil,
this peak largely disappears (dashed line
and open squares). Figure adapted from
Figs. 5 and 6 [10].

at a linear analysis of the human accommodation mechanism by John Carter [16] was unable to reconcile the various behavioral characteristics.
This nonlinear part is an essential aspect of the accommodative mechanism.
It is placed before the linear portion because of the approximately quasi-
linear form of the sinusoidal response shown in Fig. 3b. Appendix A, written by George Zames, summarizes the mathematical approach employed in
this analysis of the lens system. Further material may be found in [7, 21,
42].

The letter symbols of Fig. 7 stand for reference signal (R), error
(E), output of nonlinear part and input to linear part (M), and controlled
variable (C). Table I lists the successive functions displayed graphically
in this paper: the experimentally measured variables, their frequency
functions; computed error and thence the open-loop Bode plots; the separation into the nonmemory nonlinear part, defined by the describing function;
and the frequency-dependent linear part; the nonlinear models; their
closed-loop and open-loop frequency characteristics.

According to the model of Fig. 7, that which we measured in the frequency-response experiment is $[C(j\omega)]/[R(j\omega)]$, while the amplitude of
$R(j\omega)$ is constant. The amplitude of E does not remain constant as the fre-
quency changes. We calculated $|E/R|$ from $[C(j\omega)]/[R(j\omega)]$ as

$$\left|\frac{E}{R}\right| = \left|\frac{1-C}{R}\right| = \sqrt{1 + G^2 - 2G \cos \varphi} \qquad (1)$$

where

$$\frac{C(j\omega)}{R(j\omega)} = Ge^{-j\phi} \tag{2}$$

In Fig. 8, E is plotted against frequency. For this calculation, phase-lag invariance for different input amplitudes is assumed.

We calculated $[C(j\omega)]/[(Ej\omega)]$ under the same assumption and used the following equation:

$$\frac{C(j\omega)}{E(j\omega)} = \frac{C(j\omega)}{R(j\omega)} \qquad \frac{E(j\omega)}{R(j\omega)} = \frac{Ge^{-j\phi}}{1 - Ge^{-j\phi}} \tag{3}$$

Table I

Name	Function	Experiment or model	Figure
Time functions	C(t), R(t)	Experimental	3
Closed-loop frequency response	C/R(f)	Initial experiments	4
Closed-loop frequency response	C/R(f)	Nonlinear experiments	5
Power spectra	G(f)	Experiments in [10]	6
Error magnitude as a function of frequency	E(f)	Calculation from nonlinear experiments of Fig. 5	8
Open-loop frequency response, Bode plot	C/E(f)	Calculation from nonlinear experiments of Fig. 5	9
Open-loop gain as a function of error magnitude	C/E(E)	Calculation from nonlinear experiments of Fig. 5	10a
Describing function of nonlinear element	M/E(E)	Calculation from nonlinear experiments of Fig. 5	10b
Linear element frequency response	C/M(f)	Calculation from nonlinear experiments of Fig. 5	11
Linear element frequency response	C/M(f)	Models 1 and 2	11
Input–output function	M(E)	Model 1	12a
Closed-loop frequency response	C/R(f)	Model 1	12b
Input–output function	M(E)	Model 2	13a
Closed-loop frequency response	C/R(f)	Model 2	13b

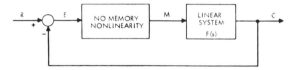

Fig. 7. Lens control system divided into no-memory nonlinear and frequency-dependent linear elements. C is controlled quantity, clear vision position; R is reference input, target position; E is error or actuating signal, position error; M is signal representing output from nonlinear element and input to linear frequency-dependent element. Table I indicates relationship between lettered variables and functions plotted in other figures.

The result, which is the open-loop frequency response or Bode diagram, is plotted in Fig. 9.

On the basis of Fig. 8 we see that the input to the nonlinear element changes with frequency. That there are three different attenuations in Fig. 9 at each frequency is the result of the nonlinear element. Since the magnitude of E at each frequency and at each R may be found from Fig. 8, we can plot $|C/E|$ vs. E for different frequencies, as shown in Fig. 10a. These curves are segments of the describing function of the nonlinear element, and they form a continuous curve when moved vertically by an amount given by the gain of the linear frequency-dependent element. This result

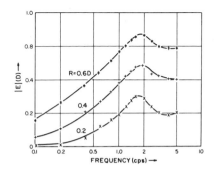

Fig. 8. Error magnitude $|E|(D)$ as a function of frequency. Symbols same as in Fig. 5.

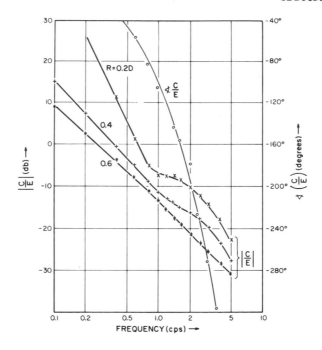

Fig. 9. Open-loop frequency response: Bode plot. Symbols as in Fig. 5. Single open-loop phase curve here implies three phase curves in closed-loop plot of Fig. 5; not widely dispersed, however.

is given in Fig. 10b which is the describing function of the nonlinearity. By using Figs. 8, 9, and 10b, the transfer function of the linear element is obtained, as shown in Fig. 11.

MODEL

We can now construct an analytical model of the lens feedback system. The transfer function of a linear element that approximates the attenuation and phase characteristic of Fig. 11 is

$$F(s) = \frac{4}{s} \cdot \frac{(1 + 0.15s)\exp(-0.1s)}{1 + 2[0.3(0.18s)] + (0.08s)^2} \tag{4}$$

and also is plotted in Fig. 11 as the dashed lines, heavy for gain and light for phase characteristics. The transfer function is written as a function of the Laplace transform complex variable, "s."

The nonlinearity can be represented by either of the two alternate input–output curves shown in Figs. 12a and 13a. The first (Fig. 12a) is

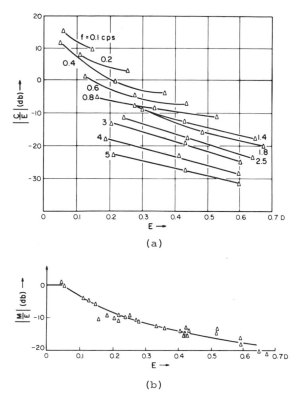

Fig. 10. (a) Open-loop gain as a function of error magnitude, parameterized by frequency. Note decreasing
gain with |E| and with f. (b) Describing function: |M/E|
as a function of |E|, obtained from (a) by separation of
no-memory nonlinearity from frequency-dependent linear function.

seen to be a simple piecewise linear element with a small–signal proportional response and flat saturation for larger inputs. It closely resembles
the data presented by Marg and Reeves [24] for the lens response of cat to
nerve stimulation. The second (Fig. 13a) is unusual with output actually
decreasing as input increases in part of its domain. However, on consideration of the fact that error signal is a blur and that a more smeared
image may well be a weaker error signal, one may find reasons for this
peculiarly shaped nonlinearity. Further, there is an experimental function,
also obtained by Hamasaki, Ong, and Marg [22], showing human static accommodative responses for various d–c inputs, which resembles the curve
of Fig. 13a.

The closed–loop frequency–response curves computed from each of
these nonlinear portions of the models, together with the model linear ele-

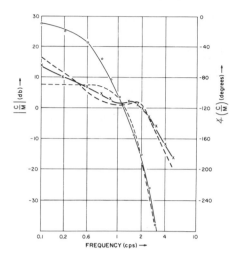

Fig. 11. Linear-element transfer function |C/M|(f). Computer results from experiment are shown in continuous lines; heavy (crosses) from gain, and thin (open circles) for phase. Model gain (thick interrupted line) and phase (thin interrupted line) from Eq. (4) (A) are also shown for comparison. Note good gain agreement; poor phase agreement at low frequencies suggesting a prediction operator, e^{+st}, effective in that range.

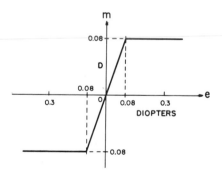

Fig. 12a. Model nonlinear relationship between input, e, and output, m: small-signal proportional region and large-signal flat saturation region. This nonlinearity would have a describing function very similar to Fig. 10b, which is modified from Fig. 1 of Ref. [24].

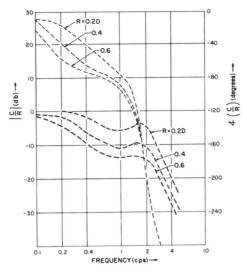

Fig. 12b. Closed-loop frequency responses
for input signals R of ±0.2, ±0.4, and ±0.6 diopters. Gain: heavy dashed lines, phase:
thin dashed lines. Compare with experimental
curves of Fig. 5.

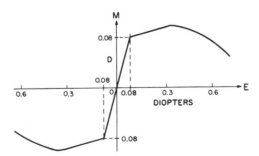

Fig. 13a. Model nonlinear relationship between input, e, and output, m: small-signal
proportional region, moderate signal saturation
region, and even larger signal region shows reduction of output. Describing function would
have less marked early decreasing slope and
more marked later decreasing slope than in
Fig. 10b. Similar to Fig. 1 of Ref. [22].

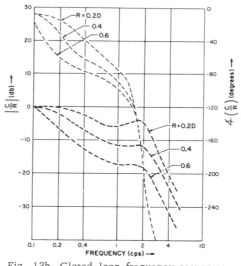

Fig. 13b. Closed-loop frequency responses
for input signals of ±0.2, ±0.4, and ±0.6 diopters. Gain: heavy dashed lines; phase: thin
dashed lines. Compare with experimental
curves of Fig. 5.

ments defined in Eq. (4), are shown in Figs. 12b and 13b for each of the
input amplitudes corresponding to the experiment of Fig. 5. Correspondence of these curves with the experimental curves of Fig. 5 is one test
of the model.

DISCUSSION

A first major discrepancy is one found between the model derived
from sinusoidal data, and the response curves in the step experiments.
The model has a time delay of $e^{-0.1s}$ whereas the step responses show a
dead time or time delay of $e^{-0.36s}$. This at once suggests that the lens system is not a simple nonlearning, nonpredicting control system as is the
iris reflex to light [32, 37]. When the lens follows simple target motions
such as a single sinusoid, it evidently is aided by a prediction operator [35,
41]. This prediction operator is an important characteristic of many neurological control systems and shows up as a relative phase advance not
present in the response of the same system to unpredictable input signals.
An unrealizable $e^{+0.27s}$ would be an adequate representation. Further evidence for a prediction operator acting under conditions of our experiments
comes from studies of initial direction of accommodative movements to
targets displaced in randomly sequenced directions [38, 39]. No odd-error
signal is present in the basic lens control system — only an even-error signal with its incumbent disadvantages in many experimental situations such
as the present one, with restriction of many usually available clues of depth
perception, size change, intensity variation of image, fusional disparity.

Thus, for the lens to have tracked without directional errors, learning of the repetitive input pattern might well have occurred. This could also be evidenced as the $e^{+0.27s}$.

A second discrepancy between the model and experiment is clearly shown in Fig. 11. The phase of the model approaches 90° lag because of the integrator, whereas the experimental phase curve approaches 0°. One hypothesis might be that prediction is more adequate at lower frequencies and that an additional variable e^{+st} provides the phase lead that makes the model disagree with the experiment.

The interesting feature of these suggestions is that direct experimental tests capable of disproving them exist (for anyone with an adequate optometer). One test would be to drive the system with pseudo-random inputs (sums of noncommensurate sinusoids) and eliminate the effect of the prediction operator. If the experimental phase remains the same, the prediction operator is disproved; if the pseudo-random experimental phase coincides with the model, the validity of the model is strengthened. In either case the model has fulfilled its function of clarifying and correlating existing information, pointing out discrepancies, and suggesting crucial experimental tests.

Many dynamic and stability features of the accommodative control system are included within our model, which therefore serves to expose, define, and interrelate these features. The borderline instability for small signals apparently shapes the noise-power spectral-density curves and accounts for the 2-cycles/sec peak. The important saturation-type nonlinearity, however, provides for the varying effect of different input amplitudes on the stability characteristics of the frequency response. In particular, the stable transient response to large or moderate steps in target position is well accounted for by the now-stabilized system. The relatively structureless low band-pass characteristics of the large-to-moderate input frequency response is now shown to be compatible with the closed-loop noise peak. The absence of this noise peak when the loop is opened, and also the increase in amplitude of the low-frequency noise, is predictable from and included within the model. The integrator, transport delay, and second-order pole configuration satisfy other physical characteristics of the system. Work is needed to correlate and assign various parameters of the model to the physical anatomic-physiological elements of the lens mechanism. For example, future study may reveal the natural frequency of the combination of elastic capsule of the lens, suspensory ligaments, ciliary muscle, and scleral restoring force to be $\omega_0 = 12.5$ radians per second (where ω_0 is the natural frequency in radians per second of the second-order differential equation), thus accounting for the complex pole pair.

Because of the increasing awareness of the importance of discontinuous control systems in biological and neurological servomechanisms [36,

45, 46], the possibility of a sampled–data mechanism playing a role in the lens system has been considered [38]. Fincham's important observation [19] that accommodation only occurred after small fixation movements in the four of his subjects who experienced enough to be aware of this aspect of their visual physiology, suggests a link with the eye movement system. Young and Stark [45, 46] have recently analyzed this latter servo and found its frequency response to have a 2.5-cycles/sec sampling peak, resulting from minimal 0.2-sec sampling periodicity. Here also an integrator is present, performing a hold function for the sampler. From the point of view of retinal physiology, small fixation saccades would serve to counteract the disappearance of the image due to stabilization on the retinal receptors [17, 30]. This would be especially true of the low–contrast, blurred, out-of-focus images.

In fact, the second nonlinear model shown in Fig. 13a incorporates a lessening sensitivity to increased blur, perhaps secondary to this adaptive aspect of retinal physiology.

The sampled–data model suggestion poses a number of crucial experiments to decide between it and the no–memory nonlinear plus frequency-dependent linear model here fitted to the present experimental data. These experiments would include both transients and steady–state inputs as exemplified in other recent studies on sampled data neurological systems [36, 45, 46]. Similarly, the phase discrepancies emphasize the importance of using pseudo–random inputs. The absence of an odd–error signal mechanism under restricted monocular viewing conditions suggests further experiments [38, 39].

The necessity for a precise and reliable direct recording optometer is thus brought home. The excellence of Campbell's apparatus is borne out by the lack of competitors, only a few other developed devices being known to us [2, 31]. New techniques for opening loops by clamping feedback gains and the use of pseudo–random inputs suggest that the questions posed by our analysis will not be long unanswered.

SUMMARY

A servoanalysis of the feedback control system for lens accommodation has been based upon experimental evidence designed to explore the nonlinear properties. Because these nonlinearities are essential to the functioning of the system, a describing function approach was employed. Computation served to obtain open–loop characteristics.

The transfer function of the linear part of Eq. (4) is

$$F(s) = \frac{4}{s} \cdot \frac{(1 + 0.15s)\exp(-0.1s)}{1 + 2[0.3(0.08s)] + (0.08s)^2}$$

and some assignments of portions of this equation to the physical elements of the lens system are suggested. The no-memory nonlinearity has general saturation characteristics and two related models, one with quite unusual aspects, are presented and independent evidence for these adduced. The combined model accounts for large and small signal responses, stability characteristics, and predicts certain noise spectral features of the open- and closed-loop lens system. Phase discrepancies, possible sampled data properties, and the even-error-signal operator put forward crucial new experiments.

APPENDIX A

The human lens accommodation mechanism is a "system" in which the lens position at any time, which we shall call the response, $C(t)$, is determined by the history of the target position, which we shall call the input $r(t)$. The mathematical counterpart of such a system is an operator, i.e., a mapping which assigns a unique output function of time r to each input function of time c.

Our object is to characterize the input vs. output relation. This operator's task is particularly simple if the operator is linear, i.e., if the superposition principle, which requires the response to a linear combination of inputs to be the same linear combination of their respective outputs, is valid. If a system is linear it is not necessary to test C with all possible inputs; instead it is enough to determine the response to a limited set of test inputs, for example, impulses of all arguments, sinusoids of all frequencies, out of which the response to an arbitrary input can be constructed by superposition.

It is especially advantageous to employ test inputs that are "eigenfunctions" of the operator, i.e., whose shape is not altered by transmission through the system and which suffer only a scale change. It is enough then to record the scale change; there is no need to record the entire response. If, in addition to being linear, the operator is also time invariant, it may be shown that complex exponentials of the type $e^{st} = \cos wt + j \sin wt$, where $s = jw = j2\pi F$, F being the frequency, are eigenfunctions. In other words, if $r(t) = R(s)e^{st}$, $R(s)$ being the complex amplitude of the input e^{st}, of arguments, then $C(t) = C(s)e^{st}$, $C(s)$ being some other complex constant. The ratio $(C/R)(s)$ which is known as the frequency response or transfer function, constitutes a complete characterization of the operator.

The lens systems is approximately linear and time invariant for small inputs, and in that range we therefore characterize it by its frequency response, $(C/R)(s)$, which is shown in Figs. 3 and 4. The lens is a feedback system (see Fig. 1). The inner part of such a system, which transforms "error" signals e into responses C is known as the "open-loop" system, while the overall feedback system is known as the "closed-loop" system. We should like to characterize the open-loop system from C/R frequency-response data which characterize the closed-loop, i.e., to determine E/R, from C/R. Provided our assumptions concerning the internal structure of this system are correct, and we actually have a feedback system of the type shown in Fig. 1, $E/R(j\omega)$ can be obtained from $C/R(j\omega)$ by means of the equation:

$$\left|\frac{E}{R}\right| = \left|1 - \frac{C}{R}\right| = \sqrt{1 + G^2 - 2G \cos \varphi}$$

In Eq. (4), the graph of the frequency response $(E/R)(s)$ (which we determined graphically from a chart known as a "Nichol's Chart") is approximated by an expression in "s."

For large target movements the lens system behaves nonlinearly, i.e., superposition is no longer valid. This nonlinearity can be accounted for, at least approximately, by a single nonlinear element in the open loop, which is "instantaneous," i.e., whose output at any time depends solely on its input at that time. We have tried to deduce the input-output characteristic of this nonlinearity from the manner in which the open-loop frequency response varies as a function of the input amplitude. (Frequency response as a function of the input amplitude is known as the describing function.) Since representation by a single nonlinear element is necessarily approximate, the exact shape of the nonlinearity is not significant — it is enough to approximate it by several straight-line segments (Figs. 12a and 13a).

We have not had to employ the frequency-response characterization for its ultimate purpose — which is to compute the response of the system to arbitrary inputs. However, from Eq. (4) an arbitrary bounded input $c(t)$ can be resolved into an integral of exponential components,

$$r(t) = \frac{1}{2\pi j} \int_{-j\infty}^{+j\infty} R(s)\, e^{st} Ds \tag{A}$$

known as "the inverse exponential (double-sided Laplace) transform," by means of the exponential transform

$$R(s) = \int_{-\infty}^{\infty} r(t)\, e^{-st} ds \tag{B}$$

In passing through the system, each exponential in the integrand in (A) is subjected to its separate simplification $(C/R)(s)$, with the result that the output can be computed by the superposition

$$C(s) = \frac{1}{2\pi j} \int_{-j\infty}^{+j\infty} R(s)\, \frac{C}{R}(s),\, e^{st}\, ds \tag{C}$$

Expressions (A), (B), and (C) have not been employed in Chapter 1, and are listed for reference only.

Chapter 2

Absence of an Odd-Error Signal Mechanism
in Human Accommodation

INTRODUCTION

The accommodative system of the eye appears to be relatively accurate in compensating for out-of-focus retinal images of a target through changes in accommodation. However, a reading of the literature indicates a difference of opinion on the means by which the system can discriminate in which direction the focus should be initially altered in order to sharpen the image.

In this chapter [39] it is proposed that:

1. The accommodative system of the eye is analogous to a closed-loop system in which blur of the retinal image serves as a stimulus to bring about an accommodative response for clearing the blurred image.

2. The accommodative system, when stripped of its connections with other clue systems, contains an even-error signal, that is, a signal giving magnitude of error, but not the direction, or sign of error.

3. The spontaneous 2-cycles/sec oscillation of the lens does not have a physiological role in converting the even-error blur signal into an odd-error signal (i.e., a signal with magnitude and sign) by a phase-sensitive demodulation operation. The oscillation may be understood as the consequence of important nonlinear characteristics of the accommodative servomechanism.

The difference between the target position and the clear vision position (*CVP*) of the lens can be thought of as a distance error which is converted into a blur by the optics of the eye. The blur is then the "error sig-

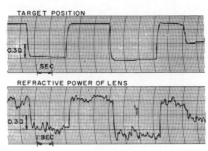

Fig. 14. The top trace shows target position as input; the lower trace shows output as refractive power of lens measured
with Campbell's optometer (infrared,
high-speed, direct-recording). Note the
large signal stability contrasted with the
low-amplitude narrow-band noise appearing as a 2-cycles/sec oscillation. That a
correct initial direction of response is seen
to occur in each of the responses above
may be attributed to the repetitive and
predictable nature of the input target motion.

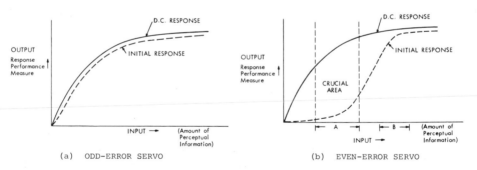

Fig. 15. Measure of amount of perceptual information as input, and similarly, measure of
response performance as output are not in any single physical dimension, but rather indicate
conceptually that performance of a system improves as more information becomes available
to it. In (a) both d-c response and initial response improved in similar fashion since an
odd-error servo is considered; and magnitude and sign information are both available. In
(b), however, d-c response may improve significantly with an amount of perceptual information which is not sufficient for initial response to perform at a better than chance level.
In lens servo with blur as input information only magnitude and not sign information is
available. D-c response can be accurate while initial response is random with this
even-error signal. When additional information, such as slight lateral movements, is
added, then input moves from region A to region B and initial response improves
markedly.

nal" that drives the accommodative servomechanism. If, as proposed, the error signal is an even-error signal, then the initial direction of corrective movement can only be random. The sensing system can then determine if the error has increased or decreased, and only then further correct the system output, or completely reverse the response if error has increased. This would be "initial-hunting" behavior, and can be noted when focusing a microscope. As in that example, an even-error signal is quite adequate to ensure eventual accurate accommodation with minimization of blur. Thus, the predicted initial-hunting behavior, that is, random initial direction of accommodation, would be evidence in favor of the even-error signal.

In order to demonstrate this, it is necessary to restrict viewing conditions to eliminate the many perceptual clues always found in normal vision. For example, the binocular disparity clue that drives convergence can be eliminated through a monocular experimental arrangement. Such an experiment would also provide a crucial test as to whether the spontaneous oscillations (at 2 cycles/sec) of the lens, described in Chapter 1, act as part of a phase-sensitive demodulation mechanism to provide directional clues for accommodation. If the oscillations did not have this functional role, then random initial directions of tracking would occur. However, if 50% erroneous initial tracking does not occur, then either (1) the oscillations play a useful role, or (2) some perceptual clue such as lateral movement, size changes, or chromatic aberration still remains, or (3) the sequence of target movements is not random and has been learned by the subject.

Figure 14 shows accommodation occurring under monocular viewing conditions without the aid of associated convergence. But in this instance, the sequence of target movements is not random and has been learned by the subject.

Figure 14 does serve, however, to illustrate both the 2-cycles/sec spontaneous oscillations that are present in the accommodation servomechanism [3, 10, 14], and quite stable large corrective movements which occur at the same time. The mechanism of generation of these oscillations has been elucidated in Chapter 1.

The design of the crucial experiment can be understood with reference to Fig. 15 which illustrates "pseudo" input-output curves for odderror and even-error servos. As input, or the amount of perceptual information increases, output, or the response performance measure, also increases. Both the final value of the d-c response and the initial response increase together, since both magnitude and *sign* of error are available. Conversely, in Fig. 15b the initial response of the even-error servo is seen to behave in a quite different manner from the d-c response, since

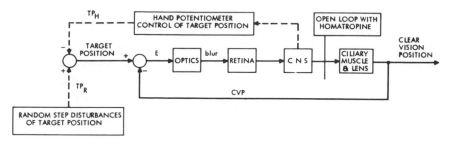

Fig. 16a. Block diagram of accommodative control system together with experimental arrangement for controlling target position. Random step of target position, TP_R; hand potentiometer contribution to target position, TP_H; clear-vision position, CVP, is the controlled output; error, E, is difference between target position and CVP; central nervous system, CNS.

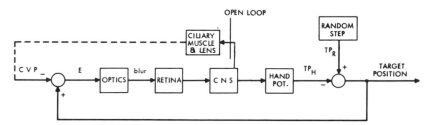

Fig. 16b. Blocks are rearranged to elucidate the experimental tracking system. Note that the CVP is fixed and considered as reference signal, rather than the controlled output variable as in (a). Homatropine opens the loop by paralyzing the ciliary muscle; thus no change in CVP occurs and no feedback from CNS output command signals result.

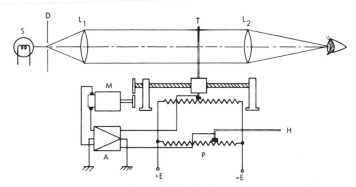

Fig. 17. Scheme of tracking experiment apparatus. Lens L_1 collimates light beam from the source, S, and lens L_2 is placed so that its second principal focus is on the pupil of the subject's eye. Target, T, is a pin and is moved back and forth by the servomechanism.

sign of error is not available. At some level of perceptual information, marked A and delimited by vertical lines about the "crucial area" on the figure, the d-c response will be accurate, while the initial response will be random; that is, it will have erroneous initial directions of tracing 50% of the time. Adding input perceptual information to the amount marked B on the figure will eventually bring the initial response to above random performance. Restricting various clues and thus reducing input information down to level A causes the initial transient response to operate at chance performance levels without significantly reducing d-c performance. This behavior, if obtained clearly, defines the system to be an even-error servo. The design of the crucial experiment then is to maneuver various perceptual clues, learning, and the 2-cycles/sec oscillation of the lens so as to: (a) produce behavior in the crucial area, identified by 50% initial errors with zero final d-c error, and (b) examine which isolated clues enable the system to escape this even-error signal behavior.

METHOD

The ideal apparatus would be a recording optometer with adequate dynamics, such as the one invented by F. W. Campbell [8] which was employed for the record of Fig. 14, and in the experiment described in Chapter 1. However, an adequate and simpler arrangement is that involving hand tracking of a target moving in the optical axis, as used by Campbell and Westheimer [12]. The accommodative system and the experimental setup are shown in the block diagrams of Fig. 16. These block diagrams are, of course, models based upon hypotheses and justified by utility. One of the assumptions, and a central feature of the experimental design, is that blur is indeed an error signal. This is considered in the discussion section. When the clear-vision position (CVP) is fed back and subtracted from the target position, an error signal, E, is obtained. This error signal is also the difference between the blurred image and the sharp image obtained in the clear-vision position. The retina operates on this blur and generates optic nerve signals. After processing by the central nervous system (CNS), motor nerve signals to the ciliary muscle change lens refraction and thus alter the clear-vision position to reduce the error. If homatropine is used to open the loop, at the point indicated in Fig. 16, by paralyzing the ciliary muscle, then the error signal may still be employed by the CNS to control the subject's hand, which rotates a potentiometer [38]. The target is mounted on an X-Y recorder pen unit and can be moved by the hand potentiometer to remove any error. Figure 17 shows a schematic of the electro-optico-mechanical apparatus. Of interest are three necessary precautions if monocular viewing is to be adequately restricted. First, the target must travel along the optical axis with no *lateral* movement. Thus, a smooth and precise mechanical system is required, as well as a bite board and headrest for the subject. Second, the convergent lens L_2 is placed so that its second principal focal place is at the plane of the pupil of the subject. This geometric disposition eliminates dependence of the angular size of the target on its position. Third, the *luminance* of the target is also kept unchanged by means of L_1, which collimates the illuminating light beam.

The relationship of normal lens accommodation to hand movement control of target in our experimental arrangement is indicated by the block diagrams of Fig. 16. In Fig. 16b, the same blocks as in Fig. 16a have been rearranged to present the experiment from a classical servoanalytical viewpoint. Note that *CVP* is fixed and is now the reference signal, while conversely, target position is the controlled variable. The subject counteracts the introduced random disturbance to reduce error. Figure 18 shows idealized signals and responses under the different experimental operating conditions. In Fig. 18a randomly changed target position, or TP_R, error, and clear-vision position of lens are shown. After a delay of 0.4 sec the error is reduced by lens accommodation, in this case with a correct initial direction. Figure 18b illustrates the same situation, but now the initial direction of accommodation is incorrect and the error increases until, after an additional delay, the direction of accommodation reverses. When the ciliary muscle is paralyzed, the hand responds, TP_H, as shown in Fig. 18c and d. For example, after a random-step generator produces an error due to a positive step change of position, the hand may produce a counter-negative step change of position of the target TP_H, to reduce the error, as in Fig. 18c. Again, if the hand makes an initially incorrect response, a second corrective response is seen after an additional delay. The response and error signals in this case are shown in Fig. 18d.

The necessity of paralyzing the ciliary muscle by means of the instillation of 1% homatropine can be circumvented, if the subject is willing to correct for step target position changes by quick responses of the hand in rotating the potentiometer. The lens reaction time is about 0.35 sec [9, 13, 16] while that of the hand movement is about one-tenth of a second less [23, 28, 35]. Thus, with the subject's eye drugged, he may correct the target position quickly enough to prevent the lens from changing the clear-vision position at all, as in Fig. 18e. However, even though the hand reacts quickly and corrects the target position, the lens system may still respond to the error that had existed during the reaction time delay of the hand. As shown in Fig. 18f, this will cause the hand to compensate for the lens change of *CVP*. However, the important information from the point of view of this study, the initial direction of tracking, will have been obtained before the lens changes.

Experiments were done with both homatropinized and undrugged subjects with quite similar results. Target position signal due to random step

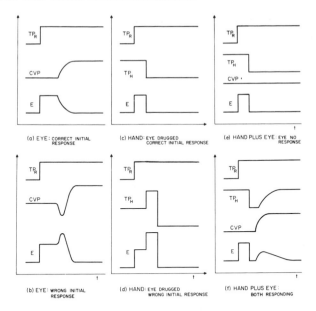

Fig. 18. Idealized signals and responses under different experimental conditions. (a) and (b) show lens accommodation; (c) and (d) hand movement when the ciliary muscle is paralyzed; (a) and (c) initial tracking in correct direction, (b) and (d) in the wrong direction. (e) and (f) show hand plus eye movements when both may occur; (e) quick hand tracking removing error signal and where no lens compensation occurs; (f) quick hand tracking but with subsequent lens compensation as well. Note secondary change of hand contribution as lens compensates.

generator, TP_R, and target position due to hand rotation of potentiometer, TP_H, are recorded separately as shown in Fig. 19. Target position is, of course, the algebraic sum of these two quantities as shown in the block diagram of Fig. 16b. Here we have rearranged the blocks to make clear the new feedback system in the hand-tracking condition.

Since it is especially important for the step disturbance of target position to be quite random, a random number table was used, and only violated on rare occasions when range saturation seemed possible.

EXPERIMENTAL RESULTS

Experiments were conducted [38] with due precautions taken to ensure random target displacements under restricted monocular viewing.

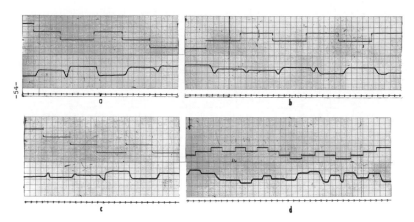

Fig. 19. Record of random step function, TP_R in upper trace; output of subject-operated potentiometer, TP_H in lower trace. Error is algebraic sum. Y-axis is in diopters, scales are equal in upper and lower traces. TP_R steps are 1.0 diopter in (a), (b), and (c) and 1.2 diopters in (d). X-axis is time in seconds marked by tics. Target was circle with a dot inside in (a) and (b), and a black pin in (c) and (d). Subject Y. T. was homatropinized in (a), (b), and (c) and subject J. B. was undrugged in (d). Note response initially in wrong direction occurring in all situations.

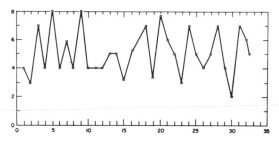

Fig. 20. Number of erroneous initial trackings in successive sets of ten trials (ordinate) vs. sequence of sets of ten trials (abscissa); 320 trials in all, mean = 5.06.

Figure 20 shows a typical result. The number of failed or erroneous initial-tracking moves in each successive ten-trial group is plotted against the serial number of the group. There is no obvious divergence from chance results; 50% of the initial movements are in error. Also, there is no tendency for decreasing initial-error percentage with time, and thus no evidence for learning or training. These experiments (10—20 in all) have been performed for several subjects (6) with their eyes either homatropinized or without drugs; the results are the same in both cases. Some statistics of higher order than the first were examined. For example, the direction of initial response was not correlated with previous stimulus or previous response.

Quantitatively, to confirm the randomness of the responses of our subjects, some statistics were obtained with respect to the internal structure of the random response process. Occurrences of uninterrupted successions of correct initial trackings were investigated. Each such uninterrupted succession has a probability of occurrence which is dependent upon its length, n. This parameter, n, is the number of correct initial responses in an uninterrupted succession. By tabulating the numbers of occurrences of sequences of length n, the experimental data shown in Fig. 21 were obtained.

If each correct or erroneous initial response occurs randomly and independently of preceding responses, then Pn, the probability of a sequence of length n, is equal to p^n. The theoretical values of Pn lie on the

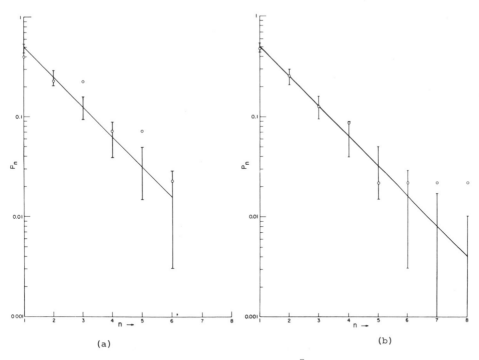

(a)   (b)

Fig. 21. Probability of n successive correct trials is Pn = p^n and is indicated on logarithmic ordinate scale as a function of n on linear abscissa scale. Points are experimental from (a) first hundred and (b) last hundred trials from a 290-trial experiment. Solid line represents theoretical probability density function for purely random events with p = 0.5. Vertical bars indicate ± standard deviation of a normal distribution from these theoretical values of Pn.

solid lines of Fig. 21, computed by assuming $p = 0.5$, close to the experimental values. The vertical bars represent ± the standard deviation of a normally distributed population about the theoretical values of Pn.

The experimental points lie within expected ranges of the theoretical curve, suggesting that the initial direction of tracking did indeed occur randomly. Furthermore, the two distributions of Fig. 21a and b were obtained for the first and last hundred trials, respectively, in a 290-trial run. It is clear that there is no difference between two distributions that might be attributed to a learning process.

Rarely were the typical results such as are shown in Figs. 20 and 21 not obtained. The required precision of lateral alignment of the moving target is considerable and technically difficult to assure. Figure 22 is an example of a run during the course of which the subject became aware of a slight lateral movement of the target. This, of course, correlated with the distance of the target, and the subject was thus able to determine the direction of movement. After the sixteenth group of ten trials she made only one erroneous initial-tracking movement. At the end of the experiment the subject indicated that the clue was indeed the slight lateral movement. When the apparatus was adjusted, she then again showed random initial tracking.

Although the importance of chromatic aberration clues has been raised in the literature [1, 12, 19], restriction of information by limiting the light frequency spectrum from white to green cannot reduce the level of performance if this is at chance level already. Figure 23 shows the results of such an experiment, which are negative, as predicted. No apparent effect was seen on switching target illumination alternately from green to white.

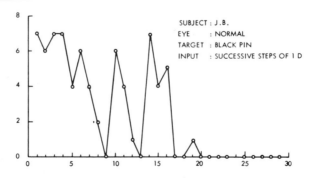

Fig. 22. Number of erroneous initial trackings in successive sets of ten trials vs. sequence of sets of ten trials as in Fig. 20. After 160 trials the subject apparently noticed lateral movement due to imperfect alignment.

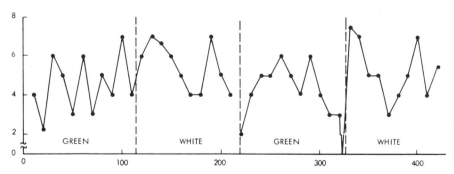

Fig. 23. Record of random target motion and tracking of a subject whose eye is not drugged. Scale as in Figs. 20 and 22. Color of target illumination was changed as indicated without apparent effect.

DISCUSSION

The servoanalytic approach to the human lens accommodation makes the question of an odd-error or an even-error signal an important one. Although such an approach requires some reorientation, the advantages of embedding this most interesting neurological feedback control system in the standard forms of nonlinear engineering control theory, hopefully, can be rewarding. Certain behavioral characteristics of this neurological response become more clearly defined and interrelated. Now that single unit studies on various levels of the visual system are producing more sophisticated information, it is particularly important to have crucial and quantitative questions to ask. The analytical and experimental material in this paper represents a frame for such "into-the-black-box" neurophysiological studies, and poses questions such as "What are the detailed neurophysiological mechanisms for operating with the even-error signal, and for interlacing the odd-error signal information derived from certain secondary clues into the basic even-error control system?"

Our main experimental result clearly shows that blur acts as an even-error signal under restricted monocular accommodative tracking. We feel that our experiments are self-consistent and that our null result is likely to be robust to a variety of experimental artifactual interferences. It is difficult to conceive of this null result, that is, 50% of initial-tracking errors, being caused other than by the system's actually operating under even-error signal conditions. Other experimental results diverging from this null result, obtained at times by ourselves as shown in Fig. 14 and further described in Chapter 3, or obtained by others in the literature [12,19], are more likely to be due to the influence of an unintentional slight perceptual clue or a less randomized input signal permitting learning of the input signal pattern.

Both of these experimental artifacts would produce a higher probability of correct initial response and lead the experimenter to a false notion of the existence of an odd-error signal under restricted monocular viewing conditions. We therefore stress three essential ingredients of the restricted experimental studies: (1) randomness of target movement; (2) absence of lateral movements; and (3) absence of size and intensity clues.

Campbell and Westheimer [12] introduced the second principal focus method to assure the third condition, but may well not have fully attained conditions 1 and 2, especially for their well-trained experienced subjects. Fincham's [19] results are consistent with the effects of not meeting conditions 1 and 3.

That blur is the main error signal was initially assumed, only to find for example, that in a recent review [4] the stimuli for accommodation considered were first "apparent size stereopsis and apparent distance." Second, such factors as "chromatic aberration, spherical aberration, astigmatism, perhaps also oscillations of accommodation and scanning movements of the eye." Blur was not considered a clue, and in fact its only mention was in connection with Fincham's [19] demonstration of the lack of accommodative response "to a target suddenly blurred but not defocused." However, other papers [1] have considered blur as a stimulus to accommodation. Perhaps one of the contributions of this paper is the evidence that even-error signals are present in the human accommodation system, thus suggesting that blur which has such an even-error characteristic is such an even-error signal. The nature of the crucial experiment is graphically explained in Fig. 15 where, by restricting clues, we can pass from region B to region A. This region A, defined by adequate d-c performance and random a-c or initial transient performance, only occurs in a system with an even-error signal, or perhaps with an even function for one region of a loop variable.

One of the important deficiencies in our experiment is the fact that the hand-movement response is a complex and indirect one. However, as will be seen in Chapter 3, these results are confirmed when hand tracking is replaced by recording of a more direct synkinetic response, i.e., the associated vergence movement of the unseeing eye when driven by an accommodative target moving along the optical axis of the seeing eye [1, 43, 44]. This asymmetrical accommodative vergence is clearly much closer to the "reflex" lens accommodation, which would of course be the ideal response to measure.

The trial-and-error procedure illustrated in Figs. 18 and 19 reflects the disadvantages of an even-error signal, the 50% incorrect initial tracking occurring under restricted monocular viewing conditions. This trial-and-error procedure cannot be considered as a straightforward phase-sensitive demodulation detector. However, it can be considered as a tactical maneuver that certainly does provide for the eventual placing of the lens servomechanism error detector on the proper slope of the convex error curve. Then the magnitude system alone can drive it to an accurate d-c response.

The second principal result of our experiment is the clear demonstration of the ineffectiveness of the 2-cycles/sec spontaneous fluctuation in providing sign information. These 2-cycles/sec oscillations might act as a phase-sensitive detection device for providing an odd-error accommodative stimulus [10, 14, 15].   These oscillations are just below the correct amplitude range for threshold sensitivity [3, 11], and appear to be absent when the lens is focused at infinity or under other special experimental conditions [10, 14, 15]. Thus, this mechanism requires an effective utilization of the nonconsciously appreciated blur at a level of 0.2 diopter. Our experiments with undrugged subjects were especially designed to measure the probability of correct initial tracking when the 2-cycles/sec oscillation of the lens was present. As indicated above, no better than chance results were obtained, thus providing direct experimental evidence against this mechanism for phase-sensitive detection to obtain an odd-error signal. These oscillations must rather be viewed as a nonfunctional indicator of the nonlinear characteristics of the accommodative servomechanism described in Chapter 1.

Effective perceptual clues include first of all binocular disparity which drives the convergence mechanism; also of importance may be monocular parallax. Size and intensity clues are clearly effective and, as shown in Fig. 22, very slight lateral movements are excellent odd-error signal clues.

Learning of a repetitive or nonrandom pattern in the signaling of target movements can also play an important role. With a long series of experimental runs done with a few very experienced subjects, this learning factor may play a particularly important and confusing role. We found it necessary, therefore, to utilize a random number table, as explained in the methods section, in order to control this factor.

A final set of clues to discuss are those of questionable effectiveness. The effects, of course, depend upon strength of the clue experience of the subject, and experimental conditions under which he is operating. Our experiments in general tend to demonstrate that these clues are of little functional significance in controlling initial direction of accommodation in normal (everyday) vision. However, these complex perceptual clues have often been proposed to have such a role and are discussed below.

Complex perceptual clues might provide an odd-error signal. Astigmatism, found in the lens of many people, could so serve [6], especially,

as demonstrated by Campbell and Westheimer [12], when exaggerated or
supplied with the use of cylindrical lenses. Its effect is, of course, quite
dependent upon image characteristics [26]. A bright self-illuminated target will accentuate this effect whereas an opaque, less brightly lit target,
such as we have used, will decrease it. Indeed, Fincham [19] has shown
that the use of small targets will often remove whatever clues have been
previously used for accommodative responses. Consideration of spherical
aberration provokes similar arguments. Annular targets have been shown
to reduce the frequency of correct responses [12]. However, these do not
seem to be the usual mechanisms for odd-error information since they are
such variable factors and are absent to any significant degree in many normal persons.

The lens control system is remarkably insensitive to chromatic aberration. Fincham [19] showed that even when the illumination was changed
from blue to red, thus requiring 1.3 diopters of positive accommodation,
none of his 55 subjects responded, even though they were subjectively conscious of blurring. In his experiment with monochromatic light and compensating lenses, some subjects responded in a constant bizarre pattern
of all correct responses to a negative lens, all absent responses to removal
of the negative lens, and all positive accommodations (wrong direction) to
a positive lens, in spite of the increased subjective blurring. Of Campbell
and Westheimer's [12] four subjects, only one experienced any difficulty in
monochromatic light, and this defect was quickly circumvented with "learning," in the next 30 trials, lasting perhaps 1−2 min. The inability of our
subjects to obtain better than random initial direction of tracking was equally unaffected by the presence or absence of chromatic aberration information (Fig. 23). It thus seems unlikely that chromatic aberration could be
the odd-error mechanism.

Fixation movement was found to be necessary by Fincham [19] for
four of the subjects who were experienced enough to control subjectively
their eye fixation. A fixation movement also has the effect of removing
that adaptation to a stationary target which causes disappearance of a stabilized image [17, 30]. If fixation movement or other intermittent process
plays a role in accommodation, then a sampled data peak should show up
in the frequency-response plot of the accommodative system [45, 46]. Indeed, there is such a peak [40] but this has been otherwise interpreted.
Another effect of a fixation movement would be to exaggerate any lateral
displacement of a target which could thus serve as a clue to accommodation.

This lack of a "simple optical" odd-error signal may be of little
functional significance in everyday vision. In such a situation a person
usually accommodates in association with convergence, while perceiving

a highly structured image with many reinforcing perceptual clues, and accommodates in a direction which may also have been predicted by the recent past history of the spatial environment.

However, it should be stressed that this restricted monocular experimental situation is by no means so abstract that accommodation or hand tracking, with or without an initial erroneous movement, does not finally eliminate the blur of the image and obtain an error-free situation, wherein clear-vision position equals target position at least to within a dead space.

SUMMARY

Experiments have been designed to demonstrate that the human accommodative system operates with an even-error signal mechanism under restricted monocular viewing conditions. Retinal blur is such an even-error input signal and thus these experimental results add to the evidence considering blur as the effective input signal in accommodation. The random or 50% erroneous initial direction of movement is a null experimental result which should be robust to a variety of experimental artifacts that may have contaminated previously published results.

The 2-cycles/sec oscillation does not have a physiological role in converting the even-error blur signal to an odd-error signal by some phase-sensitive demodulation operation. The oscillation may rather be understood as the consequence of important nonlinear characteristics of the accommodative servomechanism. Input-adaptive predictive capability of the accommodation system is related to similar capabilities in versional visual-tracking and in hand-tracking studies.

Chapter 3

Accommodative Tracking:
A Trial-and-Error Function

INTRODUCTION

Fincham [19] and Campbell and Westheimer [12] have reported that the accommodative tracking system utilized a variety of subtle clues for focus errors of less than 1 diopter. Specifically, Fincham emphasized the role of the color fringes due to the eye's chromatic aberration, and the use of changes in fixation in determining the initial direction of an accommodative effort. Later Campbell and Westheimer confirmed Fincham's findings with respect to chromatic aberration and reported that the eye's spherical aberration and astigmatism could also act as stimuli for accommodative tracking in some subjects. Fincham was unable to demonstrate the usefulness of spherical aberration as a clue.

In contradistinction, Stark and Takahashi ([39] and Chapter 2) have demonstrated that under monocular hand-tracking conditions with perceptual clues minimized, but spherical and chromatic aberration and astigmatism left unaltered, the initial-tracking error was at chance level.

Although directional clues did not exist at a level where they could be consciously perceived or utilized by any of the six subjects, they were required to make a conscious decision on which way to adjust the hand-tracking device, based on retinal blur.

In this study, by using the accommodative-convergence movements of the eye associated with changes in accommodation in the other eye, we have attempted to establish that the trial-and-error mechanism present in the accommodative system exists on an involuntary level involving no conscious decision on the part of the subject [44].

220

METHODS AND MATERIALS

Each of three subjects attempted to keep monocularly in focus the image of a target which moved either closer or farther away in a random sequence. A continuous recording of the target position and the associated accommodative-convergence movement of the occluded *fellow* eye was kept throughout the experiment. Figure 24 is a schematic representation of the experimental apparatus which is described as follows.

Target System

The target to stimulate accommodation consisted of a horizontal wire, mounted on an X-Y recorder pen unit. Its position could be varied by the experimenter by remote control. The particular position of the target at any time was registered on one channel of a Sanborn recorder. The target was moved in a stepwise fashion about an intermediate position. The length of time the target remained in either the far or near position could be varied from 200 msec to many seconds. The dioptric range of the change in accommodative stimulus was ±2 diopters. This was at times altered very slightly by a change in the gain of the X-Y recorder.

To minimize the influence of learning, successive target positions were determined from a table of random binary numbers. Illumination of the target was provided by a light source placed behind a diffusing screen. With this arrangement, luminance changes with movements of the target were negligible. Color, when used, was obtained by suitable filters.

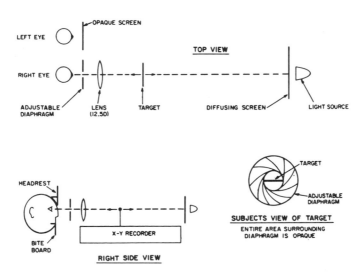

Fig. 24. Experimental apparatus. Note that the target moves in an optical axis of the right eye.

Observation System

The subject's head position was held firmly with a bite board and headrest. The subject viewed the target with his right eye through a lens whose focal point coincided with the center of his pupil. With such a system the angular size of the image of the target remained constant, irrespective of the target position.

Recording System

While the subject's right eye viewed the target, the associated accommodative-convergence movements of his occluded left eye were measured by two fixed cadmium sulfide photocells continuously monitoring the difference in reflectance of light from the iris and the sclera. The electrical signal from the two cells was fed into a Wheatstone bridge which yielded a voltage proportional to the eye position. These voltages were then amplified and fed into a second channel of the pen recorder. Convergence movements of the left eye were associated with accommodation of the right eye, and divergence movements associated with relaxation of accommodation in the right eye.

Control of Clues

Horizontal target movements was minimized by use of a horizontal line target and a variable diaphragm (see Fig. 24).

Prior to an experimental run, the subject viewed the moving target and attempted to eliminate any clues resulting from vertical target movements and difference in the size of the blur at the near and far positions. Vertical target movement was minimized by control of head position, precise initial alignment of target by means of a plastic reference grid, and by matching the symmetry of blur in extreme positions. At times, difference in size of blur at the near and far positions could be eliminated by symmetrically enlarging the step. It is important to remember that attempts to eliminate clues resulting from vertical target movement and difference in blur size were made by the subject. Thus, the results of any run could be affected by the degree of success achieved by the subject in eliminating these subtle clues. The subject wore a set of headphones which produced a relatively loud sound at 360 cycles/sec which masked auditory clues made by the target mount as it was moved.

Subjects

Three color-normal subjects with negligible ammetropia and with astigmatic error of less than 0.5 diopter were presented a series of random target positions.

RESULTS

Figure 25 illustrates a typical correct response to a step-accommodative stimulus in the near direction with overshoot of accommodative-convergence effort. Note the high signal-to-noise ratio. In Fig. 26 some incorrect responses are shown. Note the correction movement following the initial erroneous response. Figure 27 shows ten responses with five

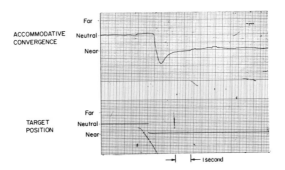

Fig. 25. A typical correct response to a step-accommodative stimulus in the near direction.

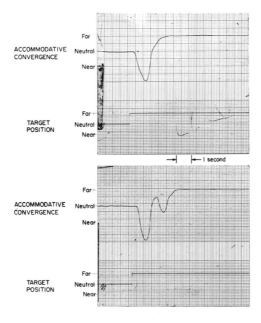

Fig. 26. Some initially incorrect responses to step-accommodative stimuli in the far direction.

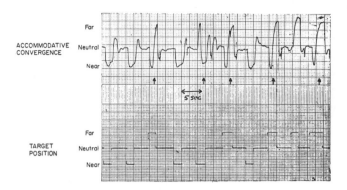

Fig. 27. Ten responses with five initial errors from a typical experiment. (Arrows indicate errors.)

mistakes taken from a typical experiment. The subject was aware of all erroneous responses as well as oscillations, which resulted in changes in the appearance of the blur.*

The results of many trials on each subject emphasized and reemphasized the difficulty involved in eliminating all perceptual clues. Of the three subjects, it appeared that only subject A was able to eliminate all perceptual clues, and he not even in all experiments.

In the first series of experiments, the target was left in each new position until the eye had correctly accommodated for it; then its position was changed. Under these conditions, with white light, subject A showed initial errors in the direction of accommodation change of 41% and 50% in two trials of about 100 stimuli. Figure 28 illustrates graphically the percentage of initial errors in every successive set of ten trials. It can be seen that the average percent error is about 50%, and that there is no indication of trends or learning.

To confirm the randomness in the direction of accommodative response, the number of successive correct responses between failures was analyzed. Figure 29 shows these intervals as a distribution function plot-

---

*Responses to the return of the target to the intermediate position were not considered responses for analysis. Since the subject always came to the correct final accommodative level, return of the target to the intermediate position represented a predictable stimulus which could be easily learned. In fact, none of the subjects ever made an error in responses to such a stimulus, emphasizing the necessity of using randomized stimulus presentation.

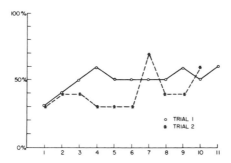

Fig. 28. Percentage of erroneous initial
tracking attempts (ordinate) in successive
ten trials vs. sequence of sets of ten trials
(abscissa).

ted with a logarithmic ordinate scale. By assuming that the occurrence of
correct initial direction of response is random, the probability of there
being a successive correct initial response is given by: $P_n = (1/2)^n$.
Therefore, if $n = 1$, $P_n = 1/2$ or 50%. If $n = 4$, $P_n = 1/16$ or 6.25%, which
means that the chances of getting four successive correct initial responses
are 16 to 1. Figure 29 illustrates the theoretical straight line resulting
from the equation. The vertical bars represent ±1 standard deviation. It
can be seen that the experimental data follow the theoretical line, further
illustrating that errors of initial tracking follow a chance or random pattern.

As expected, when the target was illuminated with monochromatic
red light, subject A again showed a similar frequency of initial errors.
While the two other subjects showed clear responses to all stimuli, they
made no initial errors, apparently indicating their inability to eliminate
all perceptual clues, and pointing out that they were not using chromatic
aberration as a clue.

In the next experiment an attempt was made to eliminate any residual
clues resulting from a still imperfect adjustment of blur by reducing the
duration of the stimulus to 400 msec. Most of the response then took place
after the target had returned to the intermediate position. Under these
viewing conditions 46% of the trials were in error in initial-tracking direction for subject A. The results for the two other subjects still indicated
that they had not made 50% errors.

DISCUSSION

In studying accommodation, it was assumed that fluctuations in accommodation were faithfully reflected by changes in accommodative con-

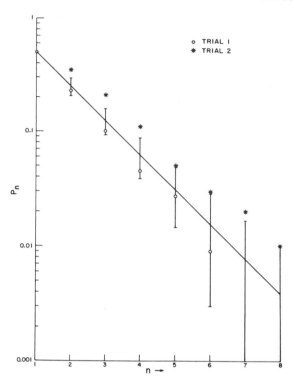

Fig. 29. Probability of getting n successive correct responses (P_n) vs. n. Straight line is predicted from theoretical equation.

vergence. Experimental and clinical studies regarding the linearity and constancy of this AC : A ratio (accommodative convergence to accommodation ratio) would appear to support this assumption [5, 20, 25, 27].

In experiments such as this and others which have been previously reported [12, 19, 29], the object is to strip the system of all clues which might play a role in indicating the sign of a given error (sign clues). It is apparent that two types of clue are available to the accommodative system: (1) target-associated sign clues, e.g., size of intensity changes, horizontal or vertical target movement, blur asymmetry due to target optics, non-random stimulus presentation, etc., and (2) eye-associated sign clues, e.g., astigmatism, chromatic, spherical, and other aberrations and small 2-cycles/sec lens oscillations. All these sign clues can be eliminated, leaving only nonsign clues, the most important of which is the gross blur of the image on the retina. This clue yields sufficient information for the system to arrive eventually at the correct final level or accommodation regardless of whether or not an initial error was made (see Chapter 2).

In the light of this and other recent work [39], eye-associated sign clues play an uncertain role in indicating the sign of the error. It is clear, however, that one subject used in the present study was unable to utilize chromatic or spherical aberration astigmatism, or 2-cycles/sec lens oscillations to gain information about the sign of the error. The two other subjects were apparently not using chromatic aberration, although for them some clue persisted which yielded sign information.

From both reanalysis of the literature and our own experiments we conclude that it is easy to attain 100% correct accommodative responses. It is only through painstaking attention to every detail of the stimulus that all clues may be eliminated and the randomness of the system appreciated. Here, too, the experience and skill of the subject is essential in isolating and eliminating all extraneous clues. Indeed, it is our contention that the failure of two of the subjects to make 50% errors resulted from their inability to eliminate all extraneous clues. This contention is based on numerous discussions with these subjects following runs of 100% correct responses, even after the most painstaking initial alignment of the apparatus. Furthermore, subject A occasionally made only 20 or 30% errors. After such runs he was frequently able to report the existence of a slight clue, for example, vertical target movement, or discrepancy in blur symmetry in the near and far positions. Finally, we have found, as has been found elsewhere [39], that the smallest clues, intentionally introduced, will result in any subject's making 100% correct responses.

The results of this study stress the importance of eliminating all extraneous clues by: (a) control of learning by use of random target presentation; (b) control of horizontal and vertical target movement; (c) control of auditory clues; (d) control of blur symmetry and size in both near and far positions; and (e) control of illumination uniformity and size clues.

SUMMARY

The directional tracking of the accommodative system of three subjects has been analyzed by following associated accommodative convergence movements. An attempt was made to eliminate extraneous clues such as apparent vertical and horizontal target movement, size change, blur asymmetry, and illumination differences and also the possibility of learning patterns of stimulus presentation. Spherical and chromatic aberrations of the eye were left intact and astigmatism was uncorrected.

For one subject the initial direction of change in accommodation was in error about 50% of the time. The gross clue of the blurred image on the retina was always sufficient stimulus to allow the system to eventually reach the correct final level of accommodation. Statistical verification of the randomness of these data led to the conclusion that it is possible for

the accommodative system to operate on an even–error signal with spherical and chromatic aberration intact, with uncorrected astigmatism, and in
the presence of normal noise oscillations of the lens.

Two other subjects made 100% correct responses even under conditions where chromatic aberration was removed as a possible clue. It is
felt that major clues as to the direction of target movement can result from
less than absolutely perfect alignment of the target on the visual axis and
from blur asymmetry.

These studies appear to minimize the importance of the eye's chromatic and spherical aberration, astigmatism, and lens noise oscillations
in determining the initial direction of accommodation. The study further
emphasized how ably the accommodative system utilizes the smallest perceptual clue and how difficult it is to eliminate completely every extraneous
clue.

REFERENCES AND FURTHER READING

1.   Allen, M. J.: The stimulus to accommodation. Am. J. Opt. 32: 422–
     431 (1955).
2.   Allen, M. J., and Carter, J. H.: An infra-red optometer to study the
     accommodative mechanism. Am. J. Opt. 37: 403 (1960).
3.   Alpern, M.: Variability of accommodation during steady fixation lev-
     els of illuminance. J. Opt. Soc. Am. 48: 193 (1958).
4.   Alpern, M.: Accommodation, in the Eye, Vol. 3, H. Davson, ed.,
     Academic Press, New York (1961), pp. 191–229.
5.   Alpern, M., and Ellen, P.: A quantitative analysis of the horizontal
     movements of the eye in the experiment of Johaness Müller. I.
     Methods and results. Am. J. Ophthalmol. 42: 289–296 (1956).
6.   Bannon, R. E., Cooley, F. H., Fisher, H. M., and Textor, R. T.: The
     stigmatoscopy method of determining the binocular refractory status.
     Am. J. Opt. and Arch. Am. Acad. Opt. 27: 8 (1950). (We wish to
     thank Professor G. Westheimer for calling our attention to this ref-
     erence.)
7.   Bower, J. C., and Schultheiss, P. M.: Introduction to the Design of
     Servomechanisms, Wiley, New York (1958).
8.   Campbell, F. W., and Robson, J. G.: High speed infrared optometer.
     J. Opt. Soc. Am. 49: 268 (1959).
9.   Campbell, F. W., Robson, J. G., and Westheimer, G.: The control of
     accommodation in the human eye. Proceedings of 2nd International
     Congress of Cybernetics, Namur, Sept. 3–10, 1958, 1: 924 (1958).
10. Campbell, F. W., Robson, J. G., and Westheimer, G.: Fluctuations
     of accommodation under steady viewing conditions. J. Physiol. 145:
     579 (1959).
11. Campbell, F. W., and Westheimer, G.: Sensitivity of the eye to dif-
     ferences in focus. J. Physiol. 143: 18 (1958).

12.   Campbell, F.W., and Westheimer, G.: Factors influencing accommo-
       dation responses of the human eye. J. Opt. Soc. Am. 49: 568 (1959).

13.   Campbell, F.W., and Westheimer, G.: Dynamics of accommodation
       response of the human eye. J. Physiol. 151: 285 (1960).

14.   Campbell, F. W., Westheimer, G., and Robson, J. G.: Significance
       of fluctuations of accommodation. J. Opt. Soc. Am. 48: 669 (1958).

15.   Campbell, F. W., Westheimer, G., and Robson, J. G.: Fluctuations
       of accommodation under steady viewing condition. J. Physiol. 145:
       579 (1959).

16.   Carter, J. G.: A servoanalysis of the human accommodative mech-
       anism. Arch. Soc. Am. Oftal. Optom. 4: 137 (1962).

17.   Ditchburn, R.W., and Ginsborg, B. L.: Vision with stabilized retinal
       image. Nature 170: 36 (1952).

18.   Eykhoff, P.: Adaptive and optimalizing control systems. IRE Trans.
       Auto. Control AC-5: 148 (1960).

19.   Fincham, E. F.: The accommodation reflex and its stimulus. Brit.
       J. Ophthalmol. 35: 381−393 (1951).

20.   Fincham, E. F., and Walton, J.: The reciprocal action of accommo-
       dation and convergence. J. Physiol. 137: 488−508 (1957).

21.   Graham, D., and McRuer, D.: Analysis of Nonlinear Control Systems,
       Wiley, New York (1961).

22.   Hamasaki, D., Ong, J., and Marg, E.: The amplitude of accommo-
       dation in presbyopia. Am. J. Opt. and Arch. Am. Acad. Opt. Mono-
       graph 192: 1−12 (1956).

23.   Houk, J., Okabe, Y., Rhodes, H., Willis, P. A., and Stark, L.: Tran-
       sient responses of human motor coordination system. Quart. Prog.
       Rept., Research Laboratory of Electronics, M.I.T. 64: 315 (1962).

24.   Marg, E., and Reeves, J. L.: Accommodative response of the eye of
       the aged cat to electrical stimulation of the ciliary ganglion. J. Opt.
       Soc. Am. 45: 926 (1955).

25.   Morgan, M. W., Jr.: The clinical aspects of accommodation and con-
       vergence. Am. J. Opt. 21: 301−313 (1944).

26.   O'Neill, E. L., and Asakura, T.: Optical image formation in terms
       of entropy transformations. J. Phys. Soc. Japan 16: 301 (1961).

27.   Ogle, K. N., and Marten, T. G.: On the accommodative convergence
       and proximal convergence. AMA Arch. Ophthalmol. 57: 702−715
       (1957).

28.   Okabe, Y., Rhodes, H., Willis, P. A., and Stark, L.: Simultaneous
       hand and eye tracking movements. Quart. Prog. Rept., Research
       Laboratory of Electronics, M.I.T. 66: 395 (1962).

29.   Rashbass, C., and Westheimer, G.: Disjunctive eye movements. J.
       Physiol. 159: 339 (1961).

30.   Riggs, L., Ratliff, F., Cornsweet, J., and Cornsweet, T.: The dis-
       appearance of steadily fixated visual test objects. J. Opt. Soc. Am. 43:
       495 (1953).

31.   Roth, H.: Electric potentials and accommodation of the rabbit eye.
       Doctoral Dissertation, University of California, Berkeley (1961).
32.   Stark, L.: Stability oscillations and noise in the human pupil servo-
       mechanism. Proc. IRE 47: 1925 (1959).
33.   Stark, L.: Environmental clamping of biological systems. J. Opt.
       Soc. Am. 52: 925 (1962).
34.   Stark, L., Campbell, F. W., and Atwood, J.: Pupil unrest: an exam-
       ple of noise in a biological servomechanism. Nature 182: 857 (1958).
35.   Stark, L., Iida, M., and Willis, P.: Dynamic characteristics of the
       motor coordination system in man. Biophys. 1: 279 (1961).
36.   Stark, L., Okabe, Y., and Willis, P. A.: Sampled-data properties of
       the human motor coordination system. Quart. Prog. Rept. Research
       Laboratory of Electronics, M.I.T. 67: 220 (1962).
37.   Stark, L., and Sherman, P.: A servoanalytic study of consensual pupil
       reflex to light. J. Neurophys. 21: 17 (1957).
38.   Stark, L., and Takahashi, Y.: Accommodation tracking. Quart. Prog.
       Rept., Research Laboratory of Electronics, M.I.T. 67: 205 (1962).
39.   Stark, L., and Takahashi, Y.: The absence of odd-error signal mech-
       anism in human accommodation. IEEE Trans. Biomed. Eng. (ac-
       cepted) and IEEE Intern. Conv. Record, Part 6: 202−213 (March 22−
       26, 1965).
40.   Stark, L., Takahashi, Y., and Zames, G.: The dynamics of the human
       lens system. Quart. Prog. Rept., Research Laboratory of Electron-
       ics, M.I.T. 66: 337 (1962).
41.   Stark, L., Young, L., and Vossius, G.: Predictive control of eye
       tracking movements. IRE Trans. Human Factors Electron. HFE-3:
       52 (1962).
42.   Termer, F. E., Harman, W. W., and Truxal, J. G.: Signals and Sys-
       tems in Electrical Engineering, McGraw-Hill, New York (1962).
43.   Troelstra, A., Zuber, B. L., Miller, D., and Stark, L.: Accommoda-
       tion tracking. Quart. Prog. Rept., Research Laboratory Electronics,
       M.I.T. 72: 262 (1964).
44.   Troelstra, A., Zuber, B., Miller, D., and Stark, L.: Accommodative
       tracking: a trial-and-error function. Vis. Res. 4: 585−594 (1964).
45.   Young, L., and Stark, L.: A sampled-data model for eye tracking
       movements. Quart. Prog. Rept., Research Laboratory of Electron-
       ics, M.I.T. 66: 370 (1962).
46.   Young, L., and Stark, L.: Variable feedback experiments supporting
       a discrete model for eye tracking movements. IRE Trans. Human
       Factors Electron. Special manual control issue. HFE-4: 38 (1963).

THE EYE

INTRODUCTION

The several years between the time I started to study the pupillary system and the time when my own concentrated research on eye tracking movements began were eventful and rewarding – both scientifically and personally. Many people – associates, students, friends old and new – were instrumental in focusing my attention on the eye and in developing scientific curiosity into active research. A discussion of these people, matched with the events, is necessary in order to establish background and a full understanding of the research described in this section.

My associate Tom Cornsweet was perhaps the first to stir my interest in eye movements when he arrived at the Psychology Department at Yale from the laboratory of Riggs in Providence, Rhode Island. His thesis research complete, he was continuing experimentation on microsaccades, and becoming quite involved in instrumentation to measure these fine eye movements. Another friend who figures prominently in this background discussion is Cyril Rashbass, whom I met while I was at University College in London. Cyril, who was then a medical student, occasionally came into our laboratory at the University to eat and chat during a luncheon break. His later work on the physiology of eye movements in London, and his important demonstration of the seemingly paradoxical behavior of the eye in tracking step-ramp targets, is gratefully remembered. More recently, Cyril became associated with Gerald Westheimer, who developed a second-order linear model for saccadic eye movements. Kurt Lion at M.I.T., Robert Cohen in Washington, and Christine Kris at Radcliffe had all used electro-oculograms, a not entirely satisfactory method for studying eye movements. However, through Dr. Kris, I met a Dr. Richter from Switzerland, who had developed a photocell system for measuring eye movements very similar to that of Torok at Illinois and Smith at Dartmouth.

Another associate, Professor Elwyn Marg, whom I had met in Stockholm in 1958, interested me in the electromyography of eye movements, based on work that he had done at Berkeley, and so in the spring of 1962, I decided to build a binocular pair of recording eyeglasses – or eye goggles – in order to obtain a more convenient and accurate measure of horizontal eye position than that which the electro-oculogram provides.

Dr. Vossius, who came to our laboratory as a visiting fellow from Frankfurt, joined Larry Young, who was just starting research for his doctoral thesis on eye movements. Taking advantage of the on-line Ramo-Wooldridge computer in our laboratory, we repeated on the eye the previous experiments on the hand movement system in order to demonstrate the predictor operator in the eye movement control system. Both Robert Payne and Allen Sandberg helped greatly with computer and instrumentation and we were able to complete this work during Vossius' two-month stay in the United States. Certain features of the transient and frequency response data suggested the presence of a discontinuous or sampled-data operator, so Larry Young went on to study this feature for his doctoral thesis. Unknown to us, three other groups – Fender and Nye in England, Dallos and Jones in Chicago, and Johnson and Fleming in Cleveland – were also working on the eye movement control system. Although these workers did not identify both the predictor and sampled-data operator, their results in general confirmed our own.

There were many other people directly involved in this background work, among them Professor David Cogan of Harvard, a world-renowned authority on the neurology of ocular movements, who provided much intellectual stimulation. Professor Cogan also introduced me to Professor Carl Kupfer and to Drs. David Miller and Ernst Meyer, who quickly realized the importance of these methods for clinical diagnosis as well as in applied research. I must also mention Professor Arne Troelstra from Dr. Bouman's laboratory in Soesterberg, Holland, who visited us for a period of a year and a half, and Professor Bert Zuber, who started work on his doctoral dissertation with us, investigating the control system for vergence eye movements. Dr. Troelstra's interest was drawn to the interaction between the various eye movement control systems, and Dr. Zuber developed the studies on saccadic suppression, one of the several effects of the intermittent motor control system on the sensory and perception side of the visuo-motor system. Not to be forgotten are Martin Lorber, Anne Horrocks, and Gerald Masek, who all contributed to these studies of the eye.

The research on eye tracking movements described in this section arose from our search for a relatively simple neurological servosystem with "class 2" characteristics – i.e., relatively linear and free from noise with sampled-data and input-signal-adaptive characteristics. Chronologically, I first studied the hand (see Section V), a class 2 computer-limited system with plant limitation superimposed on the computer limitation, and output-adaptive characteristics in addition to its plant-limited inertial features. Since the complex characteristics of the hand made its experimental definition very difficult, I looked next for a simpler example of a neurological servomechanism, and began to study the pupillary system. The pupil is a straightforward "class 1" plant-limited system which is continuous, highly nonlinear, noisy, and controlled by many inputs — as has been extensively discussed in Section II; thus, only after several years did it become clear that it would be useful to study yet another system, one with the computer-limited properties of class 2, but simpler than the hand. For this reason, we then began our research on eye-tracking movements, a classical example of a class 2 computer-limited system.

The inertia of the eyeball is essentially negligible compared with the force available in the eye muscles. Since the eye movement control system reasonably expects the eyeball to always have the same rotational inertia, there are also no output-adaptive characteristics; therefore, it does not need the ability to adapt to any such changes in eyeball inertia. The problem of eyeball inertia parenthetically raises the interesting question of the possibility of proprioceptive feedback for control of eye movements. Two facts have been well established: first, that proprioceptive spindles and fibers are present neuroanatomically in a number of animals, including man; and second, that no conscious sensation arises from stimulation of these proprioceptors. The possibility of an unconscious role for the proprioceptors in providing and in stabilizing certain phase-advancing features is contradicted by two sorts of experimental evidence. If the eyes are placed in a double open-loop situation with each eye independently following a target which is fixed to a particular portion of the retina, both eyes wander in all directions. There seems to be no clear pattern in these movements, rather a random open-loop wandering. If the proprioceptive loop had any real function, it would have helped to maintain the eyes in the straight-ahead primary position. The second experimental evidence comes from the work of Robinson, who showed that when the eyeball is loaded with an infinite impedance, the resulting pattern of force remains the same as in a normal eye movement. Any proprioceptive feedback system would certainly have produced clearly different responses in the two cases. An analysis of the trajectory of saccadic eye movements from the point of view of both optimal control theory, and of the underlying nonlinear "apparent viscosity" muscle system, again shows no evidence for proprioceptive feedback. The eye has a very adequate predictor operator or input-adaptive system which is somewhat less complex than in the hand inasmuch as the eye appears to adapt less readily to open-loop or variable feedback operating conditions. The hand, by contrast, adapts to these conditions readily.

Chapter 1 of this section is directly concerned with the prediction operator and shows behavioral data in the form of both transient and steady-state responses. The prediction operator is clearly demonstrated and these results should be compared to a similar one (Section V, Chapter 2) on the hand. Both the hand and the eye are to be contrasted with the pupillary system in Section II, which shows no such prediction operator effect.

Of considerable interest is the comparison between Figs. 12 and 13 in this chapter, which show two different Bode plots of eye frequency response. The first plot represents our best approximation from the point of view of continuous servoanalytic theory. However, when the intermittency operator with its sampled-data model seemed established on the basis of time domain studies, we reviewed and reinterpreted the same data. What previously had appeared as quite scattered results near the high-frequency end of the spectrum now looked like confirmation of the sampling mechanism in the frequency domain. The ability of a scientist to appreciate only those phenomena for which he has a model is significantly illustrated, and I would like to interject further comment on this. It has been said that if fish were to study the world, the last thing they would discover would be the sea. I believe that one of the contributions of cybernetics is to provide the biologist with a canonical set of well-knit mathematical models for common and important phenomena in experimental neurology. The behavior of these beautiful, natural neural systems — studied and described by neurophysiologists before the advent of cybernetics — became truly apparent only when the scientist could approach them armed with communication and control theory.

The crucial intermittency requiring a discrete discontinuous model is introduced in Chapter 2. Sampled-data theory was developed by Raggazini and other electrical engineers for the purpose of handling radar target tracking control systems in which intermittent signal information flow occurred. The theory also related to the complementarity of the discrete and continuous communication channel as discussed, for example, by Shannon. The z-transform permits one to handle this model analytically in a simple linear form and the luxury of being able to compute model behavior with pencil and paper will be appreciated by anyone who has had to use analog, digital, or hybrid computational systems. It is this linear-reduced model that is actually computed and compared with experimental responses and not the first, more complete, model.

Chapter 3 deals with a test of model and eye movement control system behavior under a new set of conditions — variable feedback. This provides a stringent set of requirements which strongly support the adequacy of the model. This model, of course, has its limitations since models by their very nature are too simple a representation for the so-called "real" world that lies just beyond our ken. We have discussed these limitations and are using them as guidelines for further avenues of investigation.

In the experiments discussed in Chapter 3, we used the environmental clamping technique developed for the pupillary system and subsequently used to "open the loop" on the hand system. It is one of the methods of "dry dissection" whereby we can explore and expose certain characteristics of a neurological control system without actually "wet dissecting" into the black box.

Chapter 1

Predictive Control

INTRODUCTION

The exact nature of the multiloop control system governing gross eye tracking is still largely unknown although two principal types of movement have long been distinguished [15]. The saccadic jump movement is characterized by its high velocity, the inability to control this velocity consciously, pauses of at least 100 msec between successive saccades, and possible absence of (foveal) vision during the saccadic movement. The magnitude and direction of the saccade is generally such as to correct for position error [70, 73]. Smooth pursuit movements, on the other hand, are slow (less than 20 deg/sec), seem to be continuous, are not produced consciously, occur only when the eye is tracking a smoothly moving target, and appear to keep the target image, with respect to the retina, stabilized rather than fixated [48, 74]. These two movements are integrated in any tracking task, as illustrated in Fig. 1. The important points to note are initial time delay, occurrence of saccades to correct any appreciable error, and smooth pursuit movements locking onto the target.

Previous dynamic descriptions of eye tracking behavior have not been arrived at with sufficient control of important aspects of the input signal. In particular, unpredictable inputs have not been used to clarify different properties of the control system. We have contrasted the response to predictable steps as well as simple and complex patterns of sinusoidal changes in target position [66]. In this paper the term "unpredictable input signal" denotes a pattern of target of motion which is sufficiently unconstrained so that the subject obtains little information from the past portion of the signal to permit him to predict the future motion of the target. Thus, any attempt by the subject to anticipate or predict the target movement would be in vain. In the case of discontinuous inputs, the input signal

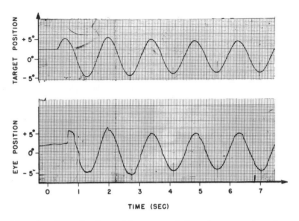

Fig. 1. A typical tracking record showing reaction time, saccadic jumps, and smooth movements.

was constructed by summing square waves of different amplitudes and incommensurate frequencies. It has been proved that similar functions constructed from irrationally related periodic sources form a process which is indeed random [8, 75].

The same argument holds for the continuous unpredictable inputs where the input function is constructed by summing several sinusoids of incommensurate frequencies. Previous experiments using these types of input patterns have demonstrated the inability of subjects to make useful predictions [60].

The experimental data to be presented illustrate the striking differences between responses to predictable input signals and to unpredictable input signals. This is true for discontinuous inputs as well as for the continuous signals discussed in the experimental results section. Our main thesis is that responses to both predictable and unpredictable inputs must be considered in any study of eye tracking movements.

METHOD

The method of measuring eye position involved detection of the difference in diffuse reflected light from the sclera (white) and iris [57, 65]. Figure 2 (top) shows the arrangement of light sources and photocells mounted on a pair of goggles worn by the human subject and Fig. 2 (bottom) shows how light is reflected from either side of the eye. It is clear that the amount of sclera illuminated by each lamp and, accordingly, the amount of reflected light received by each photocell, are functions of the angle of sight.

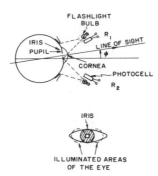

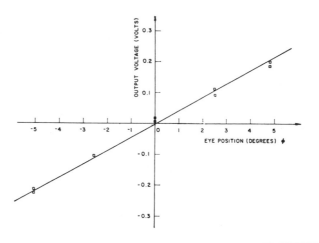

Fig. 2. Light bulb and photocell arrangement above. Front
view of eye for $\phi = 0$, below.

Fig. 3. Typical output voltage vs. eye position.

Components used were flashlight bulbs and CdS photoresistor cells.
Figure 3 shows a linearity check which is typical of those observed when
care is taken in adjusting the goggles. The bandwidth of the photoresistors
is 10 cycles/sec for high illumination and somewhat lower for dimmer
light. In all of the experiments reported in this paper, correction was
made for the small distortion attributable to the photoresistors' gain attenuation and phase lag. Since the delay in the onset of the photoresistor
response is less than 2 msec, the effect of this distortion on the experimental results obtained with discontinuous inputs is negligible.

A specially designed mirror galvanometer, manufactured by the Sanborn Co., Waltham, Mass., was used to project the target [77]. The hori-

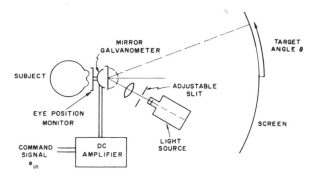

Fig. 4. Mirror galvanometer used in eye movement studies.

zontal position of the projected target was controlled by a voltage signal.
A block diagram of the apparatus as used in this investigation is shown in
Fig. 4.

The galvanometer mirror was positioned directly under the subject's
head as indicated in Fig. 4. The d-c sensitivity of the system was 80 mra-
dians/V and its linear range was ±200 mradians (±12°). By increasing the
screen-to-galvanometer distance (without altering the subject-to-screen
distance), the effective sensitivity and range, in terms of target angle as
seen by the subject, could be increased at the expense of some defocusing.
The special features of the galvanometer which make it suitable for tracking studies are its large mirror (0.7 × 4.0 cm) and ease of critical damping. The frequency response was flat out to 100 cycles/sec.

The experiments were restricted to horizontal movements. The subjects normally used binocular vision although only the left eye was monitored. Calibrations were performed and only those experiments with ac-
ceptable linearity of measurement before and after a run (about 80%) were
retained. The experiments reported herein were performed on only a few
trained subjects and care must be taken not to ascribe too much significance to the detail of the records. However, the important differences between the responses to predictable and unpredictable inputs were confirmed
in additional experiments on 20 other subjects. We did not investigate effects of significant variation in instructions to the subjects. They were
told to look at the target at all times but not directed to attempt to anticipate or not to anticipate target movement. Trained subjects were ac-
customed to the apparatus, but of course could not tell which experiment
was about to be run. The difference in response to the predictable and unpredictable signals was not notably influenced by the experience of the subject.

The target was a fairly bright projected vertical slit 0.5 × 6.0 cm,
which stood out clearly in the darkened room. The screen was 185 cm

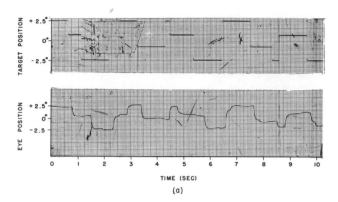

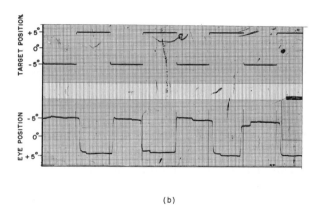

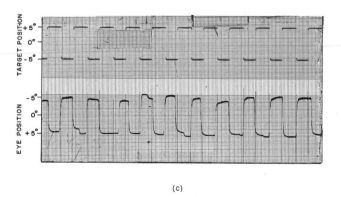

Fig. 5. Records of horizontal eye movements during tracking of random step and square-wave targets. (a) Eye movement during tracking of unpredictable steps. (b) Square wave 0.4 cycle/sec. (c) Square wave 1.0 cycle/sec.

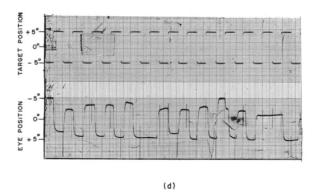

(d)

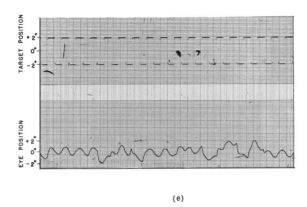

(e)

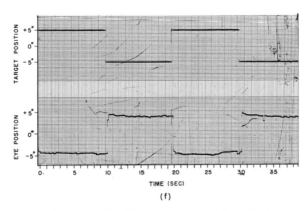

(f)

Fig. 5. (Continued). (d) Square wave 1.5 cycles/sec.
(e) Square wave 2.0 cycles/sec. (f) Square wave 0.05
cycle/sec.

from the subject. The light used for measurement of eye position was in the peripheral field and did not obscure target view or interfere subjectively with our experienced subjects.

EXPERIMENTAL RESULTS

Response Times to Unpredictable and Predictable Steps

Response time is used in this chapter to denote the length of time by which a response may precede a stimulus, as well as the latent period or reaction time, in the normal meaning of these terms. It is considered positive when response follows stimulus and negative when response occurs before stimulus. The response of the eye to a series of unpredictable step changes of position is shown in Fig. 5a and it will be noted that the movements are saccadic in nature except for some low velocity drifting. Small saccadic movements serving to correct fixation error due to drifting, overshoot, or undershoot can also be seen. These are simular to those observed under steady fixation conditions [12, 52]. The histogram of time delays in response to unpredictable steps shows their statistical distribution function in Fig. 6a.

Often the eye movement preceded the change of target position. Figure 5b—f shows the results from five square-wave experiments. At very low repetition rates, prediction is minimal; for square-wave frequencies around 0.4—1.0 cycle/sec, prediction is maximal. A prediction of 100 msec might well have the eye on target when the target appears, after allowing for the time of actual movement of the eye. We therefore define "prediction" as eye movement latencies ranging from 150 msec delay (normal response time) to 100 msec anticipation. In some cases anticipation may be so great that the eye has a chance to see itself in error. Anticipation greater than 100 msec we term "overpredictions." The eye still may not correct itself, but rather await the expected change of target position.

With moderate repetition rates (0.4—1.0 cycle/sec) the pattern seems to be a rapid buildup of prediction to the limit of overprediction and then a drop back to small prediction or a time delay [48]. At higher repetition rates (1.3—1.7 cycles/sec) there seems to be less significant prediction. Occasionally an entire step may be missed, as shown in Fig. 5d. An interesting phenomenon at still higher frequencies (around 2.0 cycles per sec) is the occurrence of apparently continuous movements rather than clear saccades (see Fig. 5e).

In Fig. 6c, the histogram of time delays at a moderate repetition rate (1.0 cycle/sec) demonstrates the high frequency of prediction when the eye follows a regular pattern. Figure 6b is a histogram of time delays for a slow repetition rate (0.4 cycle/sec) and shows a greater occurrence of overprediction. The absence of overprediction at the higher frequency of 1.5 cycles/sec is demonstrated in Fig. 6d.

The sequential pattern of time delays and predictions is of interest and may be displayed as a plot of response times in sequential order, as in Fig. 7. Here, starting with the first appearance of a square-wave pattern whose characteristics are initially unknown to the subject, the time delays, subsequent predictions, overpredictions, and reversions to time delays are exhibited. Figure 7c (1 cycle/sec) shows the quick development of prediction, but with minimum overprediction. Figure 7b (0.4 cycle/sec) exhibits the phenomenon of overprediction with occasional transitions back to prediction and delay.

Predictive behavior as a function of repetition rate is summarized in Fig. 8 where the lack of prediction at very slow and very fast rates is contrasted with the prediction and overprediction seen at intermediate rates.

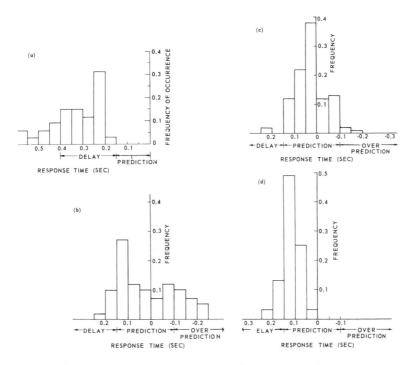

Fig. 6. Histograms of the frequency of occurrence of eye movement response times for target motions of irregular steps and regular square waves of different frequencies. (a) Irregular steps. (b) Square waves 0.4 cycle/sec. (c) Square waves 1.0 cycle/sec. (d) Square waves 1.5 cycles/sec.

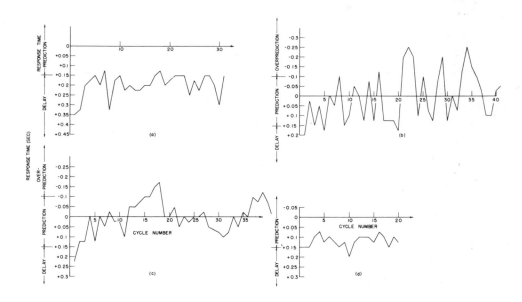

Fig. 7. Typical sequential patterns of time delays and predictions starting with the onset of random step and square-wave target motions. (a) Sequential response times — irregular steps. (b) Square wave 0.4 cycle/sec. (c) Square wave 1.0 cycle/sec. (d) Square wave 1.5 cycles/sec.

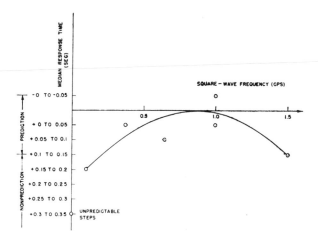

Fig. 8. Median response times in tracking target motion of unpredictable steps and square waves of different frequencies.

Response to Single Sinusoids and Unpredictable
Continuous Signals

All of the preceding results pertain to the saccadic movements. When a single sinusoid is tracked, the "smooth pursuit" system plays an important role, as shown in Figs. 1 and 9. Once accurate tracking of the target is established, the occurrence of saccades becomes less frequent. The accuracy of the tracking is so high that the amount of phase lag and attenuation at low and intermediate frequencies is quite small (Fig. 9). At higher frequencies, as shown in Fig. 10, the response becomes so erratic that it is difficult to identify the sinusoidal response. Saccadic movements are again prominent here.* Data from such predictable steady-state experiments are summarized in the Bode plot of Figs. 12 and 13, which shows that the eye moves in close synchronism with the target (0° phase lag and 0 db gain) for predictable sinusoids up to 1 cycle/sec. For random signals, however, the eye movement control system shows increasing phase lag and lower gain for increasing frequency, as might be expected of a "follow-up" servomechanism. Note the small phase advance, or phase lag, and relatively constant gain in good agreement with results in the literature [19, 67], for those frequencies at which following can be said to exist.

At frequencies above 2.5−3.0 cycles/sec the response to a single sinusoid is obviously not sinusoidal nor can any fundamental of the response at the input frequency be identified. (Since a Bode plot would obviously be meaningless for such a situation, the frequency response in

---

*Since they are in the tracking response to such unpredictable signals as the sum of four sinusoids shown in Fig. 11.

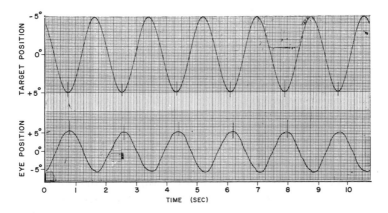

Fig. 9. Eye movement records for tracking of sinusoidal target motion at 0.5 cycle/sec.

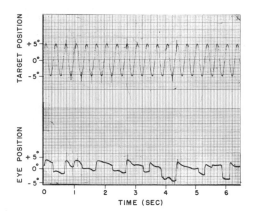

Fig. 10. Eye movement records for tracking of sinusoidal target motion at 3.0 cy-
cles/sec.

Figs. 12 and 13 is plotted only for those lower frequencies at which
sinusoidal following can be identified.)

In order to compare eye tracking movements for sinusoidal inputs
with following of continuous unpredictable inputs, it was necessary to produce a suitably random unpredictable target signal and analyze the eye
movement responses for the gain and phase of following at each frequency
present in that input. For this purpose a medium-sized digital computer
with digital-to-analog and analog-to-digital conversion equipment was used
on-line during the experiments. Whereas a perfectly random input would
consist of filtered band-limited white noise with some energy present at
all frequencies, it has been shown previously [60] that a suitable approximation for work with a human operator is a sum of three or more sinusoids
of incommensurate frequencies. In the experiments reported below, from
four to nine sinusoids were summed by the computer with the resulting
signal appearing at the analog output of the computer and forming the target signal. The eye movement response during 25 sec of following this
target motion was converted to digital form and stored in the computer.
These data were then multiplied by the sine and cosine of each frequency
which was present in the input to determine the components of eye movement in phase with and in quadrature with the components of the input.
From this information the gain and phase of the eye tracking system was
calculated by the computer for each of the input frequencies. Several
groups of test frequencies were used to cover the entire frequency range
of interest. The data shown in Figs. 12 and 13 are a composite of the results obtained with these several groups.

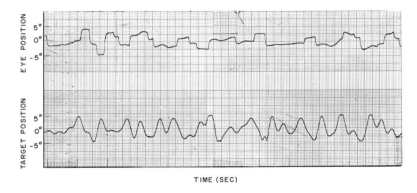

Fig. 11. A typical record of eye tracking of a sum of four sinusoids.

A typical record of the input and tracking response obtained with a random input of the type described is in Fig. 11. Smooth pursuit movements dominate the response to slowly varying portions of the continuous signal while saccadic movements are more obvious during higher velocity portions.

Gain and phase data calculated by the computer [46] from a series of experiments on unpredictable inputs are also shown in the Bode plot of Figs. 12 and 13, which show different curves fitted to the same data. Figure 12 is the graph originally published and represents the view of continuous linear

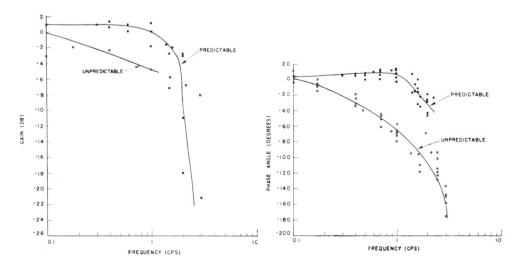

Fig. 12. Gain and phase relationships for continuous predictable and unpredictable target motions. Experimental and measurement problems often resulted in inaccurate determinations of gain characteristics. As a result, many gain data were excluded from the results shown. Phase determinations were not affected appreciably by these problems and were not excluded.

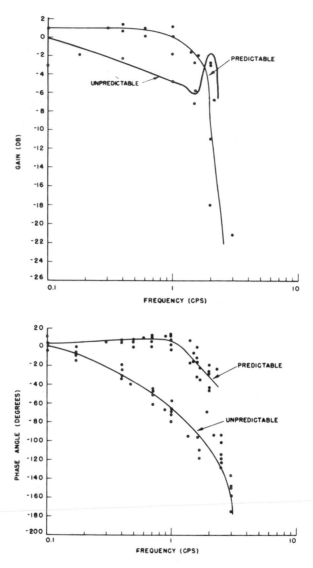

Fig. 13. Eye movement frequency response following
predictable and unpredictable targets.

servoanalytic theory. After the intermittency operator and the sampled-data model had been established on the basis of the transient input studies discussed in the next two chapters, we went back and found the data could be fitted more exactly. Figure 13 represents the best fit from the point of view of discontinuous sampled-data control system theory.

Both Figs. 12 and 13 demonstrate that there is a markedly increased phase lag for the unpredictable signal as contrasted with the steady-state sinusoidal signal which indicates that the prediction apparatus of the brain is able to compensate for almost all of the inherent phase lag of the "neurological system."

DISCUSSION

The experimental data presented above clearly substantiate the core of our main thesis — the importance of controlling the predictability of the input signal. These results are analogous to those in the manual tracking control where it has long been recognized that the nature of the control system depends upon the characteristics of the input signal [16, 35, 60]. In studying complex biological control systems it is necessary to perform experiments with a wide variety of input signals — unpredictable and complex wave forms for dissecting the underlying "neurological" feedback control system and predictable signals for evaluating system performance when the powerful predictive apparatus can be utilized.

Chapter 2

Sampled-Data Model

INTRODUCTION

The control system which enables us to move our eyes and follow a moving target has been investigated from the servomechanism point of view. The ability to identify measurable quantities as input, output, and error variables justifies such an approach: the input is the angular position of an object of interest (the target) with respect to some reference fixed to the head; the output is the angle of the line of gaze with respect to this same reference; and the error, defined as input minus output, is therefore the angular difference between the direction of the target and the direction in which the subject is actually looking. Using these measurable quantities as a basis, the analysis methods of control theory have been applied to the eye movement system to yield transfer functions for the system which should be of value to the human engineer, physiologist, and control engineer [79—81].

The function of the eye movement control system is not confined solely to nulling the error. For small targets it is satisfactory to place the image anywhere in the high resolution central field of the retina (the fovea) covering an angle of approximately 1° of the visual field. More important, however, is the realization that the purpose of the whole system is to permit a maximum amount of visual information to be extracted. To achieve this, the image must be maintained in a relatively fixed position on the retina as much of the time as possible. The dual functions of the system — centering the image on the retina and stabilizing the image — indicate the analogy to a radar tracking system which is required to acquire the target and then follow it smoothly.

The mathematical model of the eye tracking system is based upon a series of experiments in which a subject attempted to maintain fixation of

a horizontally moving target while the position of his head remained fixed.
Study of the eye movement records led to three observed characteristics
which formed the foundation of the model.

First, the system was shown to be of the input–adaptive type, exhibiting anticipation and improved tracking when the input is a predictable
wave form as shown in Fig. 1, and discussed extensively in Chapter 1.
Frequency-response curves show important differences in gain and phase
for predictable inputs of single sinusoids and nonpredictable inputs composed of many noncoherent sinusoids (see Chapter 1 [66]).

The second characteristic concerns the saccadic and pursuit movements — the two types of eye movements involved in tracking which are illustrated in Fig. 14. Saccadic movements are the rapid jumping movements used to move the eyes voluntarily from one fixation point to another,
and are particularly noticeable during reading. They are preceded by a
reaction time of 0.15 to 0.25 sec and follow a typical course of a rapid acceleration to some constant velocity which is dependent on the angle of the
total movement. There follows a rapid deceleration to rest with a small
amount of overshoot occasionally observed. The maximum velocity reaches
600 deg/sec for large saccades, and the total time of the movement is under
0.1 sec for a 10° jump. Just before the time of a saccadic jump, vision is
severely reduced. Pursuit movements, on the other hand, are slow involuntary eye tracking movements which smoothly follow a moving object of
interest in the visual field. Pursuit movements have an initial reaction
time of 0.12 to 0.25 sec and are limited to a maximum velocity of 25−30 deg

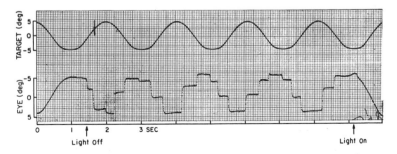

Fig. 14. Absence of velocity control branch in voluntary eye movements
without a visible target. Initial smooth tracking occurs with use of both
position and velocity control. When target light is extinguished, only
memory of the repetitive target swing guides the eyes and only position
control is available. Smooth movements return promptly when target re-
appears.

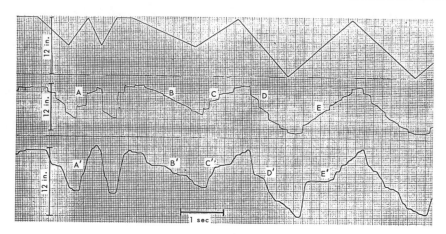

Fig. 15. A variety of eye movement characteristics during tracking of random ramp inputs shown in upper trace. At A, in middle trace, position control results in "staircase" tracking. B and E show smoother tracking with both position and velocity control operating for slower velocity ramp input. C and D show velocity control producing slower velocity segments than ramp input, and position connections are prominent. Lower trace shows simultaneous hand movement for comparison.

per sec. Functional separation of the saccadic and pursuit systems is, therefore, necessary [49]. In the model, the saccadic system must act as a position servo to direct the eye at the target, and the pursuit system must act as a velocity servo to rotate the eye at the same angular rate as the target.

The third important characteristic of the eye movement system is the discrete nature of tracking in the nonpredictive mode, requiring a discrete model for adequate description [78]. A variety of eye movement characteristics during tracking of random ramp inputs is shown in Fig. 15. The evidence leading to a sampled data control model with sampling periods of approximately 0.2 sec stems from five important experimental results.

Pulse Response

The eye movement response to a target movement pulse of less than 0.2-sec duration is a pair of equal and opposite saccadic jumps separated by a refractory period of at least 0.2 sec. The response predicted by a continuous system with a pure delay would be a pulse of the same duration as that of the target, delayed in time by an amount equal to the reaction time.

Open-Loop Step Response

Under open-loop conditions, in which the effective visual feedback is modified by addition of an external path from measured eye position to target position, the response to a target step is a staircase of equal amplitude saccades spaced approximately 0.2 sec apart.

Discrete Changes in Pursuit Velocity

During constant target acceleration, the eye velocity changes in rather discrete jumps at 0.2-sec intervals, indicating the discrete nature of the pursuit system; position errors are corrected by saccades in the direction of target motion.

Peak in Frequency Response

The frequency response of the system shows a peak in gain in the region of 2.5 cycles/sec, consistent with a sampled-data system operating at a sampling period of 0.2 sec.

Dependence of Accuracy Prediction

The inaccuracy of saccades associated with anticipation of square-wave target motion indicates that such saccadic movements cannot be modified by any visual information appearing in the 0.2-sec interval previous to the saccade.

SAMPLED-DATA MODEL

A sampled-data model for the control of eye movements during tracking of unpredictable target signals is shown in Fig. 16c. The system consists of two separate paths in the forward loop — a saccadic branch for the correction of position errors by discrete jumps and a pursuit branch for correction of velocity errors.

As indicated in the model of Fig. 16a, the error angle (e) between the desired angle of gaze (c) and the actual eye position (r) is detected at the retina. This error is sampled by an impulse modulator (M) at sampling intervals, T, where T is the average refractory period for saccadic movements (about 0.2 sec). The synchronization of the modulator must be assumed to be set to coincide with the beginning of a target motion, if the eye has made no saccadic jumps during the previous 0.2 sec. Those position errors less than approximately 0.5° are generally not corrected since the image already lies on the fovea — or central area of the retina. This is represented by the dead zone of 0.5° in the saccadic branch. Each error sample impulse is delayed by one reaction time and integrated to give a

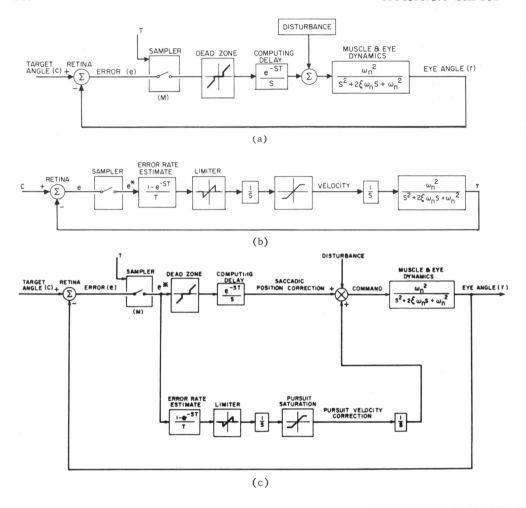

Fig. 16. Sampled-data model for (a) saccadic; (b) pursuit; and (c) combined system for eye tracking movements.

step command which is the desired change in eye position necessary to bring the eye to the previously desired position one reaction time earlier. This step is then filtered by the dynamic response of the extraocular muscles and the eye loading them to yield an actual eye position movement exhibiting finite rise time and possible overshoot.

Trajectory Dynamics

Numerical values for the muscle and eye dynamics parameters published by Westheimer [73] for his second-order model

$$\omega_n = 240 \frac{rad}{sec}, \quad \xi = 0.7$$

indicate that at the next sampling interval, 0.2 sec after the onset of a saccade, the eye position will have settled to within at least $e^{-33.6}$ of its final
value. (By final value is meant the steady-state eye position resulting from
that one saccade. If this is not the desired eye position and the error lies
outside the foveal dead zone, it will be corrected by a secondary saccadic
movement at the next sampling interval.) Since this disparity is less than
the errors in fixation resulting from miniature eye movements (designated
as disturbances in the figure), or errors in computation of the desired amplitude of the saccade, its effect on the observed error at the next sampling
instant can be neglected. (For the sake of simplicity of presentation, the
muscle and eyeball dynamics will be ignored for the remainder of this
paper; however, fuller consideration of this problem is given in the papers
on "A Model of the Human Eye-Positioning Mechanism" [68, 69].) This
being the case, it must be remembered that when the simplified model
indicates a discrete change in eye position, the actual eye position
predicted would be a typical saccadic movement.

Velocity Branch

Whereas saccadic movements serve to center the target image on the
fovea, the purpose of the pursuit branch appears to be stabilization of the
target image on the retina by keeping the angular velocity of the eye equal
to that of the target for target velocities less than 25 to 30 deg/sec. It is
therefore reasonable to describe the pursuit system as a sampled-data velocity tracker, and the block diagram of Fig. 16b represents one way in
which this branch could work. The error between the desired and actual
eye position is sampled at intervals of T seconds. An error rate sample
is estimated from the difference between the past two position error samples divided by the sampling interval, T. The integral of this error rate
impulse is a step command which is the desired change in eye velocity. A
second integration gives the desired change in eye position due to the pursuit branch. The first limiter reflects the fact that the pursuit system does
not attempt to follow the high velocity changes present in discontinuities of
the target position, and the second limiter indicates that the pursuit velocity saturates at 25 to 30 deg/sec. The output of the second integrator will be
a sequence of ramps of eye position at the desired velocity. The effect of
the muscle and eye dynamics is to smooth out the discontinuities in velocity and also to introduce a small constant steady-state error between the
desired ramp and the actual output.

LINEAR REDUCED MODEL

The block diagram of Fig. 16a may be linearized and simplified for
purposes of analyzing the transient and frequency response.

With this simplification, and the effect of disturbances and the 1.0° dead zone neglected, the saccadic system model for discrete position tracking reduces to the flow chart of Fig. 17A(a). By isolating the discrete data mode $e*$, the flow chart can be redrawn as in Fig. 17A(b), in which $K_1(z)$ represents the discrete transfer function of the open loop.

$$K_1(z) = z\left(\frac{1}{s}\right)^* \tag{1}$$

$$K_1(z) = \frac{z}{1 - z} \tag{2}$$

$$\frac{1}{1 + K_1(z)} = 1 - z \tag{3}$$

The flow chart is finally reduced to that of Fig. 17A(c), showing that the output of the saccadic system alone is

$$R(s) = C(s)* z\left(\frac{1 - z}{s}\right) \tag{4}$$

which can be recognized as a delayed zero-order hold.

Velocity

The flow chart for the simplified pursuit model with the nonlinearities removed is shown in Fig. 17B(a). As before, the flow chart is reduced by considering all inputs and outputs at the discrete data point $e*$. In Fig. 17B(b) the z transform of the open-loop transfer function is denoted by $-K_1(z)$:

$$K_1(z) = (1 - z)\left(\frac{1}{Ts^2}\right)^* \tag{5}$$

$$K_1(z) = \frac{(1 - z)}{T}\left[\frac{Tz}{(1 - z)^2}\right] \tag{6}$$

$$K_1(z) = \frac{z}{1 - z} \tag{7}$$

Since

$$\frac{1}{1 + K_1(z)} = (1 - z) \tag{8}$$

the sampled-data velocity tracker finally reduces to the flow chart of

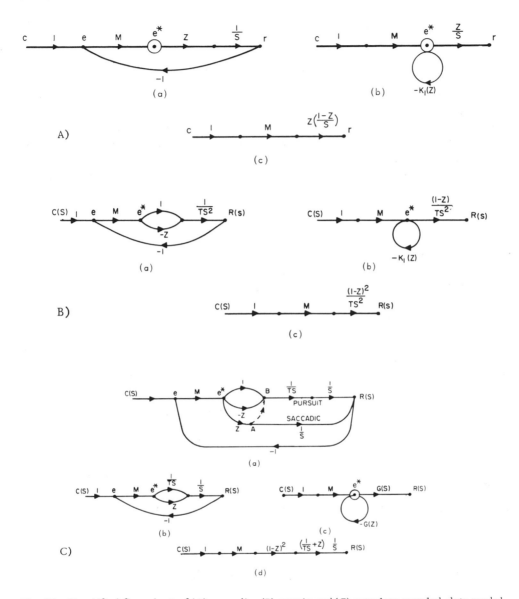

Fig. 17. Simplified flow chart of (A) saccadic; (B) pursuit; and (C) complete sampled-data model.

Fig. 17B(c). For the pursuit system alone, eye position would be given by

$$R(s) = C(s) * \frac{(1 - z)^2}{T s^2} \tag{9}$$

This equation may be rewritten to clarify the operation of the velocity tracker:

$$R(s) = C(s) * \left(\frac{1-z}{T}\right)\left(\frac{1-z}{s}\right)\left(\frac{1}{s}\right) \tag{10}$$

This equation states that the target position is sampled and its velocity is
estimated by the first difference $(1-z)/T$. Since old data are used to calculate present velocity, this term accounts for the delay in the pursuit system. The term $(1-z)/s$ is a zero-order hold, keeping the output velocity
constant between sampling instants, and the integrator yields the eye position resulting from this velocity.

Complete Sampled-Data Model

Inasmuch as the refractory period is not a deterministic function, the
simplifying assumption will be made that T is a constant sampling interval
equal to the mean experimental value for the saccadic system and also for
the pursuit system.

Because the average refractory period for velocity changes is equal
to or slightly less than that for saccadic movements for any one subject
tested, the sampling intervals will be assumed to be synchronous, both
starting with the initiation of any significant change in target position or
velocity following a quiescent period. Under these simplifying assumptions a single sampler, or impulse modulator, may be used to furnish error samples for the saccadic and pursuit tracking loops.

The pursuit loop, as a velocity tracker, should use estimates of error rate to keep the eye velocity equal to the target velocity, with corrections coming in as regularly spaced ramps of eye position. It is important
that the pursuit model not try to null out any apparent smooth error rate
resulting from a step change in error during a saccadic jump.

The function of the first limiter in the pursuit system block diagram
was to force the pursuit system to ignore all very rapid changes in the observed error. These error discontinuities may occur either from target
discontinuities or from the step changes in eye position caused by the saccadic system. This nonlinear element has been removed, and its function
retained as follows:

1. The pursuit loop in the model is considered open at the time of
   (or at the sampling instant following) any target velocity greater
   than 30 deg/sec. This artifice is easily managed in considering
   random target inputs.

2. The component of error discontinuity resulting from a saccadic movement may be prevented from stimulating the pursuit loop by introducing an extra cross-coupling branch (from A to B) between the saccadic and pursuit models.

In Fig. 17C(a), which is a flow graph representation of a linearized version of Fig. 16c, the saccadic dead zone, the pursuit saturation element, and the dynamics of the eye are neglected. (Note that z is defined as a pure delay of T sec, or $z = \exp(-sT)$. This definition of z is the inverse of the definition used by some workers in sampled-data systems.) The function of the first limiter, which was to open the pursuit loop to error changes greater than 30 deg/sec, is retained in the following manner:

1. The pursuit branch must be open during target velocities exceeding 30 deg/sec. This is handled directly for deterministic inputs and on a statistical basis for random target inputs.

2. The component of error discontinuity resulting from a saccadic movement is prevented from stimulating the pursuit loop by introducing an extra cross-coupling branch (from A to B) in Fig. 17C(a) between the saccadic and pursuit branches.

The equivalent flow chart of Fig. 17C(b) is identical to that of a purely saccadic sampled-data model except for the branch with transmission $1/Ts$, representing the contribution of the pursuit system. This is the branch that must be considered open during any high-velocity movements of the target. The forward transmission of Fig. 17C(b) is

$$G(s) = \left(\frac{1}{Ts} + z \right) \frac{1}{s} \tag{11}$$

The sampled transfer function around the closed loop is

$$G(z) = G(s)* = \left(\frac{1}{Ts^2} + \frac{z}{s} \right)*$$

$$= \frac{2z - z^2}{(1 - z)^2} \tag{12}$$

This contributes the forward loop transmission

$$\frac{1}{1 + G(z)} = (1 - z)^2 \tag{13}$$

to the graph in Fig. 17C(c). The resultant closed-loop flow chart for the complete system is shown in Fig. 17C(d).

For the integrated sampled–data model with both the pursuit and saccadic systems functioning, the predicted eye position is given by

$$R(s) = C(s)* (1 - z)^2 \left(\frac{1}{Ts} + z\right) \frac{1}{s}$$

(14)

Written in another form, the equation becomes

$$R(s) = C(s)* \left[(1 - z)z\left(\frac{1 - z}{s}\right) + \left(\frac{1 - z}{T}\right)\left(\frac{1 - z}{s}\right)\left(\frac{1}{s}\right)\right]$$

(15)

The first term in brackets, $(1 - z)z(1 - z)/s$, represents a delayed zero-order hold on the first difference of target samples. This describes the action of the saccadic system in compensating for the unit delay in velocity changes by the pursuit system.

The second term in brackets, $[(1 - z)/T] [(1 - z)/s] (1/s)$, stems from the pursuit loop. It represents the function of the velocity tracker in using the last difference of target samples to estimate a target rate, keeping the eye velocity constant by a zero–order hold circuit, and integrating this velocity to produce an eye position.

For target inputs exceeding 30 deg/sec the pursuit branch opens, and the model equations reduce to

$$R(s) = C(s)*z\left(\frac{1 - z}{s}\right)$$

(16)

which can be recognized as a delayed zero–order hold, or the saccadic system acting alone.

MODEL AND EXPERIMENTAL TRANSIENT RESPONSES
EXPERIMENTS STEP RESPONSE

Since this involves a discontinuity of target position, the pursuit loop is initially open

$$C(s) = \frac{A}{s}$$

(17)

$$R(s) = \left(\frac{A}{s}\right)^* z\left(\frac{1 - z}{s}\right)$$

(18)

$$R(s) = \left(\frac{Az}{s}\right)$$

(19)

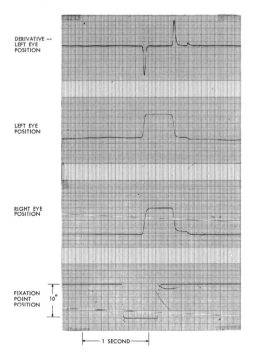

DERIVATIVE --
LEFT EYE
POSITION

LEFT EYE
POSITION

RIGHT EYE
POSITION

FIXATION
POINT   10°
POSITION

|—— 1 SECOND ——|

Fig. 18. Step response of the eye tracking
system.

The response is a delayed step, in agreement with experiments. Figure 18 demonstrates the pulse response of the eye tracking system and
shows a target motion of 10° from right to left then back again from left to
right. Approximately 0.2 sec following the initial target movement, both
the left and right eye moved to follow the target. These movements were
rapid saccadic jumps, which, in the model, are considered as discrete position-error corrections. The slight rounding of these saccades is a consequence of the finite power of the eye muscles, the inertia of the eyeball,
and the resisting force of the antagonistic muscles. Approximately 0.2 sec
following the return of the target to its initial position, the eyes made a
second jump back toward their original position. In this case, however, an
error of approximately 2° persisted, and 0.2 sec following the second saccade, a third corrective saccade occurred, again in both eyes. The top
trace of Fig. 18 shows the derivative of left eye position which indicates
the rapid velocity of the position-correcting saccade in the first and second
movements and the rather lower velocity which occurred in the third small
corrective saccade.

Pulse Response

Once again, the pursuit loop is open at the discontinuities (and the sampling instant following a discontinuity occurring between sampling instants). For a pulse width τ, with $\tau < T$,

$$C(s) = \frac{A}{s}[1 - \exp(-\tau s)] \tag{20}$$

$$C(z) = A \tag{21}$$

$$R(s) = Az\left(\frac{1-z}{s}\right) \tag{22}$$

The response is a delayed pulse of width T. Figure 19 shows the impulse response of the eye tracking system. Here the target jumped rapidly from right to left and then within 50 msec jumped back to its original position. The eyes waited their 0.2-sec or more refractory delay period before making the initial movement to follow the target. At the same time

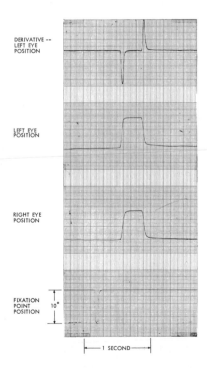

Fig. 19. Impulse response of the eye
tracking system.

this movement had been completed, there already had ensued a long enough period for the control system to realize that the target had turned to its base-line position. However, a period of at least 0.2 sec was required before the eyes could return to their original position with a second position-error correcting saccade. The requirement that the control system wait a minimum of 0.2 sec to finish its response to an impulse of target position is a crucial experiment in proving the sampled data model, indicating that there is not just a transport delay in the eye movement system, but that there is an actual refractory period before a second command can be acted upon.

Ramp Responses

The use of Eqs. (14) and (16) to describe the transient responses of the eye movement system is described elsewhere in detail [78, 79]. However, two examples of this use of the model are presented here to indicate the applicability of the sampled data model.

In Fig. 20 theoretical and experimental ramp responses are shown for the case where the ramp velocity is large enough to move the target out of the saccadic dead zone in less than one sampling period, i.e., the velocity is greater than 2.5 deg/sec. The predicted response is found by the solution of Eq. (14) with

$$C(s) = \frac{A}{s^2} \tag{23}$$

where A is the ramp velocity.

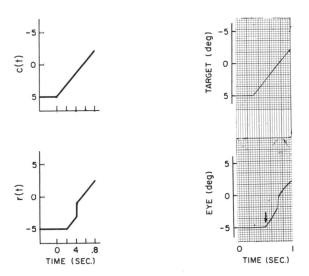

Fig. 20. Theoretical and experimental ramp response.

The predicted response transform is

$$R(s) = A\left(\frac{z}{s^2} + T\frac{z^2}{s}\right) \tag{24}$$

and the time response consists of a ramp of slope A delayed by T sec plus a step AT occurring at $t = 2T$ (Fig. 20). The experimental result shown in Fig. 20 is seen to be in close agreement with the prediction.

Ramp–Step Response

A more complex transient experiment consists of a ramp–step input, in which a constant velocity target is suddenly halted and returned to zero at time $t = KT$. The target wave form transform is

$$C(s) = \frac{A}{s^2} - \exp(-sKT)\left(\frac{A}{s^2} + \frac{AKT}{s}\right) \tag{25}$$

$$C(z) = A\left[\frac{Tz}{(1-z)^2}(1 - z^k) - \frac{KTz^k}{1-z}\right] \tag{26}$$

Since the second term in brackets represents a target position discontinuity, the eye response to this portion of the input is determined by the saccadic system alone, using Eq. (16). The complete response transform becomes

$$R(s) = A\left[\left(\frac{z}{s^2} + \frac{Tz^2}{s}\right)(1 - z^k) - \frac{KTz^{k+1}}{s}\right] \tag{27}$$

Figure 21 shows the predicted and actual transient response for this type of input. Notice that the model predicts a continuation of the eye velocity for one sample period after the input step and the occurrence of a second saccadic correction, which are observed experimentally.

Parabola Response

As noted, the system response to a parabolic input yielded evidence for the discrete nature of the velocity tracking system. The input function is

$$C(t)\frac{At^2}{2} \tag{28}$$

$$C(s)\frac{A}{s^3} \tag{29}$$

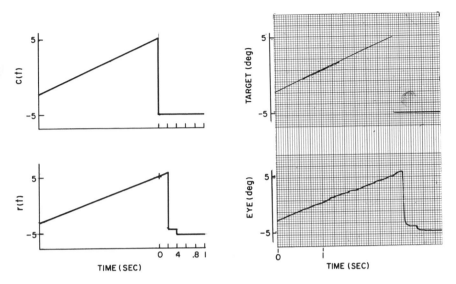

Fig. 21. Theoretical and experimental ramp-step response.

The z transform of this function can be found in a table of z transforms.

$$C(z)\frac{AT^2}{2}\frac{z(1+z)}{(1-z)^3} \tag{30}$$

$$R(s)\frac{AT^2z}{2}\left(\frac{1+z}{1-z}\right)\left(\frac{1}{Ts^2}+\frac{z}{s}\right) \tag{31}$$

The corresponding time function is plotted in Fig. 22. Notice that the predicted response consists of constant velocity segments and regularly spaced saccadic jumps in the direction of the target motion.

In addition to the model agreement with experimental transient responses, the model-frequency characteristics closely resemble the experimental Bode plots. Furthermore, experimental transient and frequency characteristics change with the effective visual feedback (varied by controlling target position from measured eye position) in a manner exactly predicted by the sampled-data model.

FREQUENCY RESPONSE: MODEL

The frequency response is not nearly as unambiguous a concept in sampled-data systems as it is in continuous systems. When a continuous linear system is driven by a sinusoidal input function, the output is a sinusoid of the same frequency which is, in general, shifted in phase and

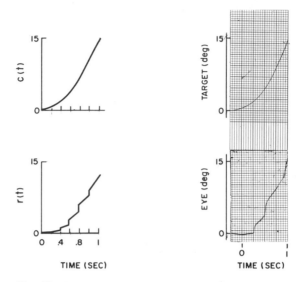

Fig. 22. Eye movement parabola response — model and experiment.

modified in amplitude with respect to the input. These gain and phase characteristics for all frequencies uniquely determine the transfer function of the system.

When a linear sampled-data system is driven by a sinusoidal input at a frequency ω, which is less than one half the sampling frequency $\omega_0 = 2\pi/T$, the output in general contains a component at the input frequency and an infinite number of higher frequency components at the input frequency plus integral multiples of the sampling frequency. This can be seen from the effect of the impulse modulator on the Fourier transforms of the input and output signals. With the modulator sampling frequency ω_0, the Fourier transform of the modulator output is [25]

$$C(j\omega) * \frac{1}{T} \sum_{n=-\infty}^{\infty} C[j(\omega + n\omega_0)] \tag{32}$$

The effect of this modulation, which is the convolution of a sampler with an input signal, is shown in Fig. 23a for a single sinusoid of frequency $\omega_1 < \omega_0/2$. It is clear that if the driving frequency is $\omega_1 = \omega_0/2$, the amplitude at ω_1 is twice that for any lower frequency. Furthermore, if the input spectrum contains energy at frequencies higher than half the sampling rate, not only is the higher frequency information irretrievably lost in the sampling process, but the lower frequency part of the sampled spectrum is con-

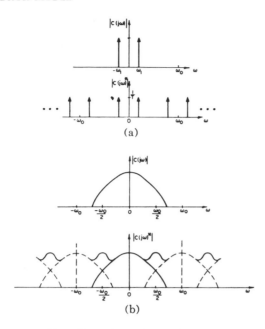

Fig. 23. Effect of impulse modulation: (a) sinusoid and (b) wide-band signal.

taminated by the interferences of the higher frequencies in the input (see
Fig. 23b). Thus, in analyzing a sampled-data system output at any fre-
quency ω_1 less than half the sampling frequency $\omega_1 = \omega_0/2 - \Delta\omega$, it must
be remembered that the output corresponds not only to the fundamental
resulting from that frequency in the input, but possibly also to the transmitted side band of an input component of frequency $\omega_2 = \omega_0/2 + \Delta\omega$. Since
the testing of the eye movement control response to unpredictable contin-
uous inputs required a wide band of input frequencies which was not always
limited to less than half the sampling frequency, this problem of ambiguity
in interpreting the output spectrum is relevant.

One possibility for defining the frequency response is to consider the
total output of the sampled-data system — not just at the sampling instants
— and to identify the frequency components of this output with the frequency response of the system. The Laplace transform of the output is given
by

$$R(s) = \frac{1}{T} \sum_{n=-\infty}^{\infty} H(s)C(s + nj\omega_0) \tag{33}$$

where $H(s)$ is the transfer function of the system from samples of the input to the actual output. Assuming that $C(s)$ has no frequency components

greater than $\omega_0/2$, then there will be no side-band interference of the type mentioned, and the frequency response can be defined as

$$\frac{R(j\omega)}{C(j\omega)} = \frac{1}{T} H(j\omega) \tag{34}$$

For the eye movement control system with only the saccadic branch opening, the model transfer function is

$$H_S(z) \frac{z(1-z)}{s} \tag{35}$$

and the frequency response becomes

$$\frac{1}{T} H_S(j\omega) = \frac{e^{-j\omega T}}{T} \left(\frac{1 - e^{-j\omega T}}{j\omega} \right)$$

$$= \frac{\sin\dfrac{\omega T}{2}}{\dfrac{\omega T}{2}} \exp\left(-\frac{j\omega T}{2} \right) \tag{36}$$

For the model including both pursuit and saccadic tracking, the transfer function is

$$H_{SP}(z) = (1 - z)^2 \left(\frac{1}{Ts} + z \right) \frac{1}{s} \tag{37}$$

$$\frac{1}{T} H_{SP}(j\omega) = \left(\frac{1 - e^{-j\omega T}}{j\omega T} \right)^2 (1 + \omega T \sin \omega T + j\omega T \cos \omega T)$$

$$= \left(\frac{\sin\dfrac{\omega T}{2}}{\dfrac{\omega T}{2}} \right)^2 (1 + 2\omega T \sin \omega T + \omega^2 T^2)^{1/2} \cdot$$

$$\exp j\left[-\omega T + \tan^{-1}\left(\frac{\omega T \cos \omega T}{1 + \omega T \sin \omega T} \right) \right] \tag{38}$$

The gain (see Fig. 24a) and phase (see Fig. 24b) of these transfer functions are plotted for the input spectrum limited to 2.8 cycles/sec.

FREQUENCY RESPONSE: EXPERIMENTAL

The experimental Bode plots for predictable and nonpredictable continuous inputs are shown in Figs. 25a and 25b. The curves marked pre-

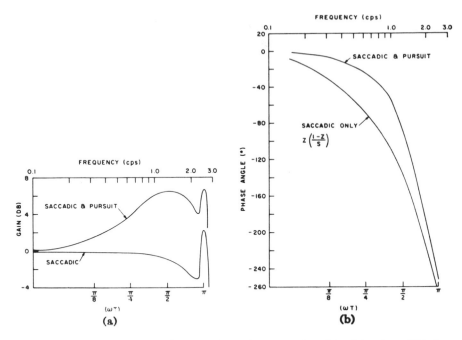

Fig. 24. Frequency response for saccadic and integrated models: (a) gain and (b) phase.

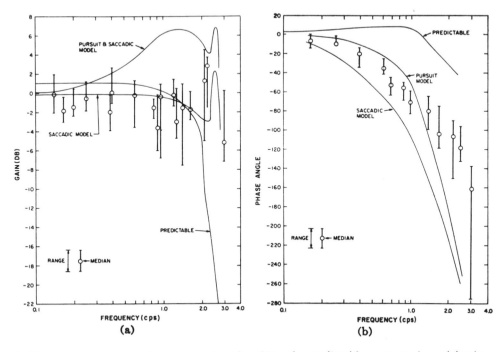

Fig. 25. Frequency responses for continuous predictable and unpredictable target motions: (a) gain and (b) phase.

dictable indicate the gain and phase of eye movement while following a predictable single sinusoid [66]. They show very little phase lag and indicate
the effect of the subject's predictive apparatus. These curves have been
matched by a linear fourth-order model with pure prediction of 0.31 sec
[78].

The frequency-response data for unpredictable inputs had to be gath-
ered in a more sophisticated manner. In order to generate the continuous
unpredictable target signal, from four to nine noncommensurate sinusoids
were summed by the RW-300 digital computer, and the resultant function
appeared at the analog output of the computer [64]. The eye movement response was stored in the computer and then analyzed for the presence of
those frequencies that made up the input.

Notice that the gain curve shows a marked peak in the region of 2−3
cycles/sec before deteriorating entirely. This peak reflects the large
amount of energy in the eye response spectrum resulting from discrete
eye movements at approximately 0.2-sec intervals during active tracking.
(The fundamental of a square-wave signal changing state every 0.2 sec lies
at 2.5 cycles/sec.)

The data in Fig. 25 represent the results of 18 runs taken on eight
different subjects. Although the range of data points is quite large, the
general characteristics of the frequency response are clear. The experimental frequency response data are seen to lie generally within the enve-
lope defined by the theoretical responses for the saccadic and integrated
models. The proximity of the median points of the experimental data to
the saccadic model reflects the predominance of saccadic movements during active tracking.

Discussion of the adequacies and limitations of the model will be post-
poned until after consideration of the valuable feedback experiments of
Chapter 3.

Chapter 3

Variable Feedback Experiments

SAMPLED-DATA MODEL

The eye movement control system may be considered as a closed-loop servomechanism with unit visual negative feedback from the eye position to the observed error at the retina. As drawn schematically in Fig. 26 (top), the controller acts on the observed error in eye position, or the difference between the angular position of the target and that of the eye. Since the object of this research was to study and describe the operation of the error sensor, controller, and load dynamics, it was designed to study the system operation in the absence of the visual feedback [80]. The feedback path is an inherent part of the system, however, since rotation of the eye displaces the target image on the retina. It could be eliminated only by physically opening the control loop; for example, by mechanically restraining the eyes from moving and observing the torque exerted by the muscles.

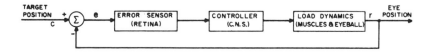

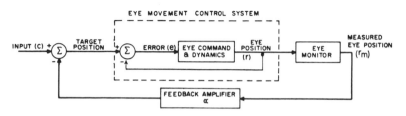

Fig. 26. Eye movement control system block diagrams. Normal visual feedback (top). Method of varying visual feedback (bottom).

Fortunately, the use of an eye movement monitor which yields an instantaneous voltage signal proportional to the eye position permits the effective visual feedback to be varied conveniently by adding an external feedback path from eye position to target position. As shown schematically in Fig. 26 (bottom), the measured eye position is amplified by an amount α and subtracted from the input command to generate the target position. Thus an eye movement Δr reduces the observed error by $(1 + \alpha)\Delta r$. By varying the sign and magnitude of α, the eye movement control system may be studied for any value of effective visual feedback. K is defined as the effective visual feedback, where

$$K = 1 + \alpha \tag{1}$$

K is thus the return difference from eye movement to observed error.

$$K = \frac{-\Delta e}{\Delta r} \tag{2}$$

In this section, the sampled-data model developed above for normal visual feedback will be logically extended to cover the variable feedback situation, and its validity will be tested by comparing predicted responses with experimental results. The purpose of effectively changing the visual feedback is not to study an abnormal situation *per se*, but to lay bare some facets of the normal system which are hidden by its closed-loop nature. Of particular interest is the performance of the control system in an open-loop situation in which the effective visual feedback is reduced to zero, and the limits of stability of the system as the effective visual feedback are varied.

The sampled-data model flow chart of the integrated system with variable external feedback is shown in Fig. 27a. In this case, a cross-coupling branch with transmission $(1 + \alpha)$ from points A to B is required to negate the effect of a saccade on the pursuit loop. The flow chart may be reduced to the form shown in Fig. 27b which has a forward transfer function

$$G(s) = \left\{ z + \frac{[1 - (1 - K)z]}{Ts} \right\} \frac{1}{s} \tag{3}$$

The sampled transform is

$$G(z) = G(s) * \frac{2z - (2 - K)z^2}{(1 - z)^2} \tag{4}$$

The system response is therefore

$$R(s) = C(s)* \frac{G}{1 + KG*} \qquad (5)$$

and the flow graph simplifies to the form shown in Fig. 27c.

For the complete system with variable visual feedback, the predicted eye position is given by

$$R(s) = C(s)* \frac{(1 - z)^2}{[1 - (1 - K)z]} \left\{ z + \frac{[1 - (1 - K)z]}{Ts} \right\} \frac{1}{s} \qquad (6)$$

When there is unity negative feedback ($K = 1$), the transfer function reduces to the normal tracking transfer function for the integrated system.

If the input velocities are outside the range of 1° to 30 deg/sec so that the pursuit system is inoperative, the saccadic tracking model must be employed as before. Its flow chart for variable feedback is drawn in Fig. 28a.

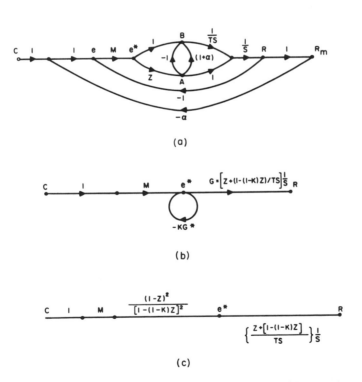

(a)

(b)

(c)

Fig. 27. Flow graphs, complete system with variable visual feedback.

The forward loop transfer function of Fig. 28b is

$$G(s) = \frac{z}{s} \tag{7}$$

yielding a sampled transform

$$G(z) = G(s)* \frac{z}{1 - z} \tag{8}$$

The resultant closed–loop flow graph for the saccadic system with variable visual feedback is shown in Fig. 28c, and predicted eye positions are given by

$$R(s) = C(s)* \left[\frac{1 - z}{1 - (1 - K)z}\right] \frac{z}{s} \tag{9}$$

When there is unit visual negative feedback, Eq. (9) reduces to Eq. (16), which is the eye position equation for the saccadic system acting alone with normal visual feedback.

Theoretical limits of stability for the sampled–data model of the eye movement system with variable visual feedback can be determined using the root locus plot in the z plane, which for sampled–data systems is entirely analogous to the root locus plot in the s plane for continuous systems. It is the locus of poles of the closed–loop transfer function in the z plane,

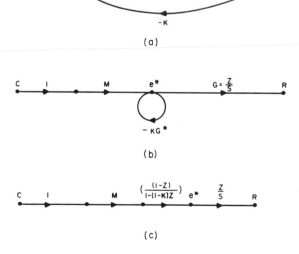

Fig. 28. Flow graphs, saccadic system with variable visual feedback.

showing how these poles migrate as the open-loop gain is varied from $-\infty$ to $+\infty$, and therefore indicating the relative stability of the system as a function of loop gain. It will be recalled that in the s plane, a pole in the right-half plane (a complex pole with a positive real part) indicates that the system is unstable, and the limit of stability is reached when a pole is on the imaginary axis. For sampled-data systems the region of instability is the interior of the unit circle on the z plane, and the unit circle represents the limit of stability. (Note that if z had been defined as $\exp(+sT)$, the region of instability would be the exterior of the unit circle.)

To determine the stability limits of the closed-loop model with variable feedback, consider the poles of the error-to-input sampled transfer functions. For the saccadic model alone

$$\frac{E(z)}{C(z)} = \frac{1 - z}{1 - (1 - K)z} \tag{10}$$

and for the pursuit and saccadic model

$$\frac{E(z)}{C(z)} = \left[\frac{1 - z}{1 - (1 - K)z}\right]^2 \tag{11}$$

The locus of roots of $1 - (1 - K)z$ is plotted in Fig. 29 as a function of K, which is indicated in parentheses. This locus represents single roots for the saccadic model and double roots for the integrated system. Thus the limits of stability are unchanged whether the saccadic branch is operating alone or in conjunction with the pursuit branch. The roots are at infinity for $K = 1$, representing a perfectly stable loop for normal tracking with no external visual feedback. Limits of stability determined from Fig. 29 are

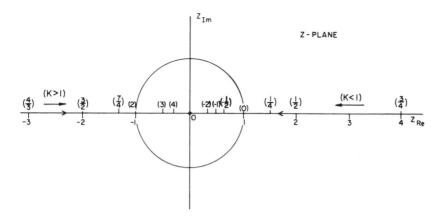

Fig. 29. Root locus of transfer function denominator.

$0 < K < 2$. The lower limit corresponds to an external visual feedback equal in magnitude and opposite in sign to the normal visual feedback ($\alpha = -1$), which is the eye movement system operating *open loop*. The upper limit corresponds to an effective visual feedback twice that of the normal eye movement system. For $K < 0$, the roots lie inside the unit circle on the positive real axis and represent unstable positive feedback. For $K > 2$, the roots lie inside the unit circle on the negative real axis, representing an unstable oscillating system resulting from the high external feedback.

OPEN-LOOP TRANSIENT RESPONSES

The variable feedback model will now be examined to test its ability to predict specific transient responses.

The open-loop case is achieved by feeding back externally with $\alpha = -1$. Under these circumstances, every eye movement causes an equal simultaneous target movement, so the observed error is independent of eye movement. If the instrumentation were much more accurate than its actual $1/4°$ noise level, this would correspond to perfect retinal stabilization, and one would expect subjective disappearance of the target.

The transfer function reduces to

$$\left[\frac{R(s)}{C(s)}\right]_{SP - OL} = M\left(\frac{z}{s} + \frac{1 - z}{Ts^2}\right) \tag{12}$$

for the saccadic and pursuit system, or

$$\left[\frac{R(s)}{C(s)}\right]_{S - OL} = M\left(\frac{z}{s}\right) \tag{13}$$

for the saccadic system alone. For the open-loop step response, the pursuit system is open at the discontinuity, and the output is yielded by the saccadic model alone.

$$R(s) = \left(\frac{1}{s}\right)^* \frac{z}{s}$$

$$= \left(\frac{z}{1 - z}\right)\frac{1}{s} \tag{14}$$

The staircase output is drawn in Fig. 30 and an experimental record showing equal amplitude steps separated by approximately 0.2-sec intervals is reproduced in Fig. 30.

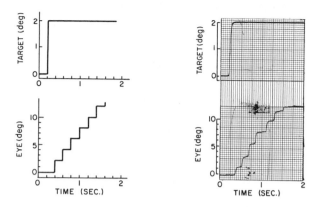

Fig. 30. Open-loop step response — model and experimental.

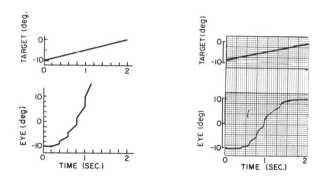

Fig. 31. Open-loop ramp response — model and experimental.

The open-loop ramp response involving the integrated system is given by

$$R(s) = \frac{Tz}{(1-z)^2}\left(\frac{z}{s} + \frac{1-z}{Ts^2}\right)$$

$$= \frac{Tz^2}{(1-z)^2}\frac{1}{s} + \left(\frac{z}{1-z}\right)\frac{1}{s^2} \tag{15}$$

The model and experimental open-loop ramp responses are shown in Fig. 31. After each refractory period, the pursuit velocity increases by a constant increment and a linearly increasing saccadic jump occurs. The experimental record shows saturation of the eye movement monitor at 10°.

VARIABLE FEEDBACK STEP RESPONSES: MODEL

The characteristics of the model step responses vary considerably with K. Some of these predictions are shown in Fig. 32. For normal feedback, the response is a delayed step (Fig. 32a). As the feedback is reduced, the model predicts series of decreasing steps (see Fig. 32b), for which $K = 0.3$. The open-loop step response (Fig. 32c) has been discussed above as representing a limit of stability, and for net positive feedback (Fig. 32d, $K = -1.0$) the model indicates an unstable series of increasing steps. For variations of K corresponding to increased effective visual feedback, the predicted model step responses indicate the appearance of more and more overshoot until instability is reached at $K = 2$. Figure 32e shows a damped alternation when the feedback is raised to 1.75. Figure 32f demonstrates that at $K = 2$ the model response to a step is a limit cycle. For further increases in feedback, the model predicts a growing alternation of response, as in Fig. 32g ($K = 2.3$).

VARIABLE FEEDBACK STEP RESPONSES: EXPERIMENTAL

By experimenting with different values of effective visual feedback, it was possible to observe the eye movement control system behavior in

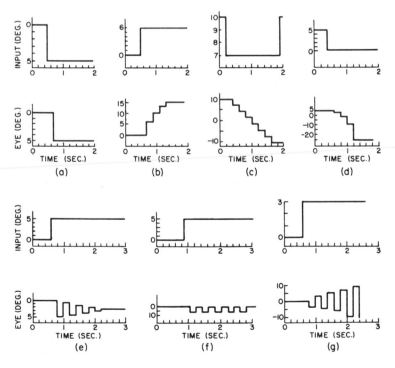

Fig. 32. Model step responses under variable feedback. (a) K = 1.0, (b) K = 0.3, (c) K = 0.0, (d) K = -1.0, (e) K = 1.75, (f) K = 2.0, (g) K = 2.3.

abnormal states which correspond to those used in the model predictions.

For normal tracking, the response to a step input is a single saccadic
movement of the amplitude of the step, following a delay of 0.12–0.25 sec.
Any error in amplitude is corrected by a small corrective saccade 0.15–
0.30 sec later. The eye then maintains steady fixation, except for miniature
fixation movements (Fig. 33a).

As positive external feedback is added, reducing the effective visual
feedback, the step response becomes a series of saccadic jumps all in the
direction of the input step. Each step is smaller than its predecessors un-

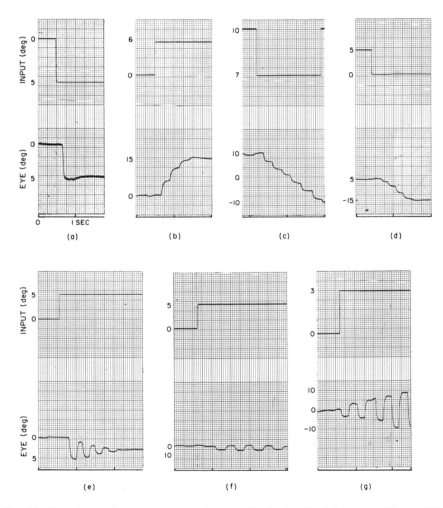

Fig. 33. Experimental step responses under variable feedback. (a) K = 1.0, (b) K = 0.3,
(c) K = 0.0, (d) K = -1.0, (e) K = 1.75, (f) K = 2.0, (g) K = 2.3.

til the eye rests in steady fixation (Fig. 33b). The number of saccades, and therefore the time taken for the eye to reach steady fixation, increases as the effective visual feedback is reduced toward zero. At $K = 0$, there is no effective visual feedback and the step response is the staircase of equal amplitude saccades already described under the open–loop condition (Fig. 33c). As more positive feedback is added, making $K < 0$, the system indeed demonstrates its instability. Each saccade is larger than the one before it, as shown in Fig. 33d.

If negative external feedback is introduced, the step response exhibits some overshoot. For $1 < K < 2$, the response is a series of saccades in alternate directions, with each saccade smaller than the previous ones, until the error between the eye position and the position toward which it was converging is less than the saccadic dead zone of approximately $\pm 0.5°$. The ratio of the amplitude of any saccade to the previous one increases as external feedback increases. This ratio is approximately the same as α, the external feedback gain. Thus in the record shown in Fig. 33e, for $K = 1.75$, the external negative feedback is 0.75, and the amplitude ratio of successive saccades is also about 0.75. When the effective visual feedback is doubled ($K = 2$), the step response is a series of constant amplitude saccades equal to the step height but alternating in direction, as in Fig. 33f. This response, which approximates a limit cycle of the amplitude of the target step and period of about 0.5 sec, may continue from 1 to more than 10 sec, and stops only when one of the saccadic jumps is significantly smaller than the previous ones, and the saccadic dead zone nonlinearity terminates the alternation.

For effective visual negative feedback greater than 2, the response is a series of alternating saccades of increasing amplitude. This unstable response continues growing until the target position reaches its saturation limits, or some other disturbance of input acts to interrupt it. A typical growing alternating response is shown in Fig. 33g.

This family of step responses is clearly predicted by the variable feedback model presented above.

LIMITS OF STABILITY: EXPERIMENTAL

The predicted region of model system stability, as shown on the root locus plot, is $0 < K < 2$. To verify that the actual system would be unstable outside the region $0 < K < 2$, eye movements were recorded for these conditions when no input signal was present. The actual target position was thus determined solely by the eye position as fed back through the external feedback amplifier. It was found that the system did indeed exhibit instability when the effective feedback was outside the range $0 < K < 2$, and that the inherent drift or small saccadic flicks in the system were sufficient to

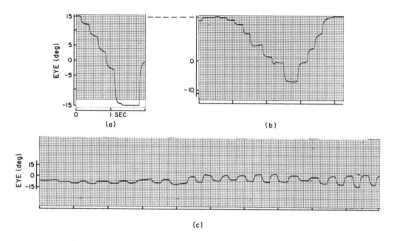

Fig. 34. Positive feedback and high negative feedback instabilities, no
input. (a) K = -0.3, (b) K = -1.0, (c) K = 2.2.

initiate a growing response. The records of Fig. 34a and b show that for
net positive feedback with no input, the eye moved in a succession of increasing saccadic jumps, until reaching the galvanometer saturation point
or being reversed by some other disturbances. In Fig. 34c, with the effective feedback slightly greater than doubled ($K = 2.2$), the system spontaneously broke into an apparent limit cycle of period 0.5 sec and about
10° amplitude.

VARIABLE FEEDBACK FREQUENCY RESPONSE: MODEL

The model frequency response may be calculated for various values
of K in the same manner as the normal tracking case. The response function for the saccadic model alone was given as Eq. (9) and its frequency
response for $K = 1$ appears as Eq. (36).

When the loop is opened, making $K = 0$, the saccadic model transfer
function becomes

$$\left[\frac{R(s)}{C(s)}\right]_{S-OL} = M\left(\frac{z}{s}\right) \tag{16}$$

and the frequency response is

$$\frac{1}{T}\,[H(j\omega)]_{S-OL} = \frac{1}{\omega T}\,\exp\,[-j(\pi/2 + \omega T)] \tag{17}$$

For the double gain case, at the other limit of stability, $K = 2$, and the saccadic model transfer function is

$$\left[\frac{R(s)}{C(s)}\right]_{\text{S - DG}} = M \frac{(1 - z)^{\text{M}}}{(1 + z)} \left(\frac{z}{s}\right) \tag{18}$$

The associated frequency response is

$$\frac{1}{T} H(j\omega)_{\text{S - DG}} = \frac{1}{2} \frac{\tan \dfrac{\omega T}{2}}{\dfrac{\omega T}{2}} \exp(-j\omega T) \tag{19}$$

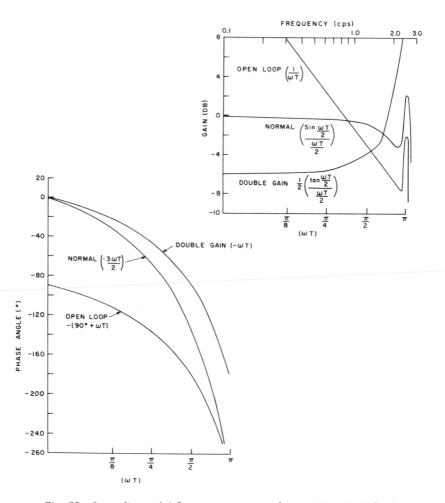

Fig. 35. Saccadic model frequency responses for variable visual feedback.

The calculated gain and phase of the saccadic model for normal, open–loop, and double gain tracking are plotted in the Bode plots of Fig. 35. These would represent the model frequency response for net target or error velocities always outside the 1 to 30 deg/sec range of the pursuit system.

If the net target and error velocities were always maintained within the range 1 to 30 deg/sec, the model frequency response would be that of the integrated (pursuit and saccadic) model. Its response function for variable feedback is given in Eq. (9), and its frequency response under normal tracking is given in Eq. (38). For the open–loop situation,

$$\left[\frac{R(s)}{C(s)}\right]_{\text{SP - OL}} = M\left[z + \frac{(1 - z)}{Ts}\right]\frac{1}{s} \tag{20}$$

and the frequency response is

$$\frac{1}{T}\left[H(j\omega)\right]_{\text{SP - OL}} = [(1 - \cos \omega T + \omega T \sin \omega T)^2 +$$

$$(\sin \omega T + \omega T \cos \omega T)^2]^{1/2}\left(\frac{1}{\omega T}\right)^2 \cdot$$

$$\exp\left[-j\left(\pi - \tan^{-1}\frac{\sin \omega T + \omega T \cos \omega T}{1 - \cos \omega T + \omega T \sin \omega T}\right)\right] \tag{21}$$

Finally, for the double gain condition, the pursuit and saccadic model transfer function is given by

$$\left[\frac{R(s)}{C(s)}\right]_{\text{SP - DG}} = M\frac{(1 - z)^2}{(1 + z)^2}\left[z + \frac{(1 + z)}{Ts}\frac{1}{s}\right] \tag{22}$$

and the fundamental frequency response associated with this function is

$$\frac{1}{T}\left[H(j\omega)\right]_{\text{SP - DG}} = \frac{1}{4}\left(\frac{\tan\frac{\omega T}{2}}{\frac{\omega T}{2}}\right)^2 \cdot$$

$$[2(1 + \cos \omega T + \omega T \sin \omega T + (\omega T)^2]^{1/2} \cdot$$

$$\exp\left(+ j \tan^{-1}\frac{\omega T \cos \omega T - \sin \omega T}{1 + \cos \omega T + \omega T \sin \omega T}\right) \tag{23}$$

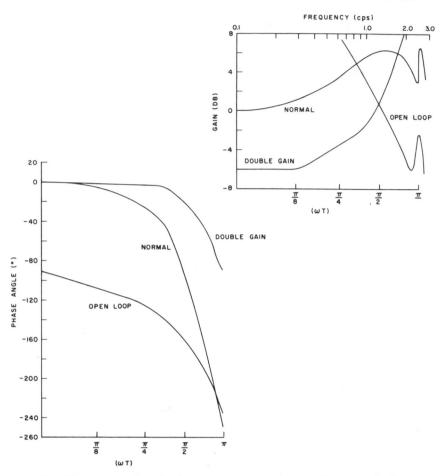

Fig. 36. Integrated model frequency responses for variable visual feedback.

Figure 36 shows the calculated gain and phase vs. frequency for the pursuit and saccadic system under conditions for normal, open-loop, and double gain tracking.

VARIABLE FEEDBACK FREQUENCY
RESPONSE: EXPERIMENTAL

In order to test these model predictions, a series of experiments with pseudo-random continuous inputs was run for different values of effective visual feedback to determine the effect of changing feedback on the eye movement control frequency response in the nonpredictive mode. As described previously, the input was generated by the RW-300 digital computer, which also recorded the eye response and calculated the gain and phase at each of the input frequencies.

The data point resulting from these experiments showed a considerable variation, with spreads of ±6 db in gain and ±20° in phase not uncommon. This variation is probably attributable to the fact that the analysis was performed only at the input frequencies, whereas the output contained considerable energy at other frequencies. The effect of the sampled-data nature of the eye control system in distorting the frequency response was discussed previously. The data were further distorted by the fact that during experiments with the effective visual feedback near the margin of instability, the target was apt to saturate for brief portions of the run.

A typical segment of record for such an open-loop tracing experiment is given in Fig. 37. Note that the input shown on the upper channel corresponds to the observed error under open-loop conditions, not to the target position (see Fig. 26). The response to the low-frequency components of the input is large, whereas the high frequencies seem to have little effect on the eye movement. As in all open-loop experiments, there is large uncorrelated drift. The tracking is heavily saccadic, with a minimum interval between saccades of 0.15−0.20 sec.

Despite the spread of the data, certain effects on the experimental frequency response as a result of varying the visual feedback were quite evident. A typical composite Bode plot for one subject is shown in Fig. 38. In comparing these experimental curves with the model predictions of the saccadic system alone (Fig. 35), or the integrated system (Fig. 36), a number of important points of agreement are apparent.

Considering first the gain vs. frequency plot of Fig. 38, it is clear that as the effective visual feedback is reduced from normal ($K = 1$) to

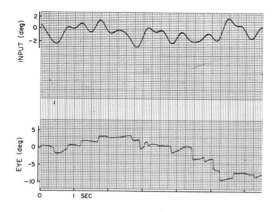

Fig. 37. Open-loop tracking of pseudo-random input.

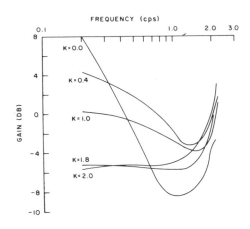

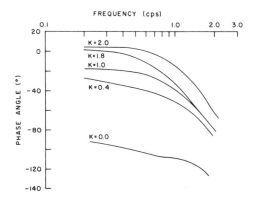

Fig. 38. Experimental frequency responses — variable feedback nonpredictive tracking.

open loop (K = 0), the low–frequency gain increases. The low–frequency slopes of the gain vs. frequency curves decrease from slightly negative for normal tracking down to about –6 db/octave for open loop. As the effective visual feedback is increased from normal up to the double gain limit of stability (K = 2), just the opposite trend is observed. The low–frequency gain is reduced with increasing feedback, down to a value of –6 db for K = 2. The slope becomes progressively less negative and then slightly positive as the effective feedback is increased, and for K = 1.8, the gain is relatively constant out to 2 cycles/sec. All of the gain curves exhibit a definite peak in the region of 2–3 cycles/sec. The height of this peak is generally about 6 db. All of these results are in good agreement with the saccadic model of Fig. 35, once again reflecting the fact that much of the tracking motion is saccadic.

The phase vs. frequency curves of Fig. 38 show that the phase lag at all frequencies increases as the effective visual feedback is reduced from normal, and decreases as this feedback is made larger. Most noticeable is the change in low-frequency phase lag as the feedback is reduced toward zero. For open-loop conditions, the low-frequency phase of the eye response is about $-90°$. These phase relationships are in good agreement with the model results, although they appear to resemble the integrated system (Fig. 36) more closely than the saccadic system (Fig. 35).

The ability of the sampled-data model to predict accurately the transient and frequency responses for a variety of unusual tracking conditions strongly supports its plausibility; since the model was developed on the basis of normal tracking data, these experiments indeed provided a test of its usefulness and generality. It is clear from the evidence presented on the open-loop step and ramp responses, variable feedback responses, limits of stability, and frequency response that the model satisfies the stringent requirements of these tests.

CONCLUSIONS

Adequacy of the Model

A useful mathematical model for the biological control system which enables us to follow a target with our eyes was developed and tested. The proposed sampled-data model is simple, analyzable, consistent with all the principal characteristics of eye tracking motions observed experimentally, and is not in conflict with any physiological evidence about the nature of the system.

The development of the model for tracking unpredictable target motions was based on recognition of the separate functions of the saccadic and pursuit systems and the necessity for describing them in terms of discrete rather than continuous control loops. The discrete nature of the system is apparent in many of the transient responses. The model adequately describes eye movements following pulses, steps, ramps, and parabolas of target motion. The general characteristics of experimental frequency responses are predicted by the model, although some details differ from the predictions.

An important test of the model's plausibility is its ability to predict accurately results of new experiments. In this paper the situation of tracking under different conditions of effective visual feedback was considered. The model adequately predicted experimental results for transient responses under open-loop conditions, step responses at a variety of feedback levels, and the limits within which feedback could be varied and system stability maintained. In addition, the model correctly indicated the effect of varying visual feedback on the experimentally determined frequency response.

Limitations of the Model

In order to develop a model for eye movements that would permit analytical treatment for the variety of test cases discussed, certain simplifications were purposely introduced.

The most general limitation involves the postulation of a deterministic model to describe the functioning of a nondeterministic biological servomechanism. A given pattern of input target motion will not always produce the same eye movement response. It was not the purpose of this research to investigate the multitude of factors which might affect the response. Instead, in order to have some basis on which to build a model, the experimental conditions were carefully controlled as to target distance, size, brightness, and background, and the experiments were repeated on more than 50 subjects to arrive at a series of typical response patterns. The occasional absence of a predicted response pattern is therefore not a contradiction of the model, whereas the regular presence of some unpredicted, nontrivial response would be a contradiction.

Another fundamental approximation concerns the use of a linear model to describe a servomechanism with many nonlinear features. One of these nonlinearities is evident in the nature of the individual saccadic movement, in which the maximum velocity increases nonlinearly with the extent of the jump. For movements less than 20°, which is the range of interest in considering eye tracking with stationary head position, a linear approximation describes the actual motion quite well. In a further simplification, the eyeball inertia, viscosity, and neuromuscular dynamics were ignored entirely in considering the overall tracking behavior of the control loop. The justification for this was that whenever the eye underwent a sudden change in position or velocity, its deviation from its final value after one refractory period was so small as to have a negligible effect on the closed-loop sampled-data system at the next sampling instant. As a result the simplified sampled-data model indicates discontinuities in eye position and velocity, whereas these movements are actually continuous rapid changes, as observed in the records in this paper and described in the literature.

The dead zone of the eye movement system for small errors in position and velocity has been established experimentally and included in the model block diagrams, but ignored in the analysis. As a result, the analytical model may predict small saccadic corrections or velocity changes, whereas the experimental record may reveal no such changes, or larger changes after a long delay during which time the error has accumulated to greater than 0.5° in position or 1 deg/sec in velocity.

The first limiter in the pursuit system, which accounts for the hypothesis that the pursuit system responds to slow movements of the target image across the retina but not to rapid position shifts, is partially accounted for in the model by a cross-coupling blanking branch from the saccadic to the pursuit loop. The other source of error discontinuities — rapid target movements — is taken into consideration in discussing transient responses by assuming that the pursuit branch of the model is opened at the appropriate times. For the frequency response, however, it is not known when the limiter is open or closed, and the predicted limits of the frequency response are given by Bode plots for both the saccadic model alone and the saccadic pursuit models.

Another pair of obvious nonlinearities is the saturation limits of eye position and velocity. The position saturation at about ±60° was never approached in the experiments since the target projection system as it was normally arranged saturated at ±20°. This latter saturation limit, however, is outside the normal range of eye movements not accompanied by head movements. The velocity saturation of 20 to 30 deg/sec was very definitely noted, and was inserted as the second limiter in Fig. 16. It was, however, neglected for the purpose of linear analysis of the model. The result is that the model output may contain velocities in excess of the saturation limit, in contradiction to the experimental results.

An important limitation of the model concerns the assumption that sampling rates for the pursuit and saccadic systems are identical and that sampling instants are synchronized with each other and with the first occurrence of an error of greater than 0.5° or of rate greater than 1 deg/sec following a quiescent period. This assumption was made to permit a model which could be handled analytically with relative ease. Judging from many experimental records taken in this investigation, the reaction time and refractory period of the pursuit system seem to be slightly less than those of the saccadic system. By ignoring this fact, the model tends to be in error in two ways. It overestimates the position error resulting from a delay in velocity correction, and therefore overestimates the amplitude of saccadic corrections. Second, the model tends to be in error by attributing too little phase advance and too much gain to the presence of the pursuit system. With a higher sampling rate for the velocity tracker, the phase lag resulting from the saccadic loop would be further reduced, and the overshoot attributable to the velocity extrapolation would be less marked. Both of these corrections would be in the direction of better agreement with the experimental data.

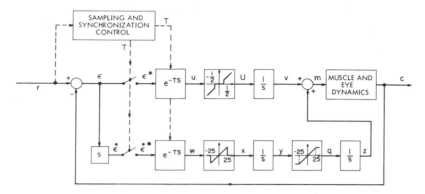

Fig. 39. Improved sampled-data model (Stark and DiStefano, unpublished).

Also, the model is not intended to describe any eye movement except gross saccadic and pursuit movements involved in horizontal tracking. Miniature movements and nystagmus are considered as disturbances, and not as control outputs of the biological servomechanism in its tracking task.

Finally, an improved model (Fig. 39) has been developed to explore a different method of velocity error estimation which is closer to the physiological or physical structure of the eye movement control system. Recent experiments have demonstrated that the sampling takes place neither at the sensory nor at the motor ends of the system, but occurs in the controller or central nervous system portion of this fascinating neurological control system.

REFERENCES AND FURTHER READING

1.   Adler, F. H.: Physiology of the Eye, C. V. Mosby, St. Louis, Mo. (1959), p. 363.
2.   Barany, E. H., and Hallden, U.: Phasic inhibition of the light reflex of the pupil during retinal rivalry. J. Neurophys. 25: 25 (1948).
3.   Bekey, G. A.: An investigation of sampled data models of the human operator in a control system. Tech. Doc. Rept. No. AST-TDR-62-36, Wright-Patterson Air Force Base (1962).
4.   Breinin, G. M.: The Electrophysiology of Extra-ocular Muscle, University of Toronto Press, Toronto (1962), p. 122.
5.   Cook, G., and Stark, L.: Dynamics of human horizontal eye movement mechanism. Quart. Prog. Rept., Research Laboratory of Electronics, M.I.T. 76: 343−352 (1965).
6.   Cornsweet, T. N.: Determination of the stimuli for involuntary drifts and saccadic eye movements. J. Opt. Soc. Am. 46: 987 (1956).

7.   Cornsweet, T. N.: Determination of the stimuli for involuntary drifts and saccadic eye movements. J. Opt. Soc. Am. 48: 808 (1958).

8.   Cox, D. R., and Smith, W. L.: The superposition of several strictly periodic sequences of events. Biometrika 40: 1 (1953).

9.   Davis, J.: Unpublished experiments. Massachusetts Institute of Technology (1962).

10. Davson, H. (ed.): The Eye, Vol. 3, Academic Press, New York (1962).

11. Davson, H.: The Physiology of the Eye, Little, Brown, Boston (1963), p. 235.

12. Ditchburn, R. W., and Ginsborg, B. L.: Involuntary eye movements during fixation. J. Physiol. (London) 119: 1 (1953).

13. Ditchburn, R. W.: Eye movements in relation to retinal action. Opt. Acta 1: 171 (1955).

14. Ditchburn, R. W.: Physical methods applied to the study of visual perception. Bull. Inst. Phys. 10: 121 (1959).

15. Dodge, R., and Cline, T. S.: The angle of velocity of eye movements. Psychol. Rev. 8: 145 (1901).

16. Elkind, J. I., and Forgie, C. D.: Characteristics of the human operator in simple manual control systems. IRE Trans. Auto. Control AC-4: 44−55 (1959).

17. Faulkner, R. F., and Hyde, J. E.: Coordinated eye and body movements evoked by brainstem stimulation in decerebrate cats. J. Neurophys. 21: 171 (1957).

18. Fender, D. H.: Personal communication (1964).

19. Fender, D. H., and Nye, P. W.: An investigation of the mechanisms of eye movement control. Kybernetik 1: 81−88 (1961).

20. Fernandez-Guardiola, A., Harmony, T., and Roldan, E.: Modulation of visual input by pupillary mechanisms. EEG Clin. Neurophysiol. 16: 259 (1964).

21. Granit, R.: Receptors and Sensory Perception, Yale University Press, New Haven, Conn. (1955).

22. Hammond, P. H., Merton, P. A., and Sutton, G. G.: Nervous gradation of muscular contraction. Brit. Med. Bull. 12: 214−218 (1956).

23. Horrocks, A., and Stark, L.: Experiments on error as a function of response time in horizontal eye movements. Quart. Prog. Rept., Research Laboratory of Electronics, M.I.T. 72: 267−269 (1964).

24. Houk, J. C., Jr.: Master's Thesis. Dept. of Electrical Engineering, Massachusetts Institute of Technology, Cambridge, Mass. (1963), p. 27.

25. Jury, E. I.: Sampled Data Control Systems, Wiley, New York (1958).

26. Kappers, C. U. A., Huber, G. C., and Crosby, E. C.: Comparative Anatomy of the Nervous System of Vertebrates, Including Man, Vol. 2, Hafner, New York (1960), p. 1079.

27. Krauskopf, J., Cornsweet, T. N., and Riggs, L. A.: Analysis of eye
     movements during monocular and binocular fixation. J. Opt. Soc.
     Am. 50: 572 (1960).
28. Kris, E. C.: A technique for electrically recording eye position.
     WADC Tech. Rept., Research Laboratory of Electronics, M.I.T. 58:
     660 (1958).
29. Latour, P. L.: Visual threshold during eye movements. Vision Res.
     2: 261–262 (1962).
30. Latour, P. L.: The neuron as a synchronous unit. In: Nerve, Brain,
     and Memory Models, N. Wiener and J. P. Schade (eds.), Elsevier,
     New York, Amsterdam (1963), pp. 30–36.
31. Lettvin, J. Y.: Student research. Quart. Prog. Rept., Research Lab-
     oratory of Electronics, M.I.T. 58: 254–258 (1960).
32. Lion, K. S., and Brockhurst, R. J.: Study of ocular movements under
     stress. AMA Arch. Ophthalmol. 46: 315 (1951).
33. Lorber, M., Zuber, B. L., and Stark, L.: Suppression of the pupil-
     lary reflex associated with saccadic suppression. Quart. Prog.
     Rept., Research Laboratory of Electronics, M.I.T. 74: 250-251 (1964);
     Nature 208: 558–560 (1965); Exptl. Neurol. 14: 351–370 (1966).
34. Matthews, P. B. C., and Rushworth, G.: The discharge from muscle
     spindles as an indicator of efferent paralysis by procaine. J. Phys-
     iol. 140: 421–426 (1958).
35. McRuer, D. T., and Krendel, E. S.: The human operator as a servo
     system element. J. Franklin Inst. 267: 1 (1959).
36. Merrill, E. G., and Stark, L.: Optokinetic nystagmus: double stripe
     experiment. Quart. Prog. Rept., Research Laboratory of Electron-
     ics, M.I.T. 70: 357–359 (1963).
37. Merrill, E. G., and Stark, L.: Smooth phase of optokinetic nystagmus
     in man. Quart. Prog. Rept., Research Laboratory of Electronics,
     M.I.T. 71: 286–291 (1963).
38. Merrill, E. G., and Stark, L.: Optokinetic nystagmus in man: the
     step experiment. Quart. Prog. Rept., Research Laboratory of Elec-
     tronics, M.I.T. 72: 269–272 (1964).
39. Michael, J., and Stark, L.: Effect of eye movements on the visually
     evoked response. Quart. Prog. Rept., Research Laboratory of Elec-
     tronics, M.I.T. 76: 332–334 (1965).
40. Miller, J. E.: Electromyographic patterns of saccadic eye move-
     ments. Am. J. Ophthalmol. 46: 183–186 (1958).
41. Mudama, E., Willis, P. A., and Stark, L.: Phototube glasses for
     measuring eye movements. Quart. Prog. Rept., Research Labora-
     tory of Electronics, M.I.T. 67: 214–220 (1962).
42. Navas, F.: Sampling or quantization in the human tracking system.
     Master's Thesis. Dept. of Electrical Engineering, Massachusetts
     Institute of Technology, Cambridge, Mass. (1963).

43.   Navas, F., and Stark, L.: Experiments on discrete control of hand movement. Quart. Prog. Rept., Research Laboratory of Electronics, M.I.T. 69: 256−259 (1963).

44.   Nelson, G. P., and Stark, L.: Optokinetic nystagmus in man. Quart. Prog. Rept., Research Laboratory of Electronics, M.I.T. 66: 366− 369 (1962).

45.   Nelson, G. P., Stark, L., and Young, L. R.: Phototube glasses for measuring eye movements. Quart. Prog. Rept., Research Laboratory of Electronics, M.I.T. 64: 214−216 (1962).

46.   Okabe, Y., Payne, R. C., Rhodes, H., Stark, L., and Willis, P. A.: Use of on-line digital computer for measurement of a neurological control system. Quart. Prog. Rept., Research Laboratory of Electronics, M.I.T. 61: 219−222 (1961).

47.   Polidira, V. J., Ratoosh, P., and Westheimer, G.: Precision of rhythmic responses of the oculomotor system. Perceptual Motor Skills 7: 247 (1957).

48.   Rashbass, C.: Barbiturate nystagmus and the mechanism of visual fixation. Nature 183: 897−898 (1959).

49.   Rashbass, C.: The relationship between saccadic and smooth tracking eye movements. J. Physiol. (London) 159: 326−338 (1961).

50.   Rashbass, C., and Westheimer, G.: Disjunctive eye movements. J. Physiol. 159: 339−360 (1961).

51.   Rashbass, C., and Westheimer, G.: Independence of conjugate and disjunctive eye movements. J. Physiol. 150: 361 (1961).

52.   Ratliff, F., and Riggs, L. A.: Involuntary motions of the eye during monocular fixation. J. Exptl. Physiol. (London) 119: 1 (1953).

53.   Richter, H. R., and Pfalz, C. R.: A propos de l'electro-oculagraphic. Confinia Neurol. 16: 270 (1956).

54.   Riggs, L. A., Cornsweet, J. C., and Cornsweet, T. N.: The disappearance of steadily fixated test object. J. Opt. Soc. Am. 43: 495−501 (1953).

55.   Ruch, T. C., and Fulton, J. F.: Medical Psychology and Biophysics, W. B. Saunders, Philadelphia (1960), p. 108.

56.   Sandberg, A. A., and Stark, L.: Model of pupil reflex to light. Quart. Prog. Rept., Research Laboratory of Electronics, M.I.T. 68: 237−240 (1963).

57.   Shackel, B.: Review of the past and present in oculography. Proc. Second International Conf. on Medical Electronics, Paris, France, Iliffe and Sons, London (1960).

58.   Smith, W. M., and Warter, P. J.: Eye movement and stimulus movement; new photoelectric electromechanical system for recording and measuring tracking motions of the eye. J. Opt. Soc. Am. 50: 245 (1960).

59. Stark, L.: Stability, oscillation and noise in the human pupil servo-
     mechanism. Proc. Inst. Radio Engrs. 47: 1925—1939 (1959).

60. Stark, L., Iida, M., and Willis, P. A.: Dynamic characteristics of the
     motor coordination system. Biophys. J. 1: 270—300 (1961).

61. Stark, L., Kupfer, C., and Young, L. R.: Physiology of the visual
     control system. NASA Rept. NASA CR-238 (1965).

62. Stark, L., and Nelson, G.: Optokinetic nystagmus. Quart. Prog. Rept.,
     Research Laboratory of Electronics, M.I.T. 64: 326—328 (1962).

63. Stark, L., Okabe, Y., and Willis, P. A.: Sampled data properties of
     the human motor coordination system. Quart. Prog. Rept., Research
     Laboratory of Electronics, M.I.T. 67: 220—223 (1962).

64. Stark, L., Payne, R., and Okabe, Y.: On-line digital computer for
     measurement of a neurological control system. Commun. ACM 5:
     567—568 (1962).

65. Stark, L., and Sandberg, A.: A simple instrument for measuring eye
     movements. Quart. Prog. Rept., Research Laboratory of Electron-
     ics, M.I.T. 62: 268—270 (1961).

66. Stark, L., Vossius, G., and Young, L. R.: Predictive control of eye
     movements. Quart. Prog. Rept., Research Laboratory of Electron-
     ics, M.I.T. 62: 271—281 (1961). Also IRE Trans. Human Factors Elec-
     tron. HFE-3: 52—75 (1962).

67. Sunderhauf, A.: Untersuchungen die Reglung der Augenbewegungen.
     Klin. Monatsbl. Augenb. 136: 837 (1960).

68. Tamler, E., Marg, E., Jampolsky, A., and Nawratzki, I.: Electro-
     myography of human saccadic eye movements. AMA Arch. Ophthal-
     mol. 62: 678 (1959).

69. Volkmann, F. C.: Vision during voluntary saccadic eye movements.
     J. Opt. Soc. Am. 52: 571—578 (1962).

70. Vossius, G.: Das System der Augenb. Z. Biol. 112: 27 (1960).

71. Wendt, G. R.: The form of vestibular eye movement response in man.
     Psychol. Monogr. 47: 311—328 (1936).

72. Wendt, G. R.: Vestibular functions. In: Handbook of Experimental
     Psychology, S. S. Stevens (ed.), Wiley, New York (1951), pp. 1191—1223.

73. Westheimer, G.: Mechanism of saccadic eye movements. AMA
     Arch. Ophthalmol. 52: 710—724 (1954).

74. Westheimer, G.: Eye movement response to a horizontally moving
     visual stimulus. AMA Arch. Ophthalmol. 52: 932 (1954).

75. Weyl, H.: Über die Gleichverteilung ven Zahlen Mod. Eins. Math.
     Ann. 77: 313 (1916).

76. Wolff, E.: Anatomy of the Eye and Orbit, W. B. Saunders, Philadel-
     phia (1961), p. 30.

77. Young, L. R.: Target projection for tracking experiments. Quart.
     Prog. Rept., Research Laboratory of Electronics, M.I.T. 62: 270—
     271 (1961).

78.   Young, L. R.: A sampled data model for eye tracking movements. Doctoral Dissertation, Dept. of Aeronautics and Astronautics, Massachusetts Institute of Technology, Cambridge, Mass. (1962).

79.   Young, L. R., and Stark, L.: A sampled data model for eye tracking movements. Quart. Prog. Rept., Research Laboratory of Electronics, M.I.T. 66: 370–383 (1962).

80.   Young, L. R., and Stark, L.: Variable feedback experiments testing a sampled data model for eye tracking movements. IEEE Trans. Human Factors Electron. HFE-4: 38–51 (1963).

81.   Young, L., and Stark, L.: A discrete model for eye-tracking movements. IEEE Trans. Military Electron. MIL-7: 113–116 (1963).

82.   Zoethout, W. D.: Physiological Optics, The Professional Press, Chicago (1947), p. 308.

83.   Zuber, B. L.: Saccadic suppression. Quart. Prog. Rept., Research Laboratory of Electronics, M.I.T. 75: 190–191 (1964).

84.   Zuber, B. L., Crider, A., and Stark, L.: Saccadic suppression associated with microsaccades. Quart. Prog. Rept., Research Laboratory of Electronics, M.I.T. 74: 224–249 (1964).

85.   Zuber, B. L., Michael, J. A., and Stark, L.: Visual suppression during the fast phase of vestibular nystagmus. Quart. Prog. Rept., Research Laboratory of Electronics, M.I.T. 73: 221–223 (1964).

86.   Zuber, B. L., Michael, J. A., and Stark, L.: Visual suppression during voluntary saccadic eye movements. Quart. Prog. Rept., Research Laboratory of Electronics, M.I.T. 74: 217–221 (1964).

87.   Zuber, B. L., and Stark, L.: Eye convergence. Quart. Prog. Rept., Research Laboratory of Electronics, M.I.T. 68: 232–234 (1963).

88.   Zuber, B. L., Troelstra, A., and Stark, L.: Eye convergence. Quart. Prog. Rept., Research Laboratory of Electronics, M.I.T. 70: 339–341 (1963).

89.   Zuber, B. L., Troelstra, A., and Stark, L.: Vergence eye movements. Quart. Prog. Rept., Research Laboratory of Electronics, M.I.T. 71: 280–282 (1963).

90.   Cook, C., and Stark, L.: Derivation of a model for the human eye-positioning mechanism. Bull. Math. Biophys. 29: 153–174 (1967).

THE HAND

INTRODUCTION

As a young neurologist, I was curious about the various neurological signs of motor discoordination evidenced by human patients with Parkinson's disease, cerebellar disease, or various types of spasticities. Nothing I had studied in basic neurophysiology seemed to explain the pathophysical mechanisms of these disorders, and discussions with such of my neurology teachers as Houston Merritt, Paul Hoefer, and Gilbert Glaser similarly indicated that no explanation was available in the neurological literature. Since these diseases seemed to be malfunctions of the neurological control systems, it seemed possible that one might make fruitful analogies with engineering control systems and their possible range of dysfunction.

I started studying engineering servomechanism theory while in the Navy by reading simple primers on fire control mechanisms written for Navy technicians, and tried to relate these studies to my basic mathematical knowledge of stability conditions for solutions of differential equations. When later I began teaching neurology at Yale University I became interested in constructing a physical model of the human arm with servomotors to maintain an artificial position or postural control system. This, then, could be examined with the usual set of neurological performance tests such as reflex hammer tapping. The internal gains and structure of the mechanical control system could then be changed so as to mimic various neurological motor syndromes. (As will be discussed later, the subsequent development of computers allowed construction of more flexible mathematical models not limited by the physical laws controlling particular physical models such as that described above.)

At the same time as the postural control studies, I was trying to develop a systems model of the physiological apparatus for movement control. My previous physiological studies had been at University College where I was interested especially in the studies of Wallace Fenn, A. V. Hill, and Douglas Wilkie on the apparent viscosity of contracting frog and human muscle, and at Columbia with Teru Hayashi investigating the force-velocity relationships of actinomycin fibers. Studies on the basic muscle spindle mechanism by Bernard Katz, Pat Merton, and Ragnar Granit (whom I visited in Stockholm in 1958) played a role in stimulating my thinking about the basic feedback properties of the postural tone mechanism. This led to the physiological system model research, started at Yale with the helpful interaction of John Atwood, which is presented in Chapter 1.

It was soon clear that more quantitative experimental evidence would be required. With hand movement, the inertial load of the hand sits on the frequency band of the control parameters, as might be expected in a system pushed by evolution to maximize performance. It was thus necessary to decide on a movement which might minimize this inertial effect; we picked wrist rotation. With the help of John Atwood, a servomechanism motor system was built to provide for an input-output systems analysis of the human hand control system.

Thus began the research reported in Chapter 2, a frequency response description of the human hand control system, done with Mitsuo Iida of Nagoya University in Japan and Paul Willis, an electrical engineer of broad interests who started working with me at Yale and later moved with me to M.I.T. One of the main findings of our first study was the importance of the prediction operator which changed the hand's performance to predictable inputs as compared with the response to unpredictable or random inputs.

Sometimes a single experiment shows important phenomena most clearly, and the "free-wheeling" experiments seemed to be of this type, indicating the open-loop nature of the stretch reflex for fast movements. Thus, our initial look at the behavior of the hand indicated that it was a much more unusual control system than the pupil servomechanism.

A further series of frequency response studies on Parkinsonian patients was carried out with Mitsuo Iida, and has been continued in Boston with the cooperation of Robert Schwab of Harvard University and the Massachusetts General Hospital. I also performed an experiment, with Geoffry Rushworth of Oxford (on the roof of the Boston City Hospital, with myself as subject), in which we used procaine to block gamma control of the spindle and thus pharmacologically ablated the postural control system. It was evident that the voluntary system, proprioceptively open loop, was competent to perform most skilled movements. The postural control system seems more of a diffuse slow system which might only be involved in clamping or damping the end of fast movements.

At M.I.T., with the help of Val Kipiniak, we developed a digital computer program, BIOSIM, both for simulating a large-scale nonlinear analog facility on the digital computer, and modeling the motor coordination system. Robert Payne, with Paul Willis, helped to develop on-line computer configurations comprising analog and digital hybrid equipment for real time control and analysis of the experiment.

A series of such experiments on the transient dynamics of the hand system were carried out with the help of Paul Willis and Yutaka Okabe (from the same department of Medicine and Neurology at Nagoya University as Mitsuo Iida). We also performed some early experiments demonstrating the sampled-data nature of the hand movement system. Despite our using special equipment and the low inertia wrist rotation movement, it was difficult to demonstrate the various discontinuities or intermittencies in the hand control system. Thus, we made the long and fruitful detour to the eye movement system described in the preceding section. In that system, since the inertia of the eyeball is not limiting, and the eye predictive capacity is less, various basic neurological features stand out more easily than in the hand system.

At the same time as the hand transient experiments were being conducted, physiological features of the dual components, postural and voluntary, of the motor control system were being extensively simulated by a student of mine, James Houk, with help from John Atwood and Jerry Elkind. In addition to the large-scale nonlinear BIOSIM model, we also developed a linear reduced model which could be placed on an analog computer. This enabled us to model various changes of state of the hand control system and to compare the behavior of the model with experiments being carried out in parallel by Okabe, Willis, and myself.

Another student, Fernando Naves, now teaching at the University of Los Andes in Bogota, Columbia, did a series of experiments with me on the intermittency operator in the hand movement control system and was able elegantly to demonstrate this, as well as a number of related facts such as the absence of a velocity control system, and the highly adaptive nature of the hand movement system even to such an extraordinary procedure as opening a loop artificially. These findings form the basis of Chapter 5.

A great deal of further work is opening up as our horizons broaden in regard to the hand movement control system. For one thing, adaptive changes may be classified into peripheral and central mechanisms for the adaptation, thus suggesting a number of experiments already designed and planned. A partitioned Markov conditional probability matrix with dummy states to increase variance can predict successive reaction times to a se-

quence of predictable step changes, as another of my students, John Bill-
heimer, demonstrated. The application of optimal control theory has already started in our laboratory with particular reference to eye movement,
and we hope to extend this to the hand. More intimate dissecting studies
using electromyograms, better designed input–output apparatus, as well as
implanted electrode stimulation in human patients with Parkinson's syndrome, have also been started in cooperation with neurosurgical groups
both in Boston and in Washington, D.C.

The bioengineering approach can provide the clinician with three in-
vestigative tools: instrumentation for an objective measurement, on–line
real–time computation for mathematical analysis and display, and elucida-
tion of these complex neurological control systems in the conceptual framework of cybernetics.

First of all, it provides the neurologist with an objective clinical record of a patient's performance. Second, by employing a computer to control and analyze the experiment, it is possible to obtain almost immediately the results of a complex analysis in an understandable form such as a
graphical frequency–response curve. Similarly, a set of parameters of a
model can be quickly fitted to the experimental data. Contrast this with
analyzing a single experiment, *in extenso*, by hand over a period of weeks,
and only then understanding the results sufficiently to feed them back to
modify the experiment.

In this section we study the human motor coordination system at two
levels bridged by a model or series of models. The elemental level tries
to understand the role that a wide variety of physiological, neurophysiological, and neurological mechanisms play in determining human hand movement. The systems level attempts, first, to measure quantitatively input–
output behavior using a variety of instrumental methods, including on–line
computers, and second, to cast the results into a servoanalytic framework
for precision and clarity of description.

The model described in Chapter 1 unites these two levels. It enables
one to organize the elemental mechanisms in explicit form and to state
definitely their complex interrelationships. Their activity in particular behavioral phenomena can then be assessed by simulation runs of the model and
analysis of its performance. A linear reduced model was also developed
to aid in "getting a feel" for the model.

Design of human behavioral experiments, both complex quantitative
ones, and at times, remarkably simple and straightforward ones, is suggested by the structure and performance of the model. Interpretation is
likewise aided.

Chapter 2 presents quantitative frequency-response data basic to a comparison of the model with the human system. It serves to define behavior from the control systems approach and lays bare the input adaptive characteristics based upon the prediction operator, which sets up a requirement for carefully selected complex inputs. Conversely, nonlinearity is minimal on first examination.

Chapter 3 turns toward the clinical Parkinson's syndrome as an example of coordination grossly impaired by defects in the basic control system. The dual-mode controller, represented in the model, and intermittently switching between postural and voluntary control, provided an hypothesis to explain the rigidity and performance failure in these patients. Conversely, further behavioral evidence from neurology buttressed the model.

Study of the time domain exposes system features complementary to those displayed in frequency domain experiments. Chapter 4 deals with behavior of hand and eye tracking and the interrelationships in response to transient inputs. There the effects of changes in postural dynamics and responses to both mechanical-impulse disturbance inputs and to visual-impulse signal inputs are compared. Finally, early findings pointing out discrete control system behavior are shown.

Chapter 5 develops in full the evidence for intermittency in the hand control system. The absence of a velocity control, a quantitative transfer function for the tracking of predictable signals, and the input-synchronized rather than clock-synchronized sampler are all demonstrated. Adaptive behavior in both open-loop and variable feedback experimental conditions is observed. The linear reduced model is expanded to include many of these findings.

Finally, in the Appendix, two adaptive sites are hypothesized and contrasted, and an experimental test is put forward.

Chapter 1

Physiological Model for Hand Movement Control System

MECHANISMS OF MOTOR CONTROL

Two mechanisms play an important role in the function and adaptive capability of the manual control system, namely, the postural control system and the voluntary control system. The postural system, a feedback system which functions to maintain the posture or position of the body and its limbs, is essentially a position servo loop with position sensors, motor elements, and some data processing. This servosystem is controlled by higher centers which both provide reference inputs and control the parameters of some of the elements of this system, as shown in Fig. 1.

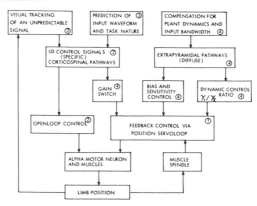

Fig. 1. Simplified block diagram for movement control.

The voluntary control system provides the mechanism for executing skilled, precise movements. It too, of course, is composed of sensory, motor, and computational components. It is quite different from the postural control system in that the controller is thought to be open-loop sampled-data which executes preprogrammed control movements. These movements are proprioceptively open loop in the sense that the postural control system appears to be at least partially disabled when voluntary control movements are being executed. The feedback is obtained from a variety of sensors, including joint position receptors and the visual system, but the information obtained from these feedback elements must be processed by higher centers before they can effect a movement. There appears to be some evidence that the voluntary control system acts like a sampled-data, input-synchronized, type 1 control system.

An important part of the control of the voluntary movements is the predictive system, which is able to extract information from signals and responses to predict the future course of events. This predictive ability is largely responsible for the input adaptive properties of manual control systems.

VARIABLE TOPOLOGY OF THE NEUROLOGICAL CONTROL SYSTEM

This section discusses some experimental responses which may correspond to three states of the neurological control mechanism, then introduces a general model for certain aspects of movement and postural coordination [91−93].

A subject was instructed to rotate a handle back and forth as rapidly as possible. A record of handle angle as a function of time was obtained, as in the top of Fig. 2. During the high-velocity portion of the movement, the antagonists (opposing muscles) were quite relaxed and limp. Because of this inactivity, despite marked stretching of the antagonists by the agonists (contracting muscles), we concluded that the stretch reflex was inoperative. Either the afferent feedback from the muscle spindle was markedly reduced, or more likely, it was functionally ineffective in exciting the alpha motor neurons (final common path). Figure 3 shows a simplified block diagram of the situation. It is important to note: (a) the back-to-back pair of control systems (agonist-antagonist) that are always present in movement because of the inability of muscle to push; (b) the open-loop mode that permits the high-frequency oscillation; and (c) the preprogrammed set of alternating contractions sent down from the brain.

The subject was then instructed to perform successive movements again, but was told that his primary object was to prevent deflection of his hand by random input disturbances. Only secondarily was he to oscillate

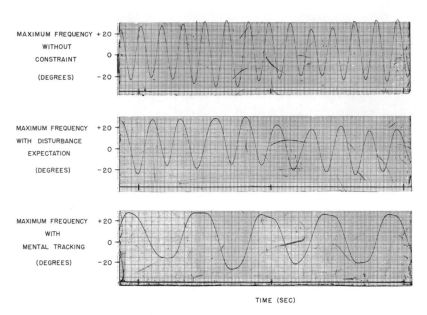

Fig. 2. Free-wheeling experiment.

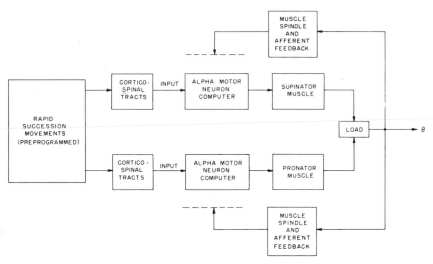

Fig. 3. Rapid-succession movements: free wheeling.

his hand at as high a frequency as possible. Figure 4 shows the neurological organization of this mode of operation. The essential features here
are: (a) the muscle spindle system, acting as a high-gain length regulator;
(b) the resultant increased stiffness or spring constant of both agonists and

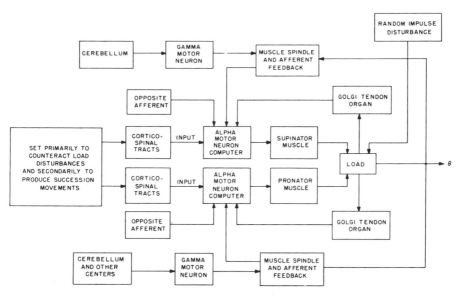

Fig. 4. Position servocontrol set primarily to counteract load disturbances.

antagonists; (c) the sharing of the alpha–motor neuron by this length regulator and the corticospinal input. As a result of these factors, the frequency of oscillations is slowed, as shown in the middle recording of Fig. 2.

This description also elucidates the essential nature of the rigidity
of Parkinson's syndrome (see Chapter 2). Here the spindle length regulator is always (except in sleep) on full gain, and thus opposes and weakens
corticospinal inputs. It might be well to mention that the diffuse anatomical organization of the gamma input makes it a highly inappropriate follow–up servosystem. In fact, it is not used for this function, but rather
for postural tone and end–of–movement damping and clamping.

In the final set of instructions given the subject, he was asked to imagine a pointer moving back and forth, to track this imaginary pointer, and
then attempt to oscillate his hand as fast as possible, as in the first mode.
The arrangement of the neurological apparatus is shown in Fig. 5. The
position afferent information is thought to feed back to a postulated visual–
motor orientation complex of the brain. Perhaps because of the necessity
of transmitting and processing all control signals through the imagery of
the mental tracking process, the oscillation is markedly slowed. This is
shown in the bottom recording of Fig. 2.

The gamma–spindle system may be utilized in this mode at the end
of each high–velocity portion of the oscillation. Evidence for this comes

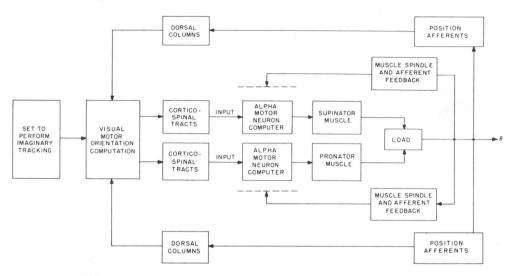

Fig. 5. Mental tracking mode configuration.

from an experiment done with G. Rushworth [104]. The gamma gain controls to the spindles were blocked with procaine, producing the so-called
cerebellar syndrome of hypotonia, astenia, ataxia, overshooting, rebound,
dysmetria, and postural drift. Skilled voluntary movements were still possible with attention. However, when sinusoidal tracking, similar in principle to mental tracking, was being performed, errors were made in a particular fashion. Occasionally, at the end of the high-velocity portion of the
sinusoid, the pencil hand would continue in the tangential direction without
halt. This indicates that the damping and clamping functions of the length
regulator (spindle) were being called for, and thus that they would be normally active in a similar manner.

In summary, we have tried to define by experiment a facet of the
complex neurological control system movement: the ability to change the
actual topology of the multiloop neural system. This introductory material
has tried to point up the interaction and indeed competition between the
postural servo loop (stretch reflex) and the proprioceptively open-loop voluntary control of the musculature.

THE QUANTITATIVE MODEL

While a general model may be useful in organizing one's notions of
the physiology of the spinal and peripheral mechanisms, it should clearly
be supplemented by a quantitative model. The BIOSIM program, providing
for the simulation of a general class of dynamical systems from specifica-
tions of their block diagram representation, has been used for this applica-

tion. It is run on a 7090 computer and contains its own Fortran-like compiler for ease of communication between user and program [57].

The expanded and modified form of the block diagrams of Figs. 3, 4, and 5 for closed-loop operation of the postural musculature system and open-loop operation of the voluntary control system is shown in Fig. 6. The experiment is carried out by instructing the subject to hold his arms fairly tense and to maintain this position despite possible deflecting forces. A force of 5 kg, lasting 0.2 sec, is applied. The subject's arm is deflected and then returns to its desired position after a slight overshoot.

The BIOSIM analog behaves as shown in Fig. 7. Tensing of the muscles is accomplished by inserting a bias signal using SPE 6 and SPE 7 (see table in Fig. 6). The disturbing force is applied by means of SPE 10, and its time function is shown. As the hand is deflected from its zero position, the error signal of the stretched muscle increases and that of the slackened

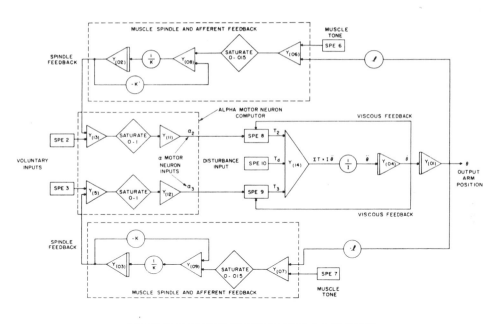

Special Function (SPE)	Definition
SPE 6 and SPE 7	Muscle Tone = 0.001 u - 1(t)
SPE 8	$\begin{cases} -Y(11) + 100\ Y(11) \bullet Y(13) & \text{if } Y(13) < 0 \\ -Y(11) + 600\ Y(11) \bullet Y(13) & \text{if } Y(13) \geq 0 \end{cases}$
SPE 9	$\begin{cases} Y(12) - 600\ Y(12) \bullet Y(05) & \text{if } Y(13) \geq 0 \\ Y(12) - 100\ Y(12) \bullet Y(05) & \text{if } Y(13) < 0 \end{cases}$
SPE 10	Disturbing Force = $5[u-1(t-0.2)]$
SPE 2 and SPE 3	Not used for this experiment

Fig. 6. BIOSIM block diagram of a simple motor coordination system, with definitions of special functions.

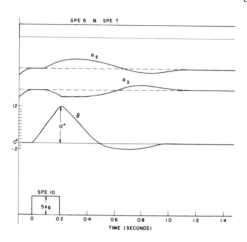

Fig. 7. BIOSIM outputs.

muscle decreases. When the disturbing force ceases, these signals, by operating on their respective muscles, drive the hand back to its zero position with a slight overshoot. The initial biasing, hand position, disturbing force, and error signals are shown as time functions in Fig. 7 and indicate a qualitative and rough quantitative agreement with the experiment. In Fig. 7, a_2 represents input to muscle-restraining initial motion; a_3 represents input to muscle slackened by initial motion; Θ represents deflection of the arm in degrees; and the dashed line indicates the bias level.

These results are encouraging, and indicate that further development in the directions of (a) better definition of the physical elements of the control system; (b) more accurate descriptions of system topology; and (c) further comparison with experiment might increase our insights into this complex system.

MUSCLE PHYSIOLOGY: BIOSIM MODEL

Muscle is unique in converting chemical energy directly into mechanical work. Limitation in the rate of this conversion acts as an apparent viscosity which plays a role in the overall dynamics of movement. This section summarizes the physiological evidence and describes the simplified mathematical model used in BIOSIM [97].

Fenn and Marsh [25] first showed that the dynamic characteristics, such as shown by dashed lines in Fig. 8, of a maximally stimulated, shortening muscle could be summarized by an exponential relationship between the load on a muscle and the maximum velocity of shortening, as shown by the dashed curve in Fig. 9. In 1938, Hill [39] showed in a thermody-

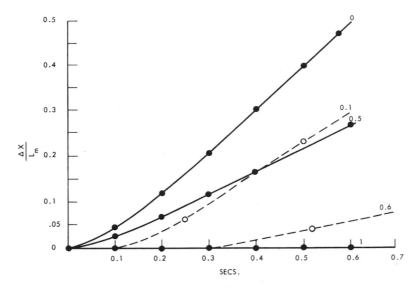

Fig. 8. Comparison of BIOSIM model with Fenn and Marsh experiments. Plots show change in length of muscle divided by length of muscle vs. time. Solid lines represent data from BIOSIM model; dashed lines show similar shape obtained by Fenn and Marsh, when normalized stretch and time are corrected to physiological scale of man. Corrections are suggested by Hill. The parameter noted on the curves is normalized load, F_{load}/F_{iso}.

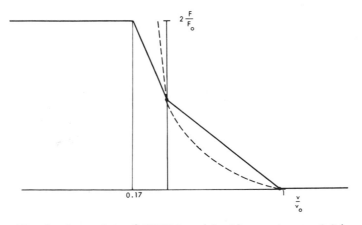

Fig. 9. Comparison of BIOSIM model with experiment. Solid curve shows force-velocity relationship of BIOSIM model; dashed curve, force-velocity relationship of frog muscle as obtained experimentally. For frog muscle $P_0 = 0.1$ kg, $v_0 = 4$ cm/sec.

namic study of muscle that the force-velocity relationship could be approximated very closely by the equation

$$(v + b)\ (P + a) = (P_0 + a)\,b = \text{constant} \tag{1}$$

where a and b are constants that are dependent on the particular muscle,
P is load, v is shortening velocity, and P_0 is isometric tension. When experiments were extended to lengthening, as well as shortening muscle,
Hill's equation failed. Katz [56] found a steep linear relationship between
muscle opposition force and lengthening velocity, as seen in the dashed
line of Fig. 9. Ten years later, Wilkie [117] repeated Hill's experiments
for the shortening muscle in the human arm by using subjects who were
directed to exert maximal voluntary effort. His results fit Hill's equation
after he corrected for the inertia of the arm. These data were very useful
in our approximation of a muscle model for digital computer simulation.

The primary purpose of building a muscle model for BIOSIM was not
to investigate the mathematical characteristics of Hill's model, but rather
to investigate the behavior of a complete agonist-antagonist muscle system. Therefore, some very simple straight-line approximations to the
physiological model were considered adequate for this work. We estimated
P_0 (the maximum isometric force) and v_0 (the maximum velocity with no
load) by using both the parameters in Wilkie's paper and our own measurements on human subjects. The slope of the straight line representing
lengthening, B_L, is made many times greater than that for shortening, B_s,
as seen in the solid lines of Fig. 7a. The following parameters were selected:

$$P_0 = 100 \text{ kg};\quad v_0 = 0.01 \text{ m/sec};\quad B_s = -10^4 \text{ kg-sec/m}$$

$$B_L = -6 \times 10^4 \text{ kg-sec/m}$$

Force velocity relationships beyond the initial velocities pose a problem.
If a muscle is shortened at a velocity greater than that at which it is capable of shortening itself, it exerts no force. Thus P is zero for velocities
greater than v_0. When a muscle is stretched more rapidly than its critical
stretching velocity, several things may happen. A phenomenon called
"slipping" or "yielding" occurs first. Hill [41] wrote: "Under a load
rather greater than it can bear an active muscle lengthens slowly, under a
considerably greater load it 'gives' or 'slips.' We can regard the first
process as 'reversible' in the thermodynamic sense, the second as largely
'irreversible.'"

Katz [56] found that contractile structures of a muscle may be dam-
aged if its velocity of lengthening is increased rapidly while the muscle is

active. Normally, the Golgi tendon organ (a tension-sensing device) re-
flexly causes the muscle to relax before this damage occurs.

It is expected that simulation will seldom operate in the critical region. Thus a compromise between complete relaxation and increased resistance caused by slipping is used. For velocities of lengthening which
are greater than critical, the BIOSIM model saturates at $2P$, as shown in
Fig. 10. When a muscle is stimulated at half maximum, we assume that
the active muscle has the same length but only one-half the cross-section
area as the fully activated muscle. Therefore, the force output, proportional to cross-section areas, is halved while the maximum velocity of
shortening, proportional to muscle length, remains constant. These relationships between force and velocity in a half-activated muscle may be obtained by applying the P_0 and v_0 values for $\alpha = 0.5$ to our original approximations; that is, straight-line relationships are shown in Fig. 10 and expressed in Eqs. (2) and (3).

$$F_{\text{muscle}} = \alpha(P_0 + B_s v) \qquad 0 < v < v_0 \qquad (2)$$

$$F_{\text{muscle}} = \alpha P_0 + B_L v) \qquad -\frac{v_0}{6} < v < 0 \qquad (3)$$

Figure 11 illustrates a model experiment conducted on the 7090 computer [97]. Figure 8 (solid lines) shows that the model response in displacement has a shape similar to that of a plot of the same variables for

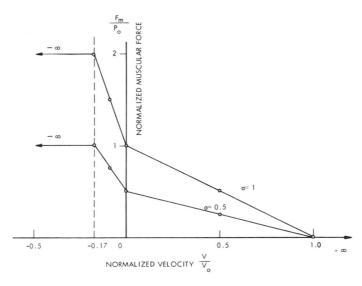

Fig. 10. Force-velocity relationship of BIOSIM model ($\alpha = 1$, maximal stimulation; $\alpha = 0.5$, half-maximal stimulation).

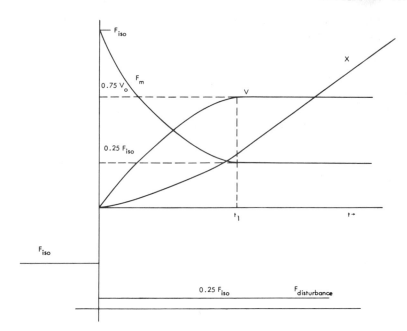

Fig. 11. Experiment performed with BIOSIM model. F_m is total force exerted by muscle, x is muscle displacement, v is muscle velocity. Excitation is maximal. Time, t_1, was necessary for model to overcome inertia
and reach constant velocity.

frog muscle in the Fenn and Marsh [25] experiments. The dead time in the
physiological studies is the time taken for the muscle isometric force to
build up to the load force. A comparison of velocity vs. time in the model
with Wilkie's data was also obtained.

The asymmetrical characteristics become very important in smoothly
terminating a rapid voluntary movement. Asymmetry increases the effective damping over that of a possible symmetrical relationship. Addition of
the postural system (muscle spindle and afferent feedback), as shown by
the middle recording of Fig. 2, further increases man's power of rapidly
damping his motion. We turn next, therefore, to an examination of the
muscle spindle.

MUSCLE SPINDLE PHYSIOLOGY: BIOSIM MODEL

The muscle spindle receptor is a differential-length receptor found
in parallel as first demonstrated by Fulton and Pi-Suner [26] with contractile fibers of muscles of many species. Its importance in human motor
coordination is great, since its positional feedback characteristics cause
the myotatic, or stretch reflex. It may also send kinesthetic information

to higher centers to help control complex motor coordination tasks. A model of this mechanism has been formulated as one component in the more complete model of human motor coordination.

The spindle is connected in parallel with the muscle contractile fibers as shown in Fig. 12a and its direct mechanical effect on the muscle is negligible. Thus, we can consider the length of the muscle, X_m, to be an input produced either through the alpha efferent nerve or through stretch by external forces. The afferent nerve carries information concerning the length of the nuclear bag to the central nervous system. Continuous signals representing a short-term average number of pulses per second are used in our model. The gamma efferent nerve excites the contractile element, or intrafusal fiber, of the spindle. It is another input that may bias the output of the nuclear bag or may, perhaps, act indirectly as an input that might control movement: this follow-up servo configuration has been suggested by Merton [68] and Roberts [82]. In our model we have simplified this input to a change in length of the intrafusal fiber, X_v, and have merely added it to X_m (Fig. 12b). The dynamics of this input may have to be considered more explicitly when we investigate coordinated movements through this input.

The response of the spindle receptor to a step input of stretch, as found by Lippold, Nicholls, and Redfearn [62] (Fig. 13) shows approximately 400% overshoot; this differential effect is called the "phasic response."

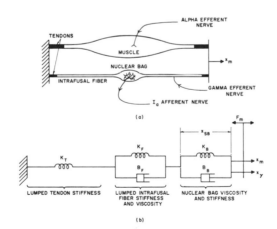

(a)

(b)

Fig. 12. (a) Diagram showing simplified muscle with one of its in-parallel spindle receptors separated for ease of illustration. (b) Mechanical model of spindle receptor. Inputs are X_m (length of muscle) and X (artificial length caused by input to intrafusal fiber); output is X_{SB} (length of nuclear bag).

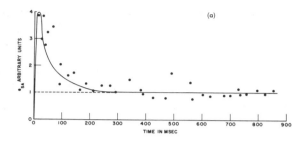

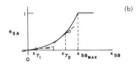

Fig. 13. (a) Step response of spindle receptor. Dots show
data points taken from Lippold [62]. Continuous curve
shows model response, $e_{SA}(t)$, in arbitrary units of pulses
per second [62]. (b) Input-output characteristics of mechanical-to-electrical transducer of nuclear bag: e_{SA} is
output of spindle receptor; X_{SB} is length of nuclear bag.
Note change in small-signal gain caused by d-c levels
of X_m or X, components of X_{SB} (Granit [29]).

After approximately 200 msec the output settles down to its steady-state
value; the steady-state response is called the "tonic response." They also
found good evidence to suggest that the phasic response is caused entirely
by mechanical factors and that the mechanical-to-electrical transducer is
a no-memory device. This information has been incorporated into our
model.

The transfer function for the mechanical model of Fig. 12b is

$$\frac{X_{SB}}{X_m} = H(s) = G_m \frac{(T_1 S + 1)}{(T_2 S + 1)(T_3 S + 1)} \tag{4}$$

where

$$T_1 = \frac{B_F}{K_F}; \quad T_2 = \frac{B_F + B_B}{K_F + K_B}; \quad T_3 = \frac{B_F B_B}{K_T(B_F + B_B)}; \quad G_m \frac{K_F}{K_F K_B}$$

We may fit this transfer function to data in Fig. 13a by choosing $T_1 =$
1/6.8, $T_2 = 1/18$, and $T_3 = 1/200$. Four estimates of the value K_F were
calculated with the use of published data. Our results showed that $K_F =$
0.1 newton/m (within one order of magnitude). Then B_F was assumed to be

Table I

	Human	Frog
K_F	0.1 newton/m	0.1 newton/m
K_B	0.4 newton/m	0.5 newton/m
K_T	1.5 newton/m	1.3 newton/m
B_F	0.015 newton sec/m	0.5 newton sec/m
B_B	0.015 newton sec/m	2.9 newton sec/m
K	0.2	27 pulses/min

approximately equal to B_B. Solving for the three remaining unknowns we obtain the value for the constants in Table I.

Inspection of Granit's data [30, 32] shown in Fig. 13b reveals that the tonic input–output characteristic of the spindle receptor is somewhat parabolic. Furthermore, other curves show that the tonic gain increases with increased gamma bias (X_v). It was found that a squaring operator fits these data quite well (Fig. 14a). The increased slope at X_v2, as compared with X_v1, shows the increased small-signal gain for an increase of gamma bias. This reconciles the opposing views of Granit [29], who claimed the gamma input is an input bias level, and Hunt and Kuffler [49, 50], who felt it has a

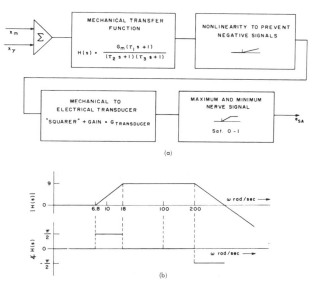

Fig. 14. (a) Complete block diagram of spindle receptor model. Inputs are muscle length, and length due to gamma input and to intrafusal fiber; output is average number of pulses per second. (b) Asymptotic Bode plots of gain and phase characteristics of H(s), transfer function of mechanical part of spindle receptor model.

role in maintaining sensitivity of the spindle to stretch. Both functions are subsumed in our model. Negative signals cannot be generated by the transducer; thus, a saturation at zero is inserted before the squarer (Fig. 14a). Also, there is an upper limit to the number of impulses that can be carried by the spindle efferent nerve; an upper-limit saturation provides this limit.

An asymptotic Bode plot of $H(s)$ in Fig. 14b illustrates the approximate frequency response of the model to small signals in normal operating ranges of bias. Note that between 7−18 radians/sec the response is differential. In this range, the spindle receptor should cause viscous damping effects, rather than spring effects, on the overall system. A sinusoidal analysis of the stretch receptors in the frog's extensor longitus digitus, IV muscle was undertaken by Houk, Sanchez, and Wells [47]. This preparation contains three to seven spindle bags with as many as three occurring on a single intrafusal fiber. No attempt was made to obtain single unit responses; rather the average response of all of the units associated with this muscle was judged to be more useful in justifying a model of this transducer suitable for simulating motor coordination. The preparation was stretched sinusoidally in position about a d-c level chosen to eliminate saturations at either maximum frequency of nerve pulses or a minimum frequency of no pulses. A sample record is shown in Fig. 15. Figure 16 shows the amplitude vs. frequency plot. Although these data are crude first runs, the break frequencies of rough asymptotes drawn on the gain plot of Fig. 16 were used to form a tentative transfer function

$$H(s) = \frac{K(30s + 1)}{(5s + 1)(0.35s + 1)}; \text{ with } K = 27 \text{ pulses/min} \qquad (5)$$

By assumption and calculation similar to those above, we obtained the values in Table I for the frog. These findings, although crude, lend

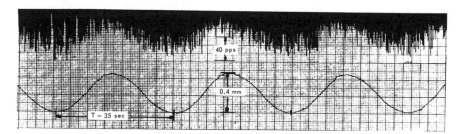

Fig. 15. Response of spindle receptor to sinusoidal drive. Length of vertical lines in top record is proportional to interval between successive nerve pulses from spindle; peak-to-peak change in instantaneous frequency, ∼40 pps. Bottom record shows sinusoidal component of spindle stretch, the input (frequency, ≈0.03 cycles/sec).

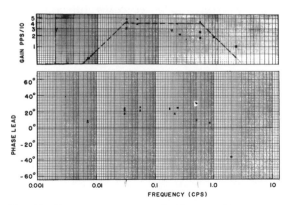

Fig. 16. Frequency response of a frog spindle receptor.
Rough asymptotes have been drawn on gain curve.

support to the model put forward and provide some feeling for the functional significance of the anatomical structure of spindle receptors.

GAMMA RATIO

The spindle system has a differentiating action over 1/2 to 3 cycles per sec, an important part of the frequency range. Since it is in the feedback path of the proprioceptive stretch reflex, or postural servomechanism loop, this obviously acts as a damping element in movement. It can be shown that in ordinary movements, the muscle apparent viscosity, which has similar dynamics, operates in a region of lesser interest below 0.25 cycle per sec. The muscle damping is the result of asymmetrical nonlinear energy conversion saturation and thus has, strictly speaking, no restricted frequency range.

However, the spindle is an ideal damping element and the gamma bias which sets its activity level and gain is well suited to modify the damping in an adaptive way.

Furthermore, Boyd [12], in a series of anatomical studies, and Matthews, in a complementary physiological study, have found another type of spindle receptor which lacks the nuclear bag. Thus, the dashpot is missing and a quite different dynamical situation results. The intrafusal muscle fiber without a nuclear bag, but with an afferent receptor, is innervated by a different gamma efferent motor fiber type — the gamma-two fiber. While this has not been worked out quantitatively, it is clear that the gamma-two is more like a pure gain element than the gamma-one, which operates as a differentiator. Thus, by altering the gamma ratio (gamma-one activity divided by gamma-two activity) the damping can be altered independently of

the loop gain. This permits even more powerful control of overall dynamics and damping, and similarly is even more suited for the adaptive role.

SUMMARY

The information presented in this chapter enables us to appreciate
the physical (anatomico-physiological) elements present in the movement
coordination system. These elements contain numerous nonlinearities as
well as fairly complex dynamic operators and require a full-scale BIOSIM
simulation for adequate computation. However, as will be shown in Fig.
56, a not too complex analytical model can be used to represent the system
satisfactorily enough for many purposes. Further reduction is possible,
since certain zeroes cancel the effect of time delays. After complete reduction, a minimum model remains, namely, a pair of complex poles whose
damping and natural frequency are set by variable gain and damping controls.

Chapter 2

Dynamic Characteristics of the Motor Coordination System in Man

INTRODUCTION

One of our aims is to describe the human motor coordination system in quantitative terms. In addition to an adequate transfer function, we hope to derive a system model which not only fits these quantitative data, but even more importantly, includes anatomical, physiological, and neurological facts in order to better define the topology of the model [30, 36, 116, 117].

This chapter [99] is limited to a consideration of the dynamic characteristics of human motor coordination obtained with experiments carried out under restricted but well-defined ranges of performance. Previous work similar to this has been done in a series of human engineering experiments reviewed by McRuer and Krendel [23, 58, 65, 66, 86, 87].

However, we are not concerned with defining human operator performance in the human engineering sense. Rather, our concern is to clarify aspects of the physiological multiple-loop control system activated during the human movement under investigation. For example, data obtained at relatively high frequencies often give the most information about the equivalent networks being studied, even though these frequencies may not be useful in terms of practical human machine-control operations.

EXPERIMENTAL APPROACH

The movement we selected to study was rotation of the wrist of an awake, cooperative human subject. The hand grasped a handle (Fig. 17)

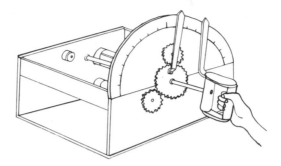

Fig. 17. Sketch showing mechanical portion of apparatus.

which could only rotate, constraining the movement to pronation and supination. Movement was further restricted by fixing the elbow. By arranging the axis of the forearm to be concentric with the shaft, the rotational inertia of the forearm was reduced. Wrist rotation was restricted to $\pm 30°$ from the vertical both to utilize similar muscles over the entire range of movement, and to stay within comfortable limits. These restrictions were probably realistic, since when the subject was instructed to rotate the handle back and forth as rapidly as possible, a mode of experiment called "free wheeling," the amplitude was generally confined to this range.

The input signal presented to the subject was the rotation of a pointer. The subject was instructed to keep the indicator attached to his handle in coincidence with the movement of the pointer. Recordings were made of the instantaneous positions of both the pointer and the handle, obtained from feedback potentiometers geared to the output shafts of their respective servomechanisms.

PREDICTABLE INPUT EXPERIMENT

In Fig. 18 is an example of a simple, predictable input experiment, in which the input pointer swings to and fro in a sinusoidal fashion. The dynamic characteristic of the hand during a steady-state predictable experiment is obtained by measuring gain and phase lag from the recordings.

Dynamic characteristics of a linear system can be described at each frequency of input signal by only two parameters, gain and phase lag. Gain and phase lag are defined in the usual manner; if a subject has a gain of 2, while the input signal pointer was swinging $\pm 10°$, then the subject's motion was $\pm 20°$. The phase difference between two sinusoids is an angle measured in degrees; if the output follows the input by one-quarter of a cycle, then the result would be stated as a 90° phase lag. Be-

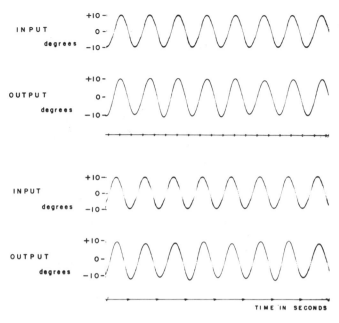

Fig. 18. Predictable input experiment.

cause of the variability inherent in biological systems, and particularly in man tracking a pointer, our analyses of gain and phase lag were based on the average response; that is, a number of consecutive performances were required to be essentially identical in gain and phase. The spread of phase lag did not become large except at frequencies which were on the limit of the ability of the human motor coordination system.

The importance of a small–signal approach in linearizing the responses of biological systems has been discussed in some detail previously [90, 106]. In order to test its applicability in this instance, we have studied the effects of changing the amplitude of the input signal [see Figs. 24–26].

COMPLEX INPUT EXPERIMENT

The block diagram of Fig. 19 indicates a further development of our experimental method. The input shown in Fig. 18 is completely predictable, and there is little useful information about the basic neurological control system to be obtained from it. It is therefore important to use an unpredictable input. Several methods of generating unpredictable signals are possible. One is the use of truly stochastic noise as the input signal, band–limited in a number of ways, if desired. This, however, requires complicated computational machinery for understanding the relationship between input and output, and more sophisticated analysis instrumentation

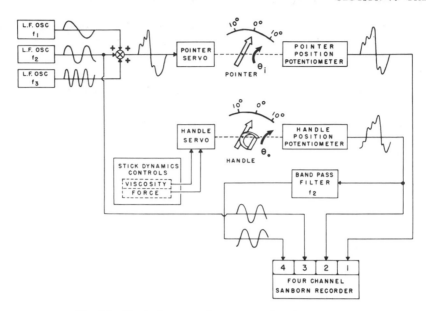

Fig. 19. Block diagram of motor coordination apparatus.

than we had available at the time. A simpler method is to sum a number
of non-low-harmonically related sinusoids to form a fairly complex input
signal. Of course, it is mathematically true that this signal is an analytic
function, and, therefore, completely predictable from its behavior in only
a small neighborhood. However, this signal is not an analytic function to
the prediction apparatus of the human subject operating instantaneously on
the pointer movement. It only takes the sum of a small number of these
sinusoids, three, if they are relatively prime to each other, in order to
make the prediction apparatus of the subject inadequate. This was demonstrated in early experiments which showed *no* reduction in phase lag or
amplitude error with considerable practice, and indeed, knowledge of the
structure of the input signal by the subject. Verification of this interesting
physiological result is contained in the analysis section under the discussion of Fig. 28.

          The principal instrumentation advantage of using a mixed sinusoidal
input is that by using a narrow band-pass filter, the output can be dissected
into various frequency components related to the frequency components of
the input. Both the single input frequency of interest and its dissected output counterpart were recorded, as shown in Fig. 19. The phase lag between channels three and four has two components: one due to the subject,
another to the pointer servo. We determined the latter by recording the
oscillator sinusoid directly and the pointer potentiometer signal through

the band-pass filter. Then the handle potentiometer signal was substituted into the undisturbed filter, as shown in Fig. 19. This allowed us to subtract out the phase lag due to the pointer servo, and eliminated any possible phase lag errors due to the filter. A band-pass filter was tuned to select the single frequency component of the output to be the central frequency of input, with -4 asymptotic gain slopes occurring on either side of this frequency. These arrangements for the construction of the complex input and the dissection of the complex output enable a simple gain and phase lag analysis to be performed just as in the predictable input experiment.

EXPERIMENTAL APPARATUS

The block diagram in Fig. 19 also indicates the operational structure of the two servosystems controlling the input pointer and the output handle. Physically, the apparatus consists of two similar 400-cycles/sec carrier servomechanisms with concentric controlled shafts [5]. Each servomechanism has a summing network, an electronic amplifier, a servomotor, and a controlled rotational element attached to the servomotor shaft. This rotational element is a pointer on one servo, and a handle on the other. Each servo also has position and velocity feedback, and these two parameters as well as the gain are individually adjustable.

This ability to select control parameters of the two servomechanisms was a great aid to us. The viscosity and stiffness of the pointer servo are permanently set to achieve a slightly less than critical damping of the pointer servomechanism. In the handle servo the control of stiffness of the output handle ("stick dynamics") proved to be a useful part of our experimental system when, in the preliminary experiments, we noted that visual tracking of the input pointer was a possible limiting factor in performance. Since this occurred at frequencies in the range of 2.5 cycles per sec and above, and it was not our aim to test the subject's visual system, we adjusted the viscosity and stick dynamics so that the response of the hand would decrease sufficiently to ensure that the subject's motor system and not his vision was limiting his performance. Calibration of this equipment is performed as follows: The instantaneous positions of both pointer and handle are recorded from their respective feedback position potentiometers on the recorder. The gain of the recorder is set such that 10° of pointer or handle movement is equivalent to 6 mm on the recording paper. This relationship is found to be linear over the range used. The dynamics of the input signal pointer servo are set to achieve an equivalent second-order system damping constant of approximately 0.6. The dynamics of the handle servomechanism were adjusted so that the subject rotated the handle against a spring force of 90 inch-ounces per radian and a viscous friction force of 1 inch-ounce-second per radian [119]. The stiffness was an important factor which acted to limit the subject's response. It, however, was not frequency-dependent and therefore did not change the form of the Bode plot. The viscous friction was a much smaller factor, even though frequency-dependent, and it did not become of equal importance to the stiffness until the frequency reached 15 cycles/sec (and then dominates above this frequency). Notice that this is a much higher frequency than that of our range of interest (see Fig. 37). It thus did not contribute significant distortion to the Bode plot. Its principal use was to stabilize the handle servomechanism. The frequency responses of both servos were obtained in the usual manner, and neither limited the experiment or experimental accuracy. For the purpose of these experiments we rendered the pointer servomechanism frequency response flat out beyond the frequency range of the experiment by prewhitening the input signals.

PREDICTABLE INPUT AND TRANSIENT EXPERIMENT

The response of a subject to a suddenly applied sinusoidal input signal from the pointer follows three successive phases of operation. In Fig. 20 such a result is shown. In the upper recording, the movement of the pointer is displayed, and on the lower recording, the movement by the subject. The first of the three phases is a reaction time delay, and is the period between numbers 1 and 2 along the time axis in Fig. 20. During this time the subject makes no movement. When the subject finally does make a movement and enters into the second, or "neurological response" phase, shown between numbers 2 and 3 along the time axis in Fig. 20, he generally starts in the proper direction to catch up with the pointer. Sometimes the initial movement of the subject is in the opposite direction, but the subject soon corrects himself and attempts to follow the pointer movement. For a variable period of time lasting approximately 0.5 to 1.0 sec during this neurological response phase, there is a phase lag in the subject's following

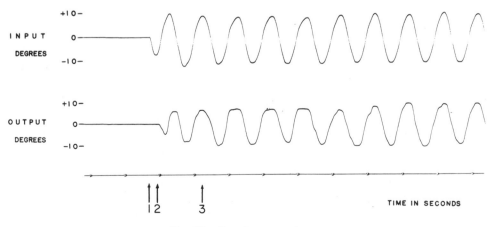

Fig. 20. Transient experiment.

movement as well as an amplitude difference, generally resulting in a
fairly sizable error. The subject, however, soon predicts the future
course of a simple sinusoidal motion and is able to phase-advance and correct his following movement so that he eventually synchronizes his response with the input signal, as shown to the right of number 3 along the
time axis in Fig. 20. It is this third or "predicted response" phase of operation which is the steady-state response of a subject to a simple predictable input. In addition to the time functions shown in the third phase
of Fig. 20, there are several samples of this steady-state predictable following shown in Fig. 19. Rarely does it take a subject longer than a second or two to get into phase three and remain there.

UNPREDICTABLE EXPERIMENT

This predicted response behavior introduces a serious and central
problem in our study. In order to handle this situation, we have been
forced to utilize nonpredictable inputs in order to prolong definitely the
second phase of response, the "neurological phase." In Fig. 21, a sample
experiment of the unpredictable signal input type is shown. The top recording shows the sum voltage of the three input sinusoids, as they appeared at the position feedback potentiometer of the signal input pointer.
The wave form as seen by the human eye across the stretch of graph paper
comprising 20 sec of time shows certain regularities. However, to the
subject who is (a) concentrating on the motion of the pointer, (b) attempt-

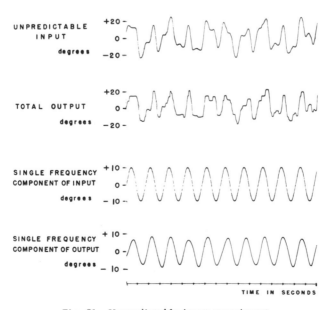

Fig. 21. Unpredictable input experiment.

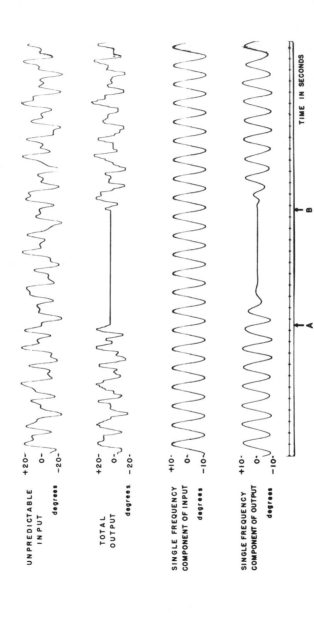

UNPREDICTABLE INPUT
+20-
0-
-20-
degrees

TOTAL OUTPUT
+20-
0-
-20-
degrees

SINGLE FREQUENCY COMPONENT OF INPUT
+10-
0-
-10-
degrees

SINGLE FREQUENCY COMPONENT OF OUTPUT
+10-
0-
-10-
degrees

A   B

TIME IN SECONDS

Fig. 22.   Control experiment showing band-pass filter transients.

ing to remain in coincidence by moving the handle (loaded with moderate viscosity and stiffness), and (c) provided with only instantaneous knowledge of pointer position, the signal in the top recording appears quite unpredictable. Evidence for this physiological result is found in the analysis of the experiments (see Fig. 28).

The second recording of Fig. 21, the total output of the subject in following this complex input signal, is seen to resemble with some degree of precision the form of the input.

The parts of the input and output signals which were analyzed are shown in the remainder of this figure. The third recording represents one frequency component of input obtained by recording directly the output of one of the input sinusoidal generators. The bottom recording is obtained from the total output by passing it through a narrow band-pass filter set to the single frequency of the input signal that we wish to study. Using the data obtained from the simplified input and output, we were able to determine gain and phase lag parameters between input and output exactly as in the steady-state response to a predictable input. We thus have a way of preventing the prediction apparatus of the human subject from dominating the physiology of the neurological response.

CONTROL EXPERIMENTS

Certain of our control experiments are necessary to demonstrate the adequacy of these methods and the difficulties involved. In Fig. 22 an experiment is shown demonstrating band-pass filter transient. In the second recording the handle movement is seen to stop suddenly when the subject is instructed to cease movement. This cessation of movement lasts for approximately 10 sec, at which time the subject is instructed to again follow the input pointer. The complex and simple inputs continue, as can be noted in the top and the third recordings.

The bottom recording shows the effect of the narrow band-pass filter which we used to simplify the complex wave form. First, the filter continues to put out a diminishing wave form after the subject has ceased to move the handle. There are certain mathematical relationships between the time and frequency representation of a function; the narrower the bandwidth of the filter, the less quickly it can respond to a transient change of conditions. In an effort to remove contributions from other frequencies from the wave form in the lower channel, we made the bandwidth of the filter quite narrow. Therefore, it takes an appreciable amount of time (approximately 1−4 sec at the frequencies we used) for the filter to reduce its amplitude as did the actual function in the total output channel. Similarly, when the subject began again to track, the filter output lagged behind the real output by several seconds. This filter action indicates the neces-

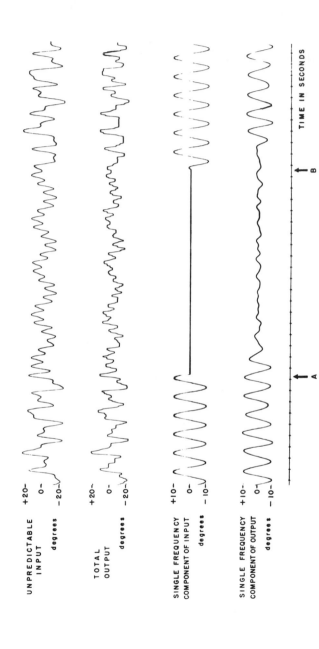

Fig. 23. Noise in signal spectrum experiment.

sity for a fairly steady-state response function in order to permit frequency analysis. We restricted our analysis to such actions of the experimental record; that is, we required the phase lag and amplitude of ten successive wave forms to be fairly closely grouped about their averages.

Figure 23, a noise in signal spectrum experiment, shows the efficiency of the band-pass filter in rejecting all but its central frequency. Again, the top recording, unpredictable input, shows the complex summation of three input wave forms up to the point marked "A." At this point, the single frequency component of the input to which the band-pass filter is tuned (shown in the third recording), was set to zero. This change is reflected in both the unpredictable input and total output channel recordings. The fourth recording, the single frequency component of output, is obtained by passing the total output shown in channel two through the band-pass filter. Notice that the ringing phenomenon of the band-pass filter is apparent from point A until a relatively small noise level is reached. This noise level continues until a bit past B, when the frequency to which the band-pass filter is tuned is again present in both the total input and the total output. This experiment demonstrates that the so-called remnant response [66, 95] of the subject is small in comparison with the signal swings and probably represents the effects of transient motions of the subject on the band-pass filter output. The experiment shown is representative of results obtained over many frequencies with many subjects.

Figure 24 shows the effect of such a transient signal injected into the experimental situation. The subject responds to the superimposed step. This response contains all frequencies, as demonstrated by a Fourier series expansion, of a step function. Only frequencies in the band-pass of the filter are accepted. The response thus appears as a brief ringing at the central frequency of the filter. This control experiment further indicates the necessity for requiring consistent steady-state relationships between input and output before accepting the data.

Another type of control experiment, and in fact an essential portion of the main experimental design, is a study of the variation in subject performance over the range of input amplitudes. We want to utilize linear system analysis; therefore we need a check on the linearity of the subject over the ranges of parameters of our experiment. As a further check on both the linearity of the experiment and the effects of the instrumentation, we performed the experiment of Fig. 25, a control experiment showing effect of different input amplitudes. Again, the four recordings are labeled as they were in the previous two figures. The central frequency component of the unpredictable input was mixed at three pointer amplitude levels, $\pm 10°$, $\pm 5°$, and $0°$. The output's central frequency component, obtained in each of these conditions, is demonstrated in Fig. 25. Notice again the

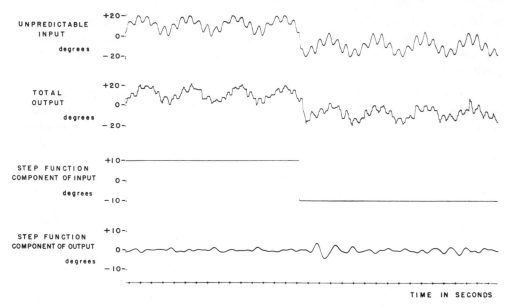

Fig. 24. Control experiment showing effect of step input through band-pass filter.

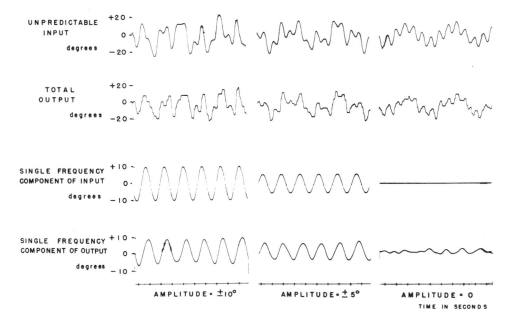

Fig. 25. Effect of different input amplitudes.

presence of the remnant response due to subject motion which transiently shocks the filter. This is most apparent in the 0° recording, and its amplitude is considerably smaller than the signal obtained under normal experimental conditions.

Experimental data to justify and substantiate that the linear approach is a feasible one are displayed in Fig. 26, a linearity control experiment. This is a plot of the output amplitude in degrees vs. the input amplitude in degrees, for three frequencies commonly used in our experiments. These frequencies were the central component frequencies of a complex unpredictable input for three different sets of input triads of frequencies. Notice that the subject's response at any one frequency is linear over a very wide range of input amplitude. The differences in slope values for the three frequencies are merely indicative of some of the information obtained in the Bode plot of the subject; it is easier for the subject to follow 0.6 cycle/sec than 1 cycle/sec, and easier to follow 1 cycle/sec than 2.5 cycles/sec. Hence, his gain is higher at 0.6 cycle/sec than at one of these higher frequencies. The data used in the experiment of Fig. 26 were obtained from experiments designed to eliminate trend effects. A rough estimate of the variance of this type of experiment is indicated by the scatter of the points in this figure.

FREE-WHEELING EXPERIMENT

One further type of experimental mode remains to be described. In this situation the pointer is not used as an input signal, but the subject is asked to oscillate the handle back and forth as rapidly as possible with varying amplitudes of oscillation being requested.

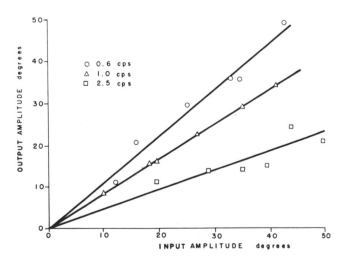

Fig. 26. Linearity control experiment.

A set of free-wheeling mode experiments is shown in Fig. 2. These experiments illustrate an interesting fact concerning the human motor coordination system. The upper recording shows a set of oscillations with a rather high mean frequency. This was obtained by asking the subject to rotate the handle as fast as possible, picking a comfortable amplitude of swing. It was noted that the subject always let his muscles go quite lax when doing this experiment. This means that the system might well be operating as an open-loop "bang-bang" control system [17].

In the second recording, the subject was requested to rotate the handle as rapidly as he could, provided always that his hand was stiff enough to receive a blow without being deflected very much from its course. In this condition, the subject kept his agonist and antagonist contracted against each other to a considerable degree. Naturally, this produced much slower oscillations. (This is an example of impaired voluntary movement performance similar to slowness in Parkinsonian patients.)

The lowest recording of this figure indicates still another and very interesting operating mode. The subject is told to keep the frequency as high as possible, in the mode of the first part of the figure. However, he is instructed to imagine the pointer oscillating back and forth as fast as he can possibly track it, and is then instructed to track this imaginary pointer. Perhaps, because of the necessity of transmitting and processing all control signals through the imagery of the mental tracking process, the oscillation is markedly slowed.

MAIN SERIES OF EXPERIMENTS

We have now discussed each type of experiment either as an example of an experiment type to be utilized in the main series below, or as a control experiment.

We arranged in random order a large number of frequencies and proceeded to study the responses of some fairly well-trained subjects over a period of weeks. At each frequency the responses to both predictable and unpredictable inputs were obtained.

Each frequency triad of the unpredictable input was arranged so that the ratio of the amplitude of the central frequency component to the amplitudes of the outer frequency components was either 1.0, 0.5, or 0.0, corresponding to $\pm 10°$, $\pm 5°$, and $0°$, respectively. A gain and phase lag analysis of the experiments was done immediately following each run and strict criteria, previously discussed, were adhered to in order to obtain the reproducible results described below. A set of free-wheeling experiments was performed; these represent limits of the physical capacity of the hand in any type of oscillation, whether coherent with an input signal or not. Together these make up the experiment analyzed in the next section.

ANALYSIS

The basic relationships which will be displayed here are those of gain and phase lag as functions of frequency. A composite Bode plot, that is, the frequency response, of the data is shown in Fig. 27. The response to the predictable input has a gain of 1 for almost the entire range of frequencies studied. At approximately 3 cycles/sec the gain decreases somewhat until the end of the range. However, the data are somewhat scattered in this high-frequency end of the behavioral range of the subject.

To delineate further the actual frequency range of the mechanical behavior of the human hand in our experiment, condition one of the free-wheeling experiment is also plotted on Fig. 27. There is, of course, no "gain" value possible for this free wheeling, because there is no physical input signal to the system; therefore, the amplitude is plotted on Fig. 27 in magnitude figures relative to the driven responses. It can be seen that an almost vertical line becomes less steep at low free-wheeling oscillation amplitudes, and this line apparently represents the maximum performance of the hand in amplitude and frequency. It was mentioned earlier that when the subject tracks a mental image of the pointer, his free-wheeling frequency decreases. It is believed that several factors may similarly reduce the response of the subject in tracking physical inputs.

The curve showing the response to an unpredicted signal over this range of frequencies is markedly different from the other two. This unpredicted or neurological response shows a gain of 1 up to 1 cycle/sec and

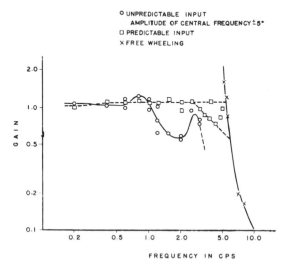

Fig. 27. Effect of signal predictability on frequency-response gain data.

then a gradual reduction at a −1 slope with increasing frequency except for
two suggestive peaks, at approximately 1 cycle/sec and 3 cycles/sec.
There is a large area, beyond 1 cycle/sec, between the gain curves of predicted and unpredicted responses. This region represents increased performance of the system due to the prediction operation which enables the
subject to correct phase and gain error in his response, and so reduce the
total error.

    In Fig. 28 the phase data are shown as a function of frequency cor-
responding to the gain data of Fig. 27. The upper curve shows the phase
response to the predicted signal input. The phase changes from 0° to a
small amount of phase advance. Above 1 cycle/sec while the mean phase
relationship continues to be one of slight phase advance, the scatter of the
phase increases markedly. The cone-shaped limits at the high-frequency
end of the upper curve are crude variance limits of this phase behavior in
the predicted experiments.

    The lower curve in Fig. 28 represents the phase relationship obtain-
ed with an unpredictable signal input. Several interesting points are seen
in this phase plot; for one, the phase lag is steadily increasing with frequency, an increase more rapid than is possible for a minimum phase lag
system. This means the human servo has a nonminimum phase lag element or elements and it is possible to analyze the phase curve into nonminimum and minimum phase components. Another aspect of the phase
curve is its several peaks and notches, showing irregularities correspond-

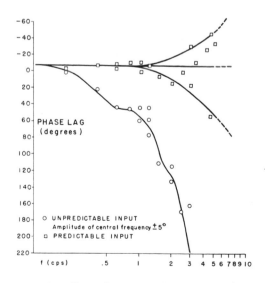

Fig. 28. Effect of signal predictability on frequency-response phase data.

ing to peaks of the gain curve and confirming the presence of these features in the gain curve. This is expected due to the close relationship of the gain and phase components of the generalized complex transfer function of a system. It is important to realize that the definite and consistent phase lags found with the "unpredictable" signal clearly show the inability of the subject to predict the complex signal.

Figures 29 and 30 show the gain and phase plots of the unpredictable experiment for two amplitudes of the central frequency components of the input triad. The two gain and the two phase lag curves are very similar, not only in their general patterns but also in much detail. There are small but significant differences between the responses to the two amplitudes; however, these differences in the gain and in the phase curves are similar in direction. For example, the high-frequency peak in the gain curve of the $\pm 5°$ amplitude is at a lower frequency than the high-frequency peak of the $\pm 10°$ amplitude gain curve. It can be seen that the phase crossover frequency is at a lower frequency in the $\pm 5°$ phase curve than in the $\pm 10°$ phase curve. The above correlations show the data to be consistent within themselves and suggest that the detailed features are reliable observations. The effect of amplitude change demonstrates the presence of some non-linearities in the system. This is similar to the experiment reported by Stark and Baker [94] in which the transfer function characteristics of the pupil servo control system were changed and the frequency of the high gain oscillation which represented its resonant peak was shifted in a parallel manner.

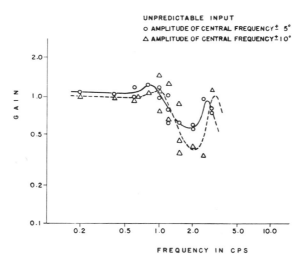

Fig. 29. Effect of amplitude on signal frequency-response gain data.

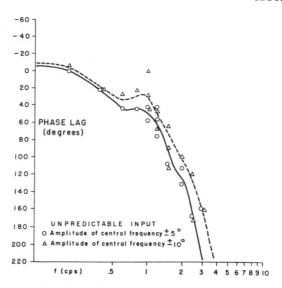

Fig. 30. Effect of signal amplitude on frequency-
response phase data.

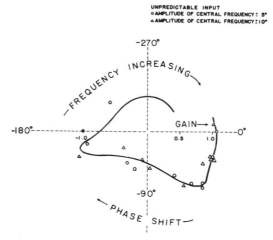

Fig. 31. Polar plot of frequency data response
(closed loop).

In Fig. 31 is shown a polar plot of gain and phase data from the unpredictable experiment. This vector plot of gain as a function of phase lag
has frequency as a monotonically increasing, but not regularly increasing,
function clockwise around the curve. The Nyquist curve is often used to
display system behavior because of the ready availability of gain and phase

stability margins of the system. This experimental curve also shows the two peaks in gain that were shown in Fig. 27.

DISCUSSION

In the Introduction we stated that this experiment was designed as a part of a complex of studies. The aim of the present paper is to describe carefully our method of obtaining quantitative information about the behavior of the control system for motor coordination and its underlying neurological servomechanisms. We have tried to give a full and careful description of the experimental methods, results, analytic methods, and faults of our present approach. We have presented evidence which shows how the prediction apparatus and the neurological response system for human motor coordination have been separated. These aspects of the physiological system have been defined in terms of their dynamic response.

Even in looking at such crude data as we have been able to accumulate, we have been impressed by the diversity of problems that are opened up by such analytic studies as described. The possibility of embedding the diverse and beautiful phenomena of human motor coordination in a sophisticated and economical mathematical representation is exciting.

SUMMARY

The dynamic characteristics of the human motor coordination system differ depending upon the physiological state. By using complex inputs, the predictive apparatus is eliminated and steady-state experiments can be used to study the neurological response. A series of linearity and other control experiments elucidate the limitations and adequacies of the methods. Some of the features shown are a resonant peak at 3 cycles/sec and both minimum and nonminimum phase elements.

Chapter 3

Dynamical Response of the Movement Coordination System of Patients with Parkinson's Syndrome

INTRODUCTION

Parkinson's syndrome is a common affliction, especially of elderly persons, well known in the neurological literature for the past hundred years. The muscles and nerves of Parkinsonian patients are generally not at all affected by the disease (until late secondary changes occur). Evidence for this comes from observation both of patients able to walk normally when sleepwalking and of the modification of this syndrome by certain brain operations. Approximately 35 years ago, it was demonstrated that interruption of the stretch reflex markedly reduced the rigidity which is a prominent sign of this syndrome [78, 112]. Recently, as a result of studies of the neurological organization of the control system of movement of normal subjects, a model of much of this system was put forward [75].

Here it is suggested that the voluntary control for moderate and rapid movements operates in an open-loop fashion, whereas postural tone and end-of-movement damping and clamping operates through a feedback path through the muscle spindles. The diffuse anatomical organization of the gamma input is well suited for this postural feedback system but argues against its role in a follow-up servosystem suggested by Merton [68].

The following experiments performed on a series of 20 Parkinsonian patients provide further evidence concerning both the nature of the defect of Parkinson's syndrome and the organization of the normal control system [93, 98].

EXPERIMENTAL METHOD

The apparatus illustrated in Fig. 17 was designed by Stark and Atwood in order to be able to perform a wide variety of tests approximating
a motor clinical neurological examination on patients with Parkinson's
syndrome or on normal subjects. It consisted of two servocontrolled units,
a pointer and a handle. The handle servo was sufficiently powerful to set
the "stick dynamics" by adjusting electrical parameters in the position,
feedback, or in the velocity feedback path. In this way, spring constant and
viscous resistance, respectively, could be arbitrarily selected.

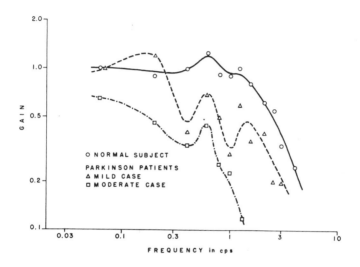

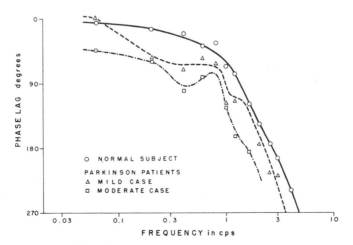

Fig. 32. Effect of Parkinson's syndrome on frequency-response
gain and phase data.

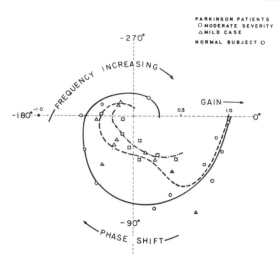

Fig. 33. Polar plot of frequency-response data.

The ability of the patient to maintain a posture was first tested. In this situation spontaneous tremors would appear as illustrated in the experimental results section. Next, mechanical impulses equivalent to stretch jerks could be administered by sticking a projection from the handle axis behind the apparatus.

Tone could be carefully measured by asking the patient to hold the handle but not to resist a programmed to-and-fro movement of the handle. An electrical signal to the input of the handle servomotor control system caused the handle to oscillate sinusoidally or to move clockwise and counterclockwise at constant velocity, thus producing a triangular or ramp input. The position of the handle and the force required to drive it were recorded continuously on the hot-stylus pen recorder.

Strength of the subject was measured in a straightforward manner by having the subject rotate the handle against a strong spring balance. Successive movements or "free wheeling" were next requested and position of the handle recorded.

Performance of skilled movements was valuable in a number of ways. First, the subject was required to maintain his attention in a rigorous fashion, and any departure from this could be instantly noted. Since it was an interesting task, there was no boredom-caused tone diminution resulting in lack of instantaneous attention. Second, by using multiple sinusoids assembled into an unpredictable complex excitation function, a quantitative frequency response could be obtained which was quite reproducible from day-to-day in any one subject. The important predictive ap-

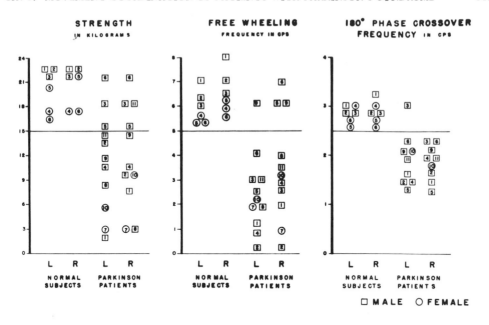

	STRENGTH IN KILOGRAMS		FREE WHEELING FREQUENCY IN CPS		180° PHASE CROSSOVER FREQUENCY IN CPS	
PARKINSON PATIENTS	LEFT	RIGHT	LEFT	RIGHT	LEFT	RIGHT
1. C.M.	1.8	7.3	1.2	2.0	1.7	1.5
2. J.B.	14.1	9.1	0.3	0.3	1.6	1.7
3. S.S.	18.2	18.2	3.0	2.7	3.0	2.3
4. J.K.	10.8	10.0	1.0	2.9	1.6	2.0
5. R.L.	15.4	15.4	2.5	6.0	1.5	1.4
6. A.S.	22.5	22.5	4.0	7.0	2.3	2.3
7. A.C.	2.7	2.7	2.0	1.0	---	---
8. J.P.	8.4	2.7	2.0	4.0	---	---
9. H.B.	10.9	14.5	6.0	6.0	2.1	2.1
10. S.B.	5.6	9.1	2.3	3.3	2.1	1.8
11. J.A.	14.6	18.2	3.0	3.5	2.0	2.0
Average	11.4	11.8	2.5	3.5	2.0	1.9
NORMAL SUBJECTS						
1. M.I.	22.5	22.5	7.0	7.8	2.9	3.2
2. P.W.	22.5	22.5	6.3	7.0	2.8	2.8
3. D.W.	22.0	22.0	6.0	6.4	2.8	2.8
4. T.A.	17.0	17.0	5.7	5.9	2.9	2.9
5. M.M.	20.0	22.0	5.3	6.0	2.6	2.7
6. J.B.	16.0	17.0	5.2	5.5	2.7	2.6
Average	20.0	20.5	5.9	6.4	2.8	2.8

Fig. 34. Effect of Parkinson's syndrome on strength, free-wheeling frequency, and frequency at which phase lag is 180°.

paratus of the human brain could be made ineffective and the basic neurological response could be exposed. A previous paper has fully described the electromechanical apparatus, the signal generation, and analysis filtering operations as well as the mathematical method [99]. Figure 21 shows

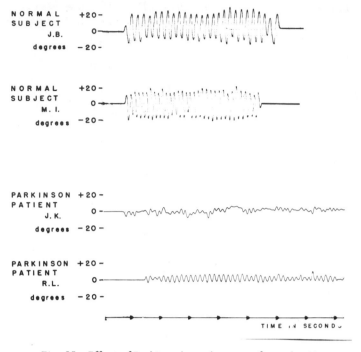

Fig. 35. Effect of Parkinson's syndrome on free-wheeling.

typical input (upper tracing) and output (second tracing) curves recorded
on a four–pen recorder. The input is the sum of three sinusoids of incommensurate frequency; the isolated middle frequency is also shown in the
third tracing.

EXPERIMENTAL RESULTS – FREQUENCY-RESPONSE DATA

By performing a series of unpredictable tracking experiments, the
frequency response of the human subject can be studied. It is important,
as shown by Stark, Iida, and Willis [99], to eliminate clearly the predictive
ability of the brain if consistent data relating to the "neurological control
system" are to be obtained. An example of such frequency–response data
is shown in Fig. 32 (top and bottom). Here, the gain and phase lag are
plotted as a function of frequency. The responses of three subjects are
shown: a normal person and two patients with mild and moderate degrees
of Parkinsonian rigidity. It can be seen that the bandwidth of the patients
is severely restricted. This is consistent with the model of a higher gain
postural feedback system operating against voluntary control of the muscles.

The polar plot shown in Fig. 33 is another display of these same data and shows in even more striking form the marked reduction in perform-ance associated with Parkinsonism.

VOLUNTARY STRENGTH

Although, as mentioned above, there seems to be no muscular pathology in Parkinson's syndrome, a marked reduction of voluntary strength exists, as shown in Fig. 34 (bottom). The model predicts this as a consequence of the opposing forces of agonist and antagonist muscles activated in a con-tradictory fashion by permanently switched-on postural servomechanism.

FREE WHEELING

The decrease in performance also affects the ability of the patient to perform rapid successive movements. In Fig. 35 the contrast between two normal subjects and two patients is marked. Later, it will be shown that another aspect of Parkinson's syndrome, tremor, may also influence this behavioral loss.

GROUP COMPARISONS

A number of Parkinsonian patients and normal subjects were studied, and from the entire experimental measurement procedure several critical parameters were extracted: the strength in kilograms obtained with vol-untary wrist rotation and the free-wheeling frequency in cycles per second obtained from the frequency-response data described above. These param-eters are presented in Fig. 34 (bottom) and also displayed in graphic form in Fig. 34 (top). From Fig. 34 (top) it can be seen that all three pa-rameters show marked correlation, as well as a striking difference be-tween normals and patients. The 180° phase crossover frequency seems to show fewer false negatives. (In fact, the left hand in patient No. 3 was considered asymptomatic by clinical neurologists.)

TONE EXPERIMENTS

When "tone" of a subject's muscles was tested, as described earlier, we found that care had to be exercised to ensure that the subject was co-operating as well as possible. Experimental records of simultaneously measured hand displacement and resultant resistive force are shown in Fig. 36. We use both sinusoidal and triangular inputs of displacement be-cause of our desire to compare our results with those of previous workers [11, 114]. The most apparent result is the larger amount of resistive force obtained from the patient group. This "rigidity" has long been known as an important part of the neurological description of Parkinson's syndrome.

NORMAL SUBJECT

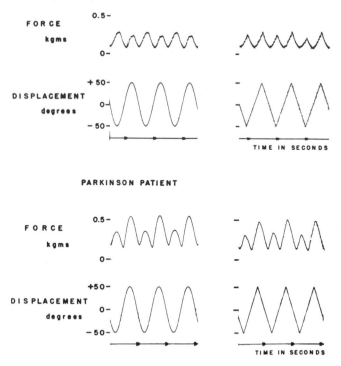

Fig. 36. Effect of Parkinson's syndrome on tone. Note in-phase relationship of displacement and resistive force thus identifying Parkinsonian rigidity as an elastic force.

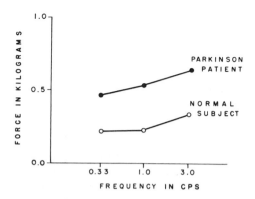

Fig. 37. Minor effect of test frequency on resistive force, or tone.

Previous theories have ascribed this either to a viscous resistance in the muscles or to an analogous viscous resistivity arising from some neurological reflex reaction. From inspection of Fig. 36 it can be seen that the force is in phase with the displacement. Thus, rigidity is entirely an increase in elastic force. This is also true for the smaller resistive forces found in normal subjects.

Further evidence for this comes from the experiment shown in Fig. 37 in which it is seen that the rigidity force is not highly frequency-dependent as would be expected if it had a viscous component.

The model for the control of normal movement and posture discussed in this report and its explanation of Parkinson's rigidity require this resistive force to be an elastic one. This is so because according to our hypothesis, a postural reaction of the stretch reflex to a displacement error in the length servo results in a force proportional to error and thus to displacement.

The elastic force clearly demonstrated in these tone experiments constitutes evidence in favor of the model.

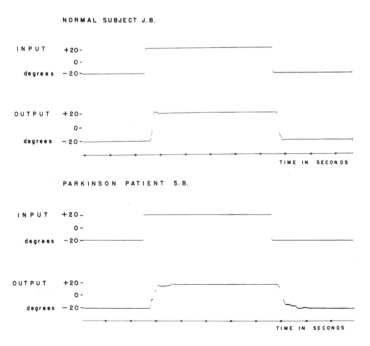

Fig. 38. Effect of Parkinson's syndrome on step responses.

STEP RESPONSES AND TREMOR

Certain time-function records, although they have not yet been used in further analytic studies, are of value for a qualitative description of Parkinson's syndrome. Figure 38 shows responses of a normal subject and a Parkinsonian patient to a step input. The slowed irregular response of the patient correlates with the diminished frequency response and lower free-wheeling frequency described above.

The normal subject is able to shut off his postural servomechanism when making a voluntary movement. This occurs so rapidly that its site of action is most likely at or near the alpha motor neuron. During voli-tional movement the Parkinsonian patient, unable to reduce the gain of his postural servo, must share control of the muscles with contradictory con-trol of the same muscles by the postural servomechanism.

Parkinson's syndrome is a more complex disorder than can possibly by explained with our present knowledge [85]. For example, tremor still eludes the neurophysiologist, although the presence of this clear oscilla-

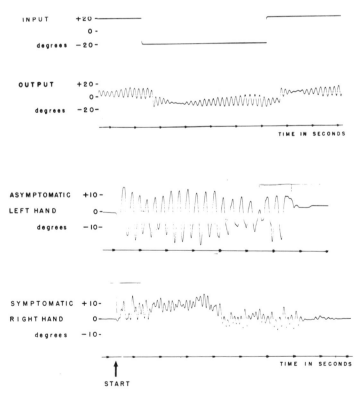

Fig. 39. Interaction of tremor with step response (upper) and tremor development with free wheeling (lower).

tion is tantalizing. In addition to the presence of the rest tremor, the interaction of this tremor with volitional movements is unpredictable and erratic. Figure 39 (top) shows an instance of tremor superimposed on a step response. The response of the patient is slowed and inaccurate. Surprisingly, after the movement is completed, the tremor dies down for approximately 1/2 sec. On the other hand, as shown in Fig. 39 (bottom), tremor can develop during the performance of a voluntary movement. Here is contrasted free-wheeling of the patient's asymptomatic left hand with the development of tremor in the right hand at approximately twice the frequency of the initial abortive free-wheeling attempts. When the subject no longer attempts to free wheel, the tremor dies down. Professor Robert Schwab [85] of the Massachusetts General Hospital has shown further difficulties in initiating complex movement which are in addition to the defects ascribable to simple rigidity.

CONCLUSIONS

The set of experiments described here attempts to study the defect in movement coordination present in patients with Parkinson's syndrome from the viewpoint of servomechanism theory. It is shown that by taking advantage of known anatomical, physiological, and neurological information, and by organizing and interrelating this information from the viewpoint of servoanalysis, one can obtain a consistent model that predicts these important and central phenomena: (a) increased elastic resistance, known as rigidity; (b) decreased frequency response of voluntary movement demonstrable in tracking experiments; (c) deficit in performance of succession movements; and (d) decrease in voluntary strength.

Chapter 4

Transient Response Dynamics of Motor Coordination System in Man

INTRODUCTION

The physiological model of human operator adaptive control of two competing control mechanisms has been suggested by reviewing the neurological and neurophysiological evidence. One mechanism is the postural control system, containing the muscle, motor nerves, the proprioceptive spindle stretch receptor, and bias control of spindle receptor. Competing with this is the proprioceptively open–loop voluntary control system with discontinuous properties (when the input signal is unpredictable). The simulated model, programmed on a digital computer, has been presented in Chapter 1 of Section V.

Previous quantitative descriptions of the human motor coordination rely to a considerable extent upon frequency–response data [23, 58, 99] both because of the ability in those steady–state experiments to maintain the control system in a quasi–stationary and quasi–linear state and because of the power of analytic mathematical methods in the frequency domain [90]. Our aim is, however, not only to approximate accurately some analytical expression of the behavioral characteristics of the system, but also to dissect and understand the underlying neurological and physiological mechanisms that are the component elements of human movement control. The time domain often shows more clearly and rapidly the changing characteristics of the movement control system as it adapts to different experimental operating conditions. Also, certain interesting discontinuous or sampled data properties become apparent [69, 100, 123] and time domain analysis is often preferred.

We [101] therefore used a transducer modified so as to handle transient visual and mechanical inputs, to quantitate the steady tension level,
and to record both responses in the form of wrist rotation and as directional angle of horizontal gaze of the eyes [46, 79, 105]. Predictability of
the target position was controlled, and eye and wrist movements were compared in order to elucidate their respective roles in visual motor tracking.
The use of visual and mechanical impulses enabled us to contrast the voluntarily controlled movement with the damping following it. This following
phase was shown as similar to the position regulation response to mechanical impulse disturbances.

EXPERIMENTAL METHOD

The method of studying the rotation of the wrist of an awake cooperative subject has been described in an earlier paper [99]. Certain modifications were made in the apparatus as shown in Fig. 40 in order to adapt
it for rapid transient visual inputs, and to permit mechanical impulses to
be injected [46]. Accordingly, a mirror galvanometer was employed for
visual target generation, which has a frequency response flat to 80 cycles per
sec when set for critical damping and which was linear over a range of
±25° (Fig. 41 top and bottom). A wide range of stimuli could thus be employed. Impulses, pulses, and steps, all with unpredictable times of occurrence, were contrasted with successions of regular square waves of
target position, in addition to single sinusoids, and sums of incommensurate sinusoids, the pseudo-random signal previously described [99]. The
response of the subject, rotation of the wrist, was measured by a precision

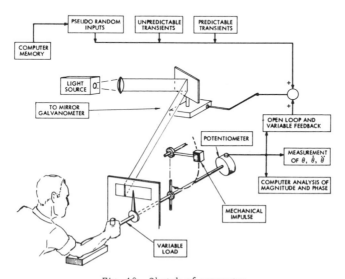

Fig. 40. Sketch of apparatus.

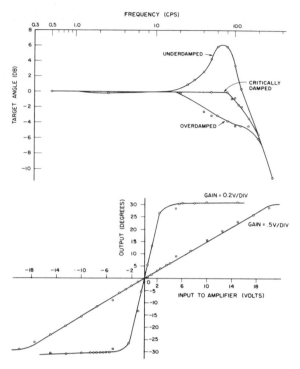

Fig. 41. Calibration curves for mirror galvanometer:
(top) frequency response, (bottom) linearity range.

continuous (infinite resolution) potentiometer, and, in addition to this angle
e, we could obtain $e^{\cdot}$ and $e^{\cdot\cdot}$ by differentiating. Figure 42 shows some
sample records. Most of the experiments were recorded on two- or four-
channel pen-writing units and analyzed by hand, although in recent experiments an on-line digital computer has been utilized [103].

A further addition to the apparatus was a mechanical pendulum whose
position was monitored by another potentiometer. It delivered impulses of
torque to the handle shaft and thus to the subject's wrist. Travel and rebound angles, length and weight of the pendulum, and elasticity of impact
determine input energy injected into the system.

For measuring voluntary tension, height of the mercury column controlled by a sphygnomanometer cuff attached to the forearm appeared to
be superior to integrated electromyogram and was used in experiments
reported here. With this objective measure, the subject could control his
voluntary tension to any one of five levels with ease and reproducibility.

Since visual targets are used and features of eye movement control
system are known to be characteristic of a sampled-data control system

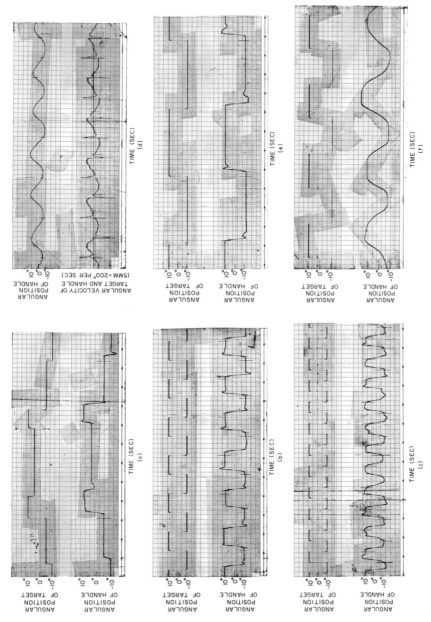

Fig. 42. Responses to random steps and regular square waves.

[120, 122, 123], a number of experiments comparing eye and hand movements were performed. Direction of gaze of the subject was monitored by
a pair of lamps and photocells connected in a bridge so as to produce an
output voltage proportional to differential area of the sclerae on either side
of the iris. The method has been described in several papers on tracking
movements [79, 100, 105]. Frequency-response characteristics of the
photocells limited exact dynamical description of eyeball movement; this
was not an important factor in the present experiments. Even so, these
frequency limitations could be removed by changing the photocell type from
solid state photoresistors to vacuum tube photoelectric cells [102].

EXPERIMENTAL RESULTS

Square-Wave Experiments

The importance of sufficiently controlling those characteristics of
the input signal that make it either predictable or unpredictable has been
experimentally demonstrated for both the hand and the eye tracking control systems [99, 107]. The following experiments were designed to demonstrate this phenomenon further and to study some of the interactions
between hand and eye movement.

The experimental records such as those of Fig. 42 show a subject
responding to an unpredictable step change in target position. In addition
to time functions as shown in Fig. 42, several other displays of system
behavior were obtained. These include sequential patterns of response
times as in Fig. 43, histograms of response times in Fig. 44, and median
response times as a function of frequency of input square wave seen in
Fig. 45.

With unpredictable steps, or with slow (0.5 cycle/sec) regular
square-wave patterns when the period between steps is apparently too long
for predictions to be attempted, response times show an average value of
approximately 0.25 sec [Figs. 42a, e, and 44 (upper and middle)] although

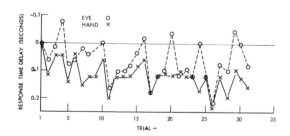

Fig. 43. Sequential order of response time delays
for both eye and brain in simultaneously recorded
experiment with input at 0.5 cycle/sec.

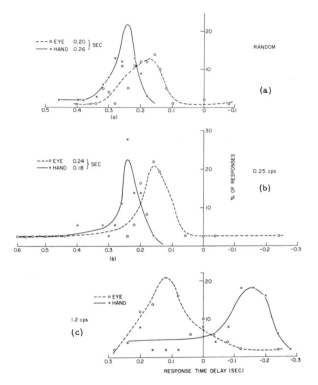

Fig. 44. Histograms of response time delays for hand and eye: (top) random target, (middle) regular square waves at 0.25 cycle/sec, (bottom) regular square waves at 1.2 cycles/sec.

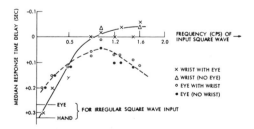

Fig. 45. Dependence of median response time delays upon input square wave. Solid line shows wrist tracking; broken line, eye tracking.

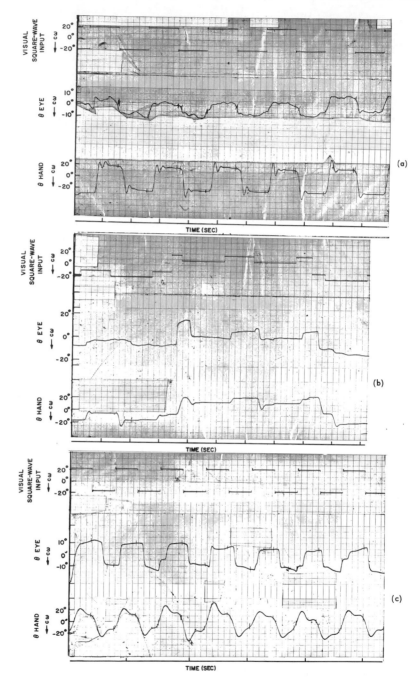

Fig 46. Eye and hand steps responses to random (b) and regular target waves (a) 0.5 cycle/sec and (c) 2.5 cycles/sec.

this value fluctuated considerably with different subjects and in different runs. When time interval between target jumps resulting from the repetitive square-wave patterns is not too long, and if dynamical characteristics of the mechanical system — muscles, leads, apparatus — permit rapid following movements, then the responses often show effects of prediction. Prediction can be observed most simply as a diminution of individual response times as in Fig. 42b, c, and f, as well as in the median response times, shown in Fig. 44 (bottom).

The subject may so anticipate the target jump that his response has a negative response time. Further, the negative response time may be large enough to permit the subject to arrive at the new target location before the target does, even after allowing for time of actual movements; this is termed "overprediction" [107]. The sequential pattern of response times shows that there is no consistent trend effect, but rather an irregular series of delays (positive response times), reduced delays, and slight anticipations (predictions), and larger anticipations (overpredictions). The dependence of median response time on frequency of repetitive square-wave pattern is shown in Fig. 45. At the higher frequencies prediction is the rule. Eventually, a point is reached where the dynamics of the wrist and unloaded apparatus prevent accurate determination of response time even with the use of derivative of input and output as demonstrated in Fig. 42d.

While performing these experiments, similar ones were being carried out in our laboratory on eye movements [107, 120], and it appeared that the hand system could operate in a higher frequency range than the eye movement system, suggesting that eye tracking is not a prerequisite for hand tracking. Observation of the subject's eyes during hand tracking indicate that eye movements occur when the hand is tracking a square-wave target motion with a frequency of 1.3 cycles/sec and an amplitude of ±10° or ±20°. However, the hand is still able to track at 2.5 cycles/sec, while the eye appears stationary.

EYE-HAND COMPARISONS

Because of the large variation in response of both the hand and the eye systems with different subjects, with minor modifications of the experimental apparatus, and with time, we decided to measure both of these system responses simultaneously.

The eye muscles have considerable power with respect to their constant load, the eyeball, and show faster rise times than the hand; especially, as shown in Fig. 46c, when tracking rapidly alternating signals. Examples of eye and hand responses both recorded simultaneously to irregular steps and slower regular steps are also illustrated in Fig. 46; for these inputs the eye has shorter response times than the hand (see also

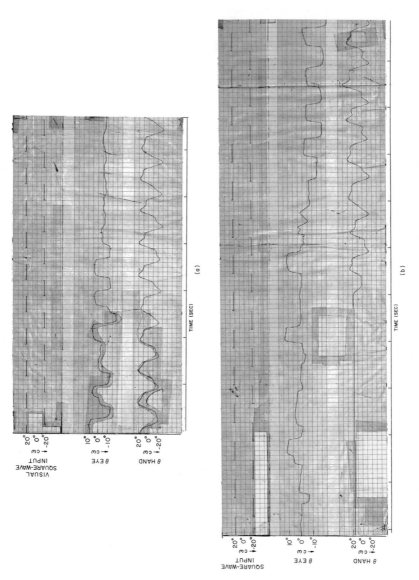

Fig. 47. Time function of target, eye, and hand position: upper shows eye tracking movement sequence spontaneously halting with no apparent effect on hand tracking; lower shows poor to absent eye tracking when hand is still and marked improvement of eye tracking association with hand tracking.

Fig. 42a, b, and Fig. 45). At moderate frequencies (0.7 to 1.0 cycle/sec) the hand develops prediction faster and to a greater extent than the eye. At higher frequencies (1.2 cycles/sec) the hand shows considerable prediction, while the median eye response time starts to lag [see Figs. 44 (bottom), 45, and 46c], in spite of the higher frequency characteristics of the actual movement dynamics of the eye.

At low frequencies, there is some correlation evident between eye and hand response times, as is shown in Fig. 46. However, at high frequencies, the eye may spontaneously stop moving without noticeably interfering with hand tracking, as clearly demonstrated in Fig. 47a. The target swing was ±20° in this experiment. The triangular points in Fig. 45 are data from wrist tracking movements while the eye was held to prevent motion, again confirming the lack of dependence of hand movement upon eye tracking movement under our experimental conditions. Conversely, hand movement clearly and consistently seems to aid the eye tracking movement control system. Figure 47b shows a typical example of improved eye movement performance when hand tracking occurs. Similarly, deterioration of performance is noted when hand tracking is stopped as a result of an instruction to the subject.

IMPULSE RESPONSE

The following impulse experiments were performed in order to attempt to quantitate the mechanical state of the position control system of the wrist and the ability of the voluntary tracking system to reproduce impulses of visual target motion.

The mechanical system of the pendulum, apparatus, and forearm are represented in the mobility analog of Fig. 48. Use of equivalent second-order systems for the apparatus and the wrist, with all elements in parallel, permits one to subtract the equivalent admittance values of the machine from the respective values of the total response.

The mechanical impulse response $[h_m(t)]$ of the wrist seems to be well fitted by a simple underdamped second-order equation as shown by the examples in Fig. 49. Approximate values found for the equivalent mechanical parameters of wrist rotation are: inertia: $J = -0.5 \times 10^{-4}$ newton-m-sec^2-radians^{-1}; viscosity, $B = 1.5 \times 10^{-3}$ newton-m-sec-radian^{-1}; and elasticity, $K = 2.0 \times 10^{-2}$ newton-m-radian^{-1}. The negative value for inertia is the result of subtracting two larger values and is not believed to be significant. These parameters of $h_m(t)$ will be discussed in further detail in the next section when their values as a function of tension are studied.

The characteristics of the response $[h_v(t)]$ of the wrist to an impulse of visual target movements are more complex, as shown in Fig. 49b. We postulate that $h_v(t)$ consists of two phases: the first, consisting of the time

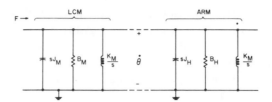

Fig. 48. Mobility analog of mechanical parameters of apparatus and hand under experimental
conditions.

delay and the first overshoot, is considered a visually driven response.
After a transition period, the second or "follow-up" movement, $h_f(t)$, is
considered controlled by the postural reflex loop attempting to bring the
hand to rest. Thus it is only this second portion, $h_f(t)$, that is comparable
to the mechanical response. However, only 30% of the $h_v(t)$ records yielded
$h_f(t)$ portions that could be reasonably described by a second-order fit, and
even these data show considerable scatter.

EFFECTS OF TENSION ON IMPULSE RESPONSES

Striking effects were obtained in free-wheeling experiments as shown
in Fig. 2 with change in set of the subject, especially with change in tension
set to control expected impulse disturbances that occurred randomly [99].
These effects agreed with the model proposed to account for the rigidity
of Parkinson's syndrome, and in order to further quantify the behavior of
the control mechanism for movement and posture, the impulse experiments
were conducted at a variety of levels of tension.

As explained in the methods section, a sphygnomanometer cuff was
used because we obtained thereby a reproducibility of tension level which
was felt to be superior to subjective open-loop estimation, or indeed to
rms value of rectified and integrated electromyographic activity. Five
different levels of tension produced smoothly graded response characteristics from most relaxed to most tense states.

The most striking effect was the decrease in amplitude of $h_m(t)$, the
mechanical impulse response with tension; this was accompanied by an increase in ringing frequency, as can be seen by comparing a and c in Fig.
49. Several hundred experiments on about six subjects showed similar effects and Fig. 50 shows a plot of the variation of the equivalent mechanical
parameters J, B, and K as a function of tension. K is a strong function of
voluntary tension, while B increases only slightly with increase in tension;
as before, values of J are small; they are the results of subtracting two
large values, and are not believed to be significant.

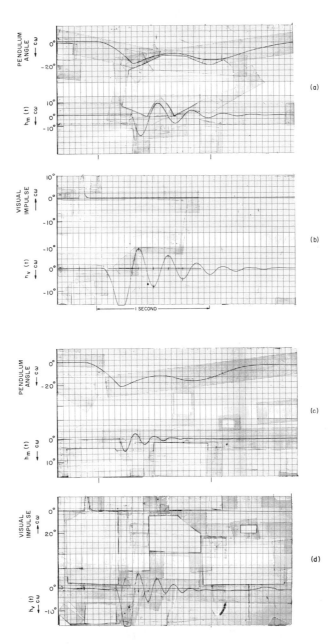

Fig. 49. Impulse experiments.

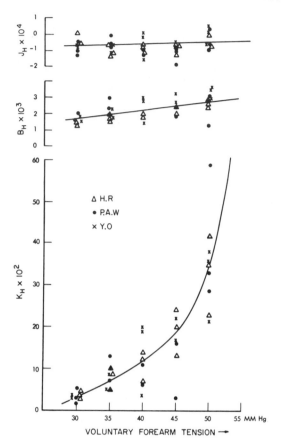

Fig. 50. Mechanical parameters of hand as functions of tension. J_H is rotational inertia in new-
ton-meters-second^2-radians^{-1}; B_H is viscosity in
newton-meters-second-radian^{-1}; and K_H spring
constant in newton-meter-radian^{-1}.

    The poles of the equivalent second-order system are at $S = -\gamma \pm j\omega_d$
where γ is real frequency and ω_d imaginary frequency in the complex plane.
The effect of tension on the system is shown in Fig. 51 by the path of migration of upper half plane poles. The principal effects of an increase in
voluntary tension on the $h_m(t)$ equivalent second-order system are an increase in absolute bandwidth and a slight decrease in stability margin.

    As discussed in the preceding section, the response $h_v(t)$ of the eye-
forearm rotation system to a visual impulse is considerably more complex
than the response $h_m(t)$ to a mechanical impulse; accordingly we have separated the visual driven initial portion from the follow-up portion $h_f(t)$.
However, Fig. 49b and d shows that the same qualitative changes appear

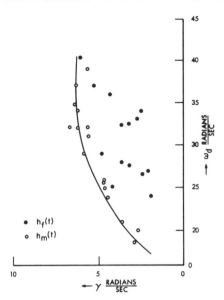

Fig. 51. Root locus of positive pole of a
complex pair fitted to $h_m(t)$ and $h_f(t)$.
Scales are imaginary, frequency ω_d, and
real frequency in radian-second^{-1}.

with increase in tension. When the $h_v(t)$ curves are parameterized as to
(a) number of ringing cycles, (b) average ringing frequency, (c) amplitude
ratio, and (d) time delay, all as a function of tension, it can be noted (see
Fig. 52) that similar changes (b), or lack of change (a) and (d), occur with
both $h_m(t)$ and $h_v(t)$ responses. Time delay is, of course, zero for $h_m(t)$.

When J, B, and K parameters are determined for $h_f(t)$ portions of
$h_v(t)$ and the pole configuration computed, it is seen that the pole path as a
function of tension is rather similar for $h_f(t)$ and $h_m(t)$ (Fig. 51). Considerable scatter in $h_f(t)$ poles can be noted. The principal correlation of the
$h_f(t)$ equivalent second-order system with increase in voluntary tension is
increase in absolute bandwidth. Little change in amplitude or shape of the
initial portion of $h_v(t)$ or in the time delay is seen as a function of tension.
It was felt that unpredictable changes of target position were not fully accomplished in this experiment, as the target pulse, although irregular in
time of appearance, was always of constant width, amplitude, and duration.
When these restrictions were removed, certain quite interacting phenomena appeared relating to earlier studies on psychological refractory period [108] and to sampled-data approaches to these neurological control
systems.

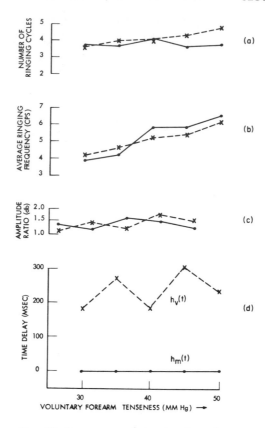

Fig. 52. Parameters of visual and mechanical
impulses as function of tension.

SAMPLED-DATA CHARACTERISTICS

Recent studies have suggested that the eye movement tracking system can be treated as a sampled-data system [120, 122, 123]. For the purpose of this paper it is sufficient to realize that the discrete nature of
sampled-data systems is in some sense similar to the notion of a refractory period, a common phenomenon in neurophysiology. Other discontinuous control systems, such as quantized systems, have properties in common with sampled-data systems, and indeed the hand movement system
demonstrates characteristics of quantization as well [100].

When slowly moving ramps are used as input target signals, and the
subject is tracking this input with a device that offers very small mechanical impedance, a response that is similar to that of Fig. 53 is seen. The
output response consists of steplike changes in position which occur with
irregular intervals and amplitudes. The output signal rotary motion trans-

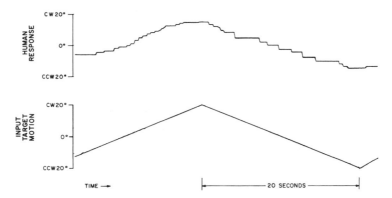

Fig. 53. Discontinuous change of hand position to slow target ramps.

ducer had an effectively infinite resolution, and neither friction nor any other component of the mechanical impedance of the transducer was of sufficient magnitude to play an important role in the dynamics of the system. Early studies utilizing this input suggested the possibility that the known sampled-data properties of the eye tracking system were possibly related to these steplike responses, but recent experiments have failed to confirm this conjecture.

When pulses of varying width are used as input target signals and are presented irregularly in time, a response that is similar to that shown in Fig. 54a is obtained. There are delays of approximately 150–250 msec before the rapid response motions occur for both leading and trailing edges of the input target motion. This delay in response has several components in addition to nerve conduction time and is sometimes called the "psychological refractory period" [108]. When a pulse of extremely narrow width is supplied unpredictably as an input target motion, a normal delay occurs before the response movement to the leading edge of the input motion, as shown in Fig. 54b. However, the response motion to the trailing edge of the input motion has a much prolonged delay, 400 msec in the case illustrated in Fig. 54b. This prolonged delay can be accounted for as a normal refractory period that starts after the initial response motion, rather than being triggered by the trailing edge of the input target. This behavior would be characteristic of a sampled-data system. Contrast this widened response to a short pulse with the rather narrow responses of Fig. 49b and d. Here the unpredictability of the pulse was only in time of occurrence; the subject knows that any disturbance would be a short pulse and accordingly preprogrammed a double sequence of movements to match the short pulse when triggered by the appearance of a disturbance. Similar evidence has been shown for the eye movement control system, in which, because of

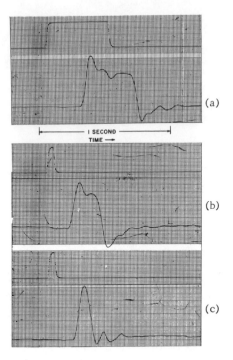

Fig. 54. Responses to unpredictable
wide pulse, unpredictable narrow pulse,
and predictable narrow pulse showing,
respectively, delay, refractory period,
and input-adapted preprogramming.

the relative power of the eye musculature compared with the low mechanical impedance of the eyeball, the load and muscle dynamics are negligible
compared to the shaping of the response wave formed by the sampled-data
control dynamics [120, 122, 123].

For a motor coordination sampled-data system with the transient responses shown in Figs. 53 and 54, the frequency response to wide bandwidth inputs could demonstrate a peak at one-half the sampling frequency,
or at approximately 2−3 cycles/sec, as well as an absence of coherent
tracking characteristics at frequencies greater than this peaking frequency. There exists very little experimental data that clearly support the occurrence of this suggested peaking frequency, since most of these experiments have used input bandwidths with either much lower frequencies than
3 cycles/sec or else too little power at these higher frequencies to obtain
effective responses necessary to demonstrate the peak. In fact, Bekey, in
the most complete study of sampled-data properties of the manual track-

ing control system, was able to show peaking at sampling frequencies only in spectral curves of error [8]. However, Stark, Iida, and Willis [99] have described steady-state frequency-response experiments that were not limited by these undesirable input spectrum characteristics (see Fig. 29).

This clearly defined peak in the response spectrum supports the transient data and the idea that the human motor coordination system can be treated as a sampled-data control system.

DISCUSSION

The first steps of experiments using unpredictable and predictable step input target motions represent a further attempt to dissect the effects of the predictive apparatus from the behavior of the underlying "neurological" system; previous experiments with single and multiple sinusoidal inputs enabled a similar dissection [99].

Processing retinal information is clearly an early operation in hand tracking a visual target. This same information is necessary for eye movement control. It was therefore of interest to compare and contrast the eye movement control system with the manual one, and this was done with somewhat surprising conclusions.

First, the manual system can track rapidly alternating target movements of a repetitive predictable nature to an octave or two higher frequency (see Fig. 54), in spite of the much lower bandwidth of the output elements. The hand has a natural frequency, ω_n = 40 radians/sec, while the eyeball has a natural frequency, ω_n = 240 radians/sec; this is a reflection of the considerable power of the eye muscle with respect to the eyeball as compared with the power of the wrist and forearm muscles with respect to the more nearly matched load of the hand. In spite of this six-fold less bandwidth of output elements, i.e., muscles and load, the hand is able to track at higher frequencies than the eye. Evidently the control and predictive apparatus for manual control is more effective than that for eye movement control.

Second, eye movement is not necessary for hand movement in spite of the rather large amplitude ($\pm 10°$ and $\pm 20°$) tracking signals in our experiments. Clearly, focal fixation of the target is not required for the order of accuracy obtained in these experiments.

Conversely, the physical movement of the hand appears to reinforce the target signal input to the eye movement system so that adequate eye tracking may occur in borderline regions if the hand is tracking and may not occur if the hand is still. This evidence supports the contention that what we are studying is manual control dynamics and not some limitation secondary to eye movement control. Of course, visual processing is still an integral part of our experimental conditions; it could be eliminated by proprioceptive tracking (one hand passively moved and the other tracking this motion), or perhaps by auditory input signals. We now turn to consideration of the "neurological" control mechanism.

The neurological control system for motor coordination has been discussed in terms of two signals competing for control of the output elements [91]. These output motor elements are the "final common path" of Sherrington, the alpha motor neurons and the muscles they control. The first of the competing signals is that carried by the feedback path of the position control system, the Ia afferent neurons from the spindle receptor system. The second of the competing signals is that carried by the corticospinal pathways' subserving voluntary movement.

The two types of impulse inputs presumably test the state of the shared control of the alpha motor neuron. The mechanical impulse response, $h_m(t)$, has no associated time delay and thus tests the gain and dynamics of the postural or position servo loop. As such it is similar to the classical tendon jerk of the clinical neurological examination. This system has been studied by simulation techniques recently [96] and our representation includes two important nonlinear dynamical systems plus time delays for nerve impulse conduction and a variety of saturation and asymmetrical elements. That this can be adequately represented by an undamped, linear second-order system is indeed a tribute to the power of these simple analytic tools. The apparent spring constant of the hand is determined by the gain of the position servo, since restoring force is proportional to displacement error. An important nonlinearity of the spindle is an increase in slope of the input-output curve with increased input, either by external stretch or by gamma motor neuron stimulation of the intrafusal fibers [30, 96]. As a consequence, increased spring constant is predicted in the experiment with increasing tension.

The visual impulse response, $h_v(t)$, appears to be a time-delayed preprogrammed control signal, probably corticospinal in route, which is followed by a return to the postural control system with dynamics predictable by the $h_m(t)$ to the experiments. The similarity of the dependence of $h_f(t)$ and $h_m(t)$ on tension suggests this latter. Evidence for the ballistic preprogrammed input comes partly from the marked disparity between the second-order model of the postural system and the initial portions of the $h_v(t)$. Even more striking are the results obtained when the duration of the pulse is made unpredictable, as in the sampled-data experiments. Here a clear-cut refractory period of about 200 msec is demonstrated, certainly unrelated to the second-order dynamics.

The free-wheeling experiment shows the competition between the voluntary control and the postural servomechanism. When tension is increased in the mental set situation where the subject is awaiting a random disturbance, the frequency of oscillation in successive movements is markedly diminished; the continual activation of the postural servo means that antagonists are of course active and degrade performance. However, little evidence for reduction in amplitude of voluntary tracking or for increase in response time is noted in these somewhat predictable impulse experiments. For this to be demonstrated, carefully controlled experiments must be performed with relatively quick process identification computation available to aid the experimenter.

The sampled–data experiments are rather convincing on the basis of the prolonged pulse response, the peak in frequency response, and the irregular positional corrections for slow ramps. These lines of evidence have already been extended and will be the subject of a separate paper [69]. The present results are especially relevant to the problem of eye and hand interaction. It has been firmly established that the response to unpredictable inputs by the eye tracking control system can be well predicted by a sampled–data model [120, 122, 123]. However, such phenomena as the irregular multiple–step responses to a slow ramp are independent of visual tracking. Further, the sampled–data control system for the eye appears to be both position and velocity control while the manual tracking sampled data seems to be only a position control system [7], thus indicating rather different origins of the control signals to these two complex motor outputs.

One point of interest is the distribution of response times for hand tracking of unpredictable targets [Fig. 44 (top)], a narrow quasi–Gaussian distribution with a mean of 0.24 msec. This indicates that the sampling times are not clock–driven, but rather input–synchronized with some random deviations from mean response time [69]. If they were clock–driven one would expect a rather square distribution from one sampling time to two sampling times with perhaps some additional random variation around this square distribution. The clock–driven distribution predicted by Lemay and Westcott [59] is clearly not found under our experimental conditions.

SUMMARY

An experimental transducer similar to that used in frequency–response studies has been modified in order to define characteristics of transient response dynamics.

The importance of distinguishing between behavior with predictable inputs and with unpredictable inputs is demonstrated.

Since the visual system is part of the tracking situation, the control of eye movements and hand movements was studied by comparing simultaneous measurements. It was seen that while the load dynamics of the eyeball were less restricting than the inertia, viscosity, and elastic resistances of the wrist, the hand was able to follow repetitive square-wave patterns of high frequencies (1.5−2.5 cycles/sec) better because of its wider bandwidth control apparatus. Further, it was apparent that eye movement did not aid hand movement, but rather that hand movement helped eye movement at high frequencies.

Mechanical impulse responses, $h_m(t)$, testing the postural control system were compared with visual impulse responses, $h_v(t)$, testing the voluntary tracking system. The former is well approximated by an equivalent second-order underdamped system. The latter has two components, a visually driven initial response and a follow-up portion $h_f(t)$; only the $h_f(t)$ is at all able to be approximated by a similar underdamped second-order system.

On changing steady background levels to tension both $h_m(t)$ and $h_f(t)$ show dynamical changes in the direction of increased bandwidth with increased tension.

Finally, certain experimental evidence relating to sampled-data properties of the hand movement system is put forward.

Chapter 5

Sampling or Intermittency in the Hand
Control System

INTRODUCTION

In the studies described in this chapter we have examined the control system for human hand movement, using the approaches of several different disciplines. In studying the performance of the system, its limits, and possibilities as a human engineer would, we have used a tracking situation and complex input to obtain differing frequency response. We have utilized known facts and data from the physiological literature, and also have simulated muscle and muscle spindle dynamics to define the blocks within the total motor control system of man. As neurologists we have organized the "topology" of these physiological elemental blocks into a model in a manner that seems compatible with actual neurological organization as far as it is known. To further refine and assess our model, we have utilized studies in persons with neurological defects, such as Parkinson's disease, and have studied the effects of drugs producing temporary paralysis of gamma efferent fibers. Finally, as control engineers interested in neurological control systems, we have used special experimental techniques such as opening loops by clamping methods in order to expose certain essential characteristics of this elegant feedback control system [70, 100].

Some of the characteristic features of the model, described below, have been derived from physiological information, some from our experimental results, and some remain untested assumptions.

After a description of the experimental equipment, including transducer and analog and digital computing equipment, the experimental section describes experimental results from four types of inputs: closed-loop

369

transients, open-loop and variable feedback transients, closed-loop pseudo-random signals, and closed-loop predictable signals.

In the discussion section our model is evaluated and our work related to previously published research in this field.

Finally, in the conclusion we attempt to summarize the present status of our knowledge of the human control system for hand movement.

THE MODEL

The adequacy of representing the hand movement system as a feedback control system has been shown by various human engineering investigators [23, 58, 60, 65, 66, 110]. Further, on the basis of our own experimental results such as short pulse (Fig. 54) and ramp responses (Fig. 55) as well as the findings of others [8, 38, 113], we decided that some kind of intermittency should be included.

The model to be presented has a number of characteristic features, each of which will be discussed in detail, justified by experiment where possible, and related to results in the literature in our discussion.

For unpredictable inputs, the human tracking system acts as a position feedback control system, having important discontinuous properties in the frequency range of interest. Its intermittency is related to an interesting control of the muscle by ballistic voluntary inputs and by a stabilizing proprioceptive reflex loop. The actuating signal of the system includes a combination of error and input (pursuit tracking) which compensates for some of the inherent delays. For predictable inputs, different continuous models which include ideal predictors are applicable.

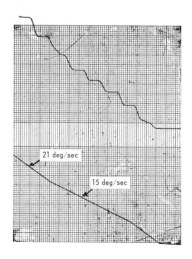

21 deg/sec

15 deg/sec

Fig. 55. Intermittent response to unpredictable ramp inputs.

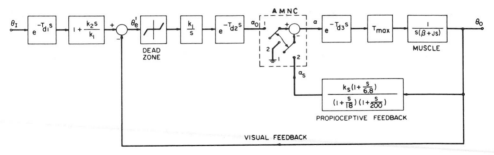

Fig. 56. Model for pursuit tracking of unpredictable inputs.

The model for pursuit tracking of unpredictable inputs is presented in Fig. 56 in block diagrammatic form. Notice (a) the visual feedback path and the integrator in the loop which makes the system a position servo; (b) the additional proprioceptive feedback path; (c) the switch providing an intermittency or quasi-sampled-data action; (d) the alternation between voluntary and reflex action for control; (e) the observation of the input allows for some delay compensation through the lead term $[1 + (k_2/k_1)s]$.

This block diagram necessarily includes elements from the visual, central nervous, and muscular systems.

EXPERIMENTAL METHOD

The experimental arrangement, shown in Fig. 40, includes the "light-coordination machine" and peripheral equipment for the analysis and generation of signals.

The light-coordination machine [46] consists of a linear mirror galvanometer which transforms an electric signal into a display of a light pencil, a screen, a pointer, and an angular potentiometer as the mechanical-to-electrical transducer. The subject tried to follow and reproduce the input by rotation of his wrist.

The handle was very light and had negligible friction; when no extra load was added, it acted as a small inertia load on the arm. Accelerations as high as 400 deg/sec^2 were recorded, and the apparatus required very small forces, much less than 100 g. This was a favorable feature since the apparent viscosity of the muscle increases with the degree of activation.

Peripheral equipment included function generators, filters, operational amplifiers set for signal transformation, recorders, measuring equipment, a pendulum for application of mechanical impulses, and a computer with digital-to-analog and analog-to-digital conversion equipment.

The computer was used for the frequency domain characterization of the
human operator, using programs developed at the Neurology Section at
M.I.T. It produced a combination of sinusoids of arbitrary magnitude and
phase, and made a Fourier analysis of the operator's response to these
signals.

EXPERIMENTAL RESULTS

Closed-Loop Transients (Unpredictable Inputs)

Closed-loop transients included visual pulses and approximations to
impulses, mechanical impulses, visual steps, and ramps. Examples of
typical performances are presented.

Responses to input pulses are shown in Fig. 57. For unpredictable
short pulses there were intervals from 0.15 to 0.30 sec between the two
halves of the response, as can be seen from Fig. 57c and d. This was no
longer true for predictable short pulses shown in Fig. 57e. Notice also
the fairly consistent overshoots in the initial correction, the oscillatory
hunting, and the decrease of velocity after the proper output level is
achieved. The addition of moderate load seems to decrease natural frequency and the acceleration of the arm.

Visual and Mechanical Impulses

Figures 58 and 49 show the responses to visual and mechanical impulses [46, 73]. The visual impulse was a very narrow light pulse of more

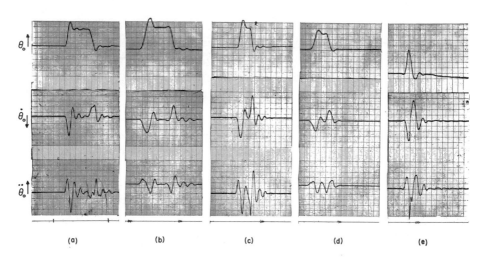

Fig. 57. Responses to pulse inputs: (a) 400-msec pulse; (b) 400-msec pulse, inertia and
friction added to handle; (c) 100-msec pulse; (d) 100-msec pulse, inertia and friction
added to handle.

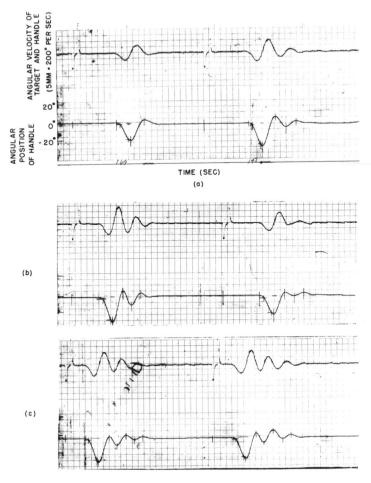

Fig. 58. Visual impulse responses with velocity indicated: (a) relaxed
condition; (b) moderately tense condition; (c) very tense condition.

than 20° and the mechanical impulse was a disturbance applied to the pointer
axis by means of the pendulum. It is of interest to fit these "impulse responses" to a simple model, a second-order system, which thus characterizes the motor end of the control system. The mechanical impulse response looks like a second-order system; the natural frequency for the
relaxed subject is 3 cycles/sec and the damping ratio is 0.15; for the tense
subject the natural frequency is 5 cycles/sec and the damping ratio is 0.20.
The visual impulse response is characterized by two high gain corrections,
and a subsequent part representing an oscillatory decay. The first part of
the response looks like an on-off action. After the first correction the
situation can hardly be called unpredictable.

Step Responses

The responses of the hand control system to unpredictable steps have most of the characteristics already noted for pulse responses, and a few others. Figure 59 shows the main types of responses whose response time varies between 200 and 450 msec. Overshoot in the initial correction is usual, and is followed by oscillations (6−8 cycles/sec). The possibility of two−mode action is clearly seen in Fig. 59b, c, e, and f. Figure 60 shows the output, its velocity and acceleration for unpredictable and predictable steps. The velocity graphs consist of a triangle followed by decaying oscillations, which for the predictable case are of minor importance.

The acceleration consists of a pulse which carries the arm to the desired position, followed by several alternating pulses during a breaking or clamping period. There is no overshoot for the predictable case, and the first acceleration pulse is followed by a deceleration pulse before reaching the desired position, thus suggesting a different action; small acceleration oscillations follow.

Studies of reaction time distribution for step responses have been done by Okabe et al. [74] as shown in Fig. 61. By a statistical argument

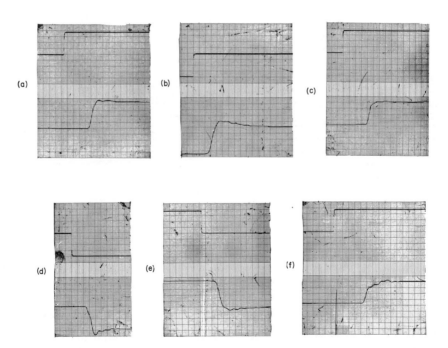

Fig. 59. Six types of step response: (a) 11 cases out of 40; (b) 7 cases; (c) 7 cases; (d) 5 cases; (e) 5 cases; (f) 3 cases. Size of steps: 20°. Time scale: 100 msec/large division.

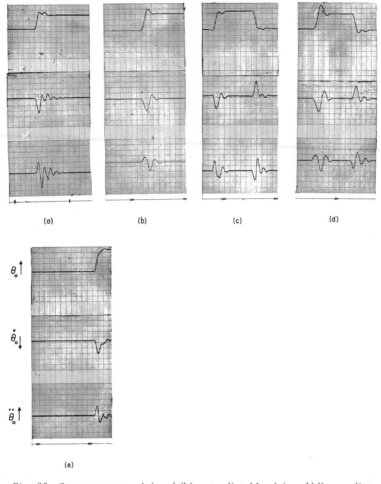

(a)   (b)   (c)   (d)

(e)

Fig. 60. Step responses: (a) and (b) unpredictable; (c) and (d) unpredictable with inertia and friction added to handle; (e) predictable response to known step after training.

these studies can shed light on intermittency in the following ways. For a sampler synchronized to an independent clock, the response time would have a uniform distribution between the delay time alone and delay time plus a sampling period. If the sampler is synchronized to the input, an impulse function–type distribution at the delay time is to be expected, and is approached by the experimental graph.

We gave considerable attention to the response of the operator to ramp inputs, since the sampling or quantization limitations of the hand control system can be determined by testing with a ramp input if the system

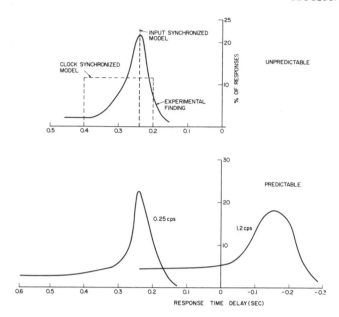

Fig. 61. Reaction time distributions for responses to both unpredictable steps and predictable square waves.

acts as a position servo. No physiological basis for a velocity servo was
known, but we were also interested in finding conditions under which the
hand could exhibit velocity servo characteristics. The ramp was presented
as part of a more complex combination to avoid prediction. Figures 53,
62, and 63 show inputs and responses to ramps ranging from 1.5 deg/sec
to 25 deg/sec. Figures 64 and 65 also present velocities and accelerations.
The main observation that can be derived from ramp experiments is evidence for samplinglike intermittency. Quantatization or dead zone features
are unimportant except for very slow ramps, less than 1 deg/sec. Other
observations include performance deterioration for fast inputs (see Fig.
62f), apparently no velocity servo, and very little interpolation between
samples, except possibly for some high–speed inputs.

Figure 66 shows the quasi–linear rise velocity of the "saccades" as
a function of input ramp velocity. Although rise velocity changes with input velocity, this does not correspond to a velocity servo. Figure 67a and
b, also derived from ramp responses, shows that the size of the saccades
is fairly linear with input velocity, and the rate of repetition of saccades
is 2.5 cycles/sec for most of the input velocity range.

Observation of velocity and acceleration wave forms for ramp responses shows that the series of saccades is like a collection of step re-

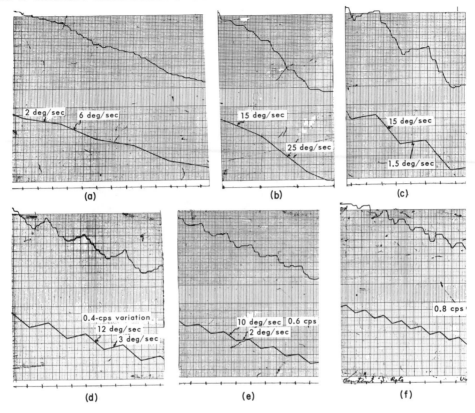

Fig. 62. Responses to unpredictable ramps, handle loaded with inertia and friction.

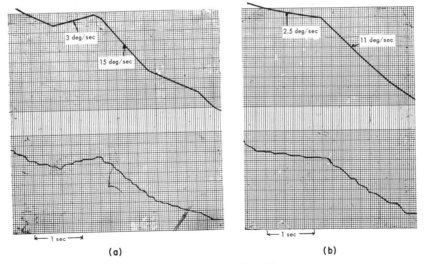

Fig. 63. Responses to unpredictable ramps.

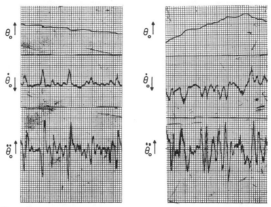

Fig. 64. Ramp responses with output velocity and acceleration indicated. Time scale: 200 msec/large
division.

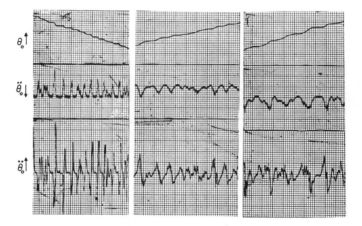

Fig. 65. Ramp responses with output velocity and acceleration
indicated. Loaded handle. Time scale: 200 msec/large division.

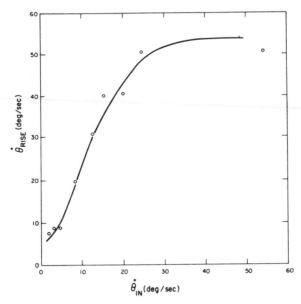

Fig. 66. Quasi-linear relationship between saccadic rise
velocity and input velocity (random ramp input).

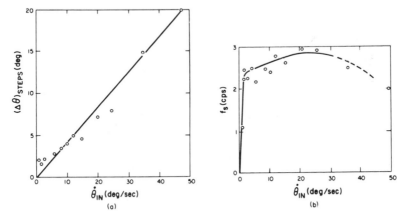

Fig. 67. (a) Size of intermittent response as a function of input velocity. (b)
Frequency of intermittency as a function of input velocity.

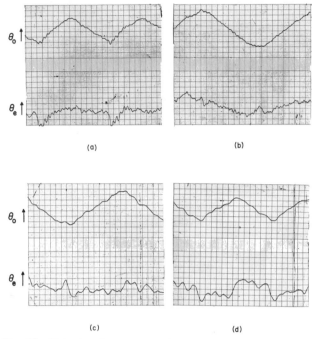

θ_o ↑

θ_e ↑

(a)   (b)

θ_o ↑

θ_e ↑

(c)   (d)

Fig. 68. Output and error, ramp input, tense conditions (a and
b), relaxed conditions (c and d).

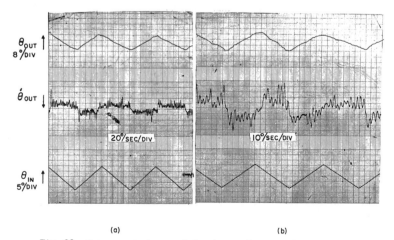

θ_{OUT} ↑
8°/DIV

$\dot{\theta}_{OUT}$ ↓

20°/SEC/DIV   10°/SEC/DIV

θ_{IN} ↑
5°/DIV

(a)   (b)

Fig. 69. Response to known 8 deg/sec and 16 deg/sec ramps.

sponses. Figure 68 illustrates the variation of these saccades from tense to relaxed conditions. Finally, Fig. 69 shows responses to predictable ramps, for contrast.

Open-Loop and Variable Feedback Transients

Open-loop experiments are very useful for the analysis of biological servosystems since the models of these systems are not unique, and an open-loop measurement can show the order of the system and its low-frequency behavior more clearly than a closed-loop measurement. Variable feedback serves to correlate open-loop and closed-loop experiments.

Figure 40 shows the experimental method for open-loop and variable feedback transients. The display becomes:

$$\theta_d = \theta_i + G\theta_o \tag{1}$$

where θ_i is input, and θ_d is hand output. For open loop $G = 1$, and for variable feedback $0 < G < 1$. The error becomes

$$\theta_e = \theta_d - \theta_o = \theta_d\left[\frac{(G - 1)}{G}\right] + \frac{\theta_i}{G} \tag{2}$$

Short pulses and steps rather than ramps were the main inputs considered. Figure 70 shows the responses to (a) 100- and (b) 200-msec pulses. From these experiments we observed that (a) the responses were entirely analogous to closed-loop step responses; (b) the initial corrections were larger than, but proportional to, pulse size; and (c) the final values were related more to pulse area, and were attained after oscillations of the same frequency as in the closed-loop responses.

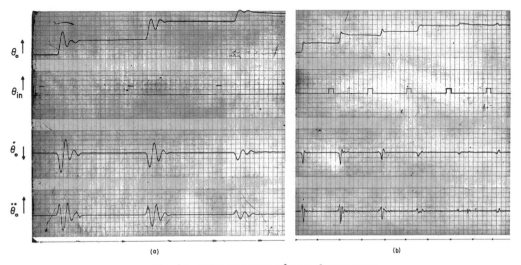

Fig. 70. Pulse responses of open-loop system.

The step responses, shown in Fig. 71, look like a staircase made of saccades at a rate of 4−5 per sec. The ratio of correction to step size is again larger than unity.

Figure 72 is an interesting correlate of these observations. The inputs are repetitive 5 cycles/sec and 2 cycles/sec positive pulses. The integral of this wave form appears as a ramp to the 3-cycles/sec sampler, and the response looks like the open-loop step response and the closed-loop ramp response.

Observations about variable feedback responses, shown in Fig. 73, are included in the discussion.

Continuous Random Inputs

As explained above, the approximations to continuous random inputs were combinations of five to ten harmonically unrelated sinusoids generated and analyzed by a computer. Short runs (20 sec) were used to avoid operator adjustments shown by previous investigators [23, 99, 110] to depend on input spectrum.

Figures 74 and 75 show various input spectra and the pertinent averages of gain and phase frequency response. Of the three different frequency responses, that of Fig. 75 is of the most interest, probably due to the inclusion of high frequencies. Important characteristics are: peaks in gain and related phenomena in phase around 1.5 cycles/sec; decrease in

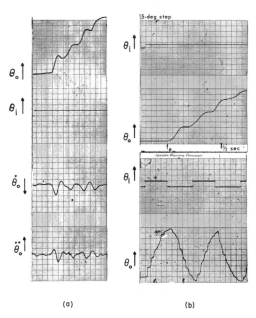

(a)   (b)

Fig. 71. Step responses of open-loop system, loaded handle.

response at high frequencies which can be fitted by a -20 decilog/decade line; increase in gain from 0.5 cycle/sec up; and low-frequency gain equal or less than unity.

Response to Predictable Inputs

In order to obtain a description of performance as a function of frequency, the subjects were asked to track single sinusoids. Magnitude and phase were obtained up to 3 cycles/sec (then the subjects began to act as independent generators). The experimental points and a fit by transfer function are shown in Fig. 76. For this fit to agree with the phase points, an ideal predictor exp(Ts) has to be included, an expected phenomenon for these predictable inputs. The transfer function was

$$F(s) = \frac{e^{+0.15}}{(1 + 0.13s)(1 + 0.053s)(1 + 0.016s + 0.0028s^2)} \tag{3}$$

Figures 57e and 60e refer also to predictable inputs, in the time domain. Figure 69 shows the response to known triangular waves; saccades are absent and only 10 cycles/sec fluctuations show up. All these phenomena will be considered in the discussion.

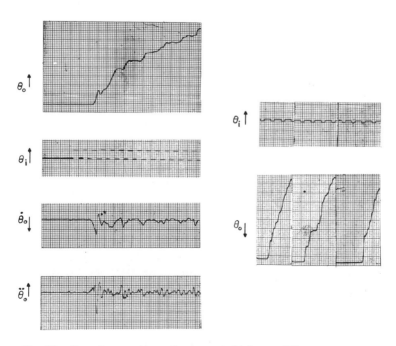

Fig. 72. Open-loop system. Responses to high-repetition square waves; analogous to step responses.

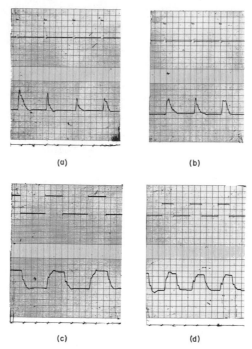

Fig. 73. (a) Pulse response with G = 1/2; (b) pulse response with G = 2/3; (c) step response with G = 1/2; (d) step response with G = 2/3.

DISCUSSION

We have postulated a feedback, position control system (Fig. 56), for the hand movement system, and included an intermittency due to alternation between voluntary and reflex control. We will now discuss the various experiments, observations, and assumptions.

Feedback System

Human engineering investigators like Tustin [110], Elkind [23], and McRuer and Krendel [58, 65, 66] have proposed feedback systems for the hand control. Indeed, the adequacy of such representation depends upon the ability of the operator to correct for disturbances and to follow a visual input up to his dynamic limits. The human operator can apply several criteria to his tracking such as minimization of disturbance or exact reproduction of the input.

Position Servo

We can explain the topology of the model in terms of the behavior characteristics of the human operator for unpredictable inputs. That the

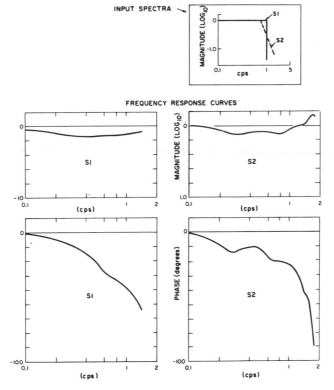

Fig. 74. Response to quasi-random continuous signals.

hand system is essentially a position servo is supported by the following arguments:

1. The final value of the error is zero for any kind of step input. This requires an integrator in the loop so that if a step (K/S) is applied:

$$\text{Final value} = \lim_{s \to 0} \frac{sK}{s} \frac{1}{1 + \dfrac{H(s)}{s}} = \lim_{s \to 0} \frac{Ks}{S + H(s)} = 0 \qquad (4)$$

2. The open-loop responses shown in Figs. 70, 71, and 72 give evidence of the integrator, since a step input produces a (discrete) ramp, and pulse input produces a step output. A discrete ramp results also from fast repetitive steps whose integral, and that of a step half as large, are equal, except for a small ripple.

3. The ramp experiments show the inability of the subject to follow the input when the ramp was unpredictable; the error did not go to zero. The ramp input (V/S^2) is precisely one order higher than the tracking system.

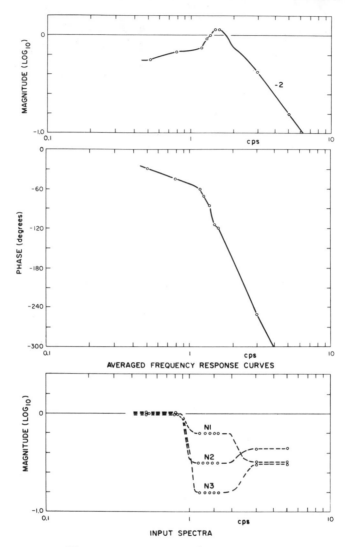

Fig. 75. Response to quasi-random continuous signals.

     Velocity servo characteristics are unimportant since very little in-
terpolation appears between saccades, for ramp inputs (Figs. 53 and 62−
65), and constant velocity sections are zero. In addition, very few corrections occur in the direction opposite to that of target when following combinations of ramps. A combination of position and velocity servo could
originate such corrections, as happens in the eye tracking system [107,
120]. Of striking interest is the inability of the hand servo to follow
smoothly an unpredictable ramp.

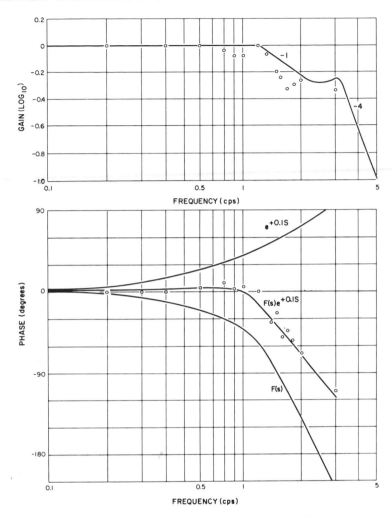

Fig. 76. Response to closed-loop predictable inputs and fit by a transfer function.

The quasi–line dependence of $\dot{\theta}_{rise}$ on $\dot{\theta}_{in}$ (Fig. 14) should not be confused with the action of a velocity servo.

Psychological intermittency and the details of the behavior of the muscle and spindle are considered later in this discussion. We will now consider other features of our model, and review previous models.

Figure 77 presents references for the discussion of compensatory and pursuit tracking configurations, as well as the models proposed by Bekey [8] and Lemay and Westcott[59] for compensatory tracking, and by

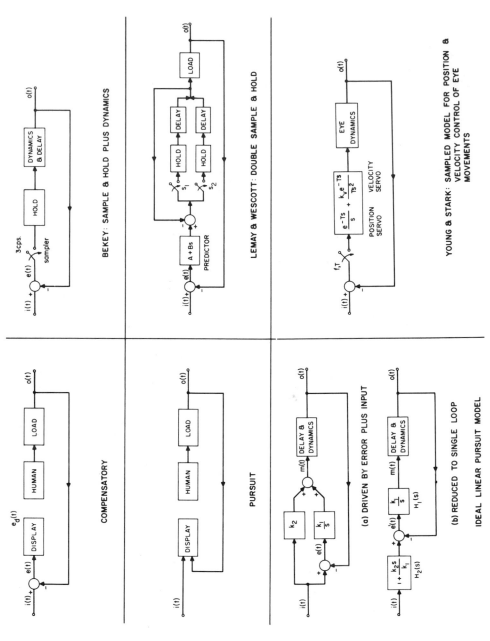

Fig. 77. Other continuous and discontinuous models for hand tracking [8, 59, 123].

Young and Stark [107, 120] for eye tracking. Smith and Cortes [17, 89] proposed an optimal bang–bang system.

In compensatory tracking the error is the only information received by the human; conversely, in pursuit tracking both input and output are perceived. Since the system is a position control system, thus requiring an integration in the loop, we choose the ideal linear pursuit model as a starting point. In such a model the actuating signal is:

$$m(t) = \theta_{\text{oct}} = k_2 \theta_i + k_1 \int \theta_e dt \qquad (5)$$

and

$$e^1(t) = \theta_e^1 = \frac{1}{k_1} \frac{dm}{dt} \qquad (6)$$

We attribute the operation k_1/s and a delay, T_{d2}, to the central nervous system (see Fig. 56). Although a more complex model including intermittency could be assumed for the delay and dynamics block, we try to use the simplest models for the less known parts of the tracking system; indeed, the integration in the central nervous system must be discrete in some unknown way. We attribute the delay T_{d1} to perception. The dead zone results from the experimentally verified impossibility of correcting extremely small errors.

The integrator followed by an intermittent element introduces a delay, approximately equal to $T_s/2$ where T_s is the "sampling" period.   Taking this and known physiological factors into account, we divide the total response delay as follows:

Visual latencies = T_{d1}	= 40 msec
Central nervous system and AMNC intermittency = $\dfrac{Ts}{2}$ (implicit)	= 165 msec
Conduction time = T_{d2}	= 15 msec
Muscular contraction time = T_{d3}	= 30 msec
	= 250 msec

The dead zone varies with the individual and with the system. It may have influenced the results for very slow ramps (see Fig. 67a and b).

Tustin [110] and Elkind [23] obtained fits of the form

$$\frac{K_c e^{-0.18s}}{1 + 0.85s}$$

for the human operator in compensatory tracking. These were optimum fits in a mean square sense, and this factor and the problem of spectrum selection may explain the absense of the integrator [60]. Bekey took Elkind's fits and added sampler and hold elements. His compensatory model thus explains the occasional peaks in the frequency response but does not explain the overshoot in the step response, the open loop and variable feedback responses, and the oscillations which were consistent except for improbable dead time resonances.

We can consider now our variable feedback experiments, ignoring intermittency since, from experimental data, it appears only to add a ripple to most outputs. We will consider the effects of delay and hand dynamics after obtaining equations which neglect these factors:

$$\theta_d(s) = \theta_i(s) + G\theta(s) \tag{7}$$

$$\theta_e(s) = \theta_d(s) - \theta_o(s) \tag{8}$$

$$\theta_o(s) = k_2\theta_d(s) + \frac{k_1}{s}\theta_e(s) \tag{9}$$

solving

$$\frac{\theta_o}{\theta_i}(s) = \frac{1}{1-G} \cdot \frac{[1 + s(k_2/k_1)]}{\left\{1 + s\left[\dfrac{1-k_2G}{(1-G)k_1}\right]\right\}} \tag{10}$$

If $G = 1$

$$\frac{\theta_o}{\theta_i}(s) = \frac{k_1}{s(1-k_2)}[1 + s(K_2/K_1)] \tag{11}$$

If $k_2 = 0$, but $G = 1$,

$$\frac{\theta_o}{\theta_i}(s) = \frac{k_1}{k_1(1-G) + s} \tag{12}$$

If $k_2 = 0$, and $G = 1$,

$$\frac{\theta_o}{\theta_i}(s) = k_1/s \tag{13}$$

If we consider Eq. 13, the open–loop impulse response is a delayed step with a rise time determined by hand dynamics. The step response is a delayed ramp approximately. We deduce from Eqs. (12) or (10) that the effect of nonunity external feedback is that the step response goes to a final value $1/(1 - k_2)$.

If $G < 1$ and $k_2 > 0$, we can expect step responses to have an initial jump of the order of $k_2/(1 - k_2G)$ and a higher final value $1/(1 - G)$. The short pulse response should die out at a high rate if G is small. Open–loop experiments show that the size of the correction is generally larger than the size of the step; this observation agrees with our assumption of $k_2 \neq 0$ in Eq. (5).

Variable feedback experiments, shown in Fig. 73, show that for pulse inputs the initial jump is independent of G, as was to be expected, and the decay rate is slightly faster for $G = 1/2$ than for $G = 2/3$, as expected. For step inputs the final value is higher than the initial value, and higher for $G = 2/3$, than for $G = 1/2$, although not as high as expected.

We have shown that our general topology and assumption of a position control system is consistent with all unpredictable transient results. Indeed, our model is similar to Young and Stark's model of the eye system [122, 123], which applies to a situation similar to compensatory hand tracking ($k_2 = 0$), and which differs from the hand system in the dynamics and in the location and type of intermittency.

PREDICTION OPERATOR

Before explaining the contributions of Smith and Cortes, and Lemay and Westcott, we recall the importance of the distinction between unpredictable and predictable tracking [91, 107].

1. Comparison of Figs. 74 and 75 with Fig. 76 shows, both in magnitude and phase, the greater ability of the subject to track continuous predictable signals. Furthermore, no peaking occurs at 1.5 cycles/sec.

2. Figure 69 shows that if ramps are predictable the subject can follow them without 3 cycles/sec intermittency and with apparent velocity servo characteristics.

3. Figures 57e and 60e show clear improvement in subject's performance when the step and pulse inputs are known beforehand.

4. Response times are reduced for predictable steps [74] (Fig. 61).

5. The hand system can perform very rapid and complex voluntary tasks, while it cannot follow simple unknown signals of frequencies higher than 3 or 4 cycles/sec.

When the signal is not entirely unpredictable, a whole range of intermediate situations can occur. The importance of considering unpredictable signals is also due to the essential uniqueness of unpredictable models, and to the large variety of predictable models as the input and learning conditions change.

Smith and Cortes [17, 89] made an extensive study of the step response with various large inertia loads, and postulated a "bang-bang" model for the human operator, which was optimal for constant force (see Fig. 78a and b). Lemay and Westcott [59] started from step response and presented a model (Fig. 77) which included a representation of the arm dynamics as a pure inertia, and a system of samplers for production of a preprogrammed

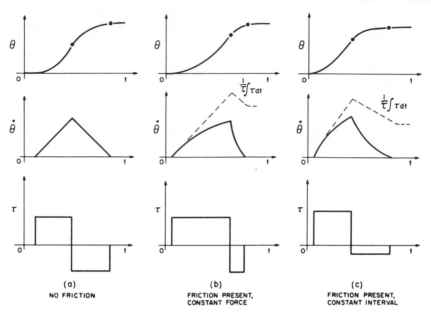

Fig. 78. Time-optimal step responses.

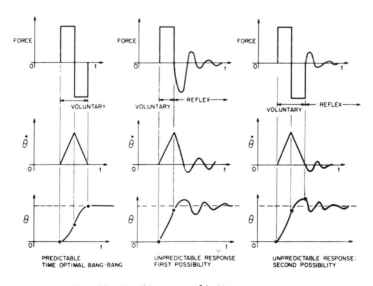

Fig. 79. Possible types of ballistic movements.

time–optimal step response for constant interval and adjustable force (see Fig. 78c). The predictor was added in order to improve the model's fit to experimental random signal responses, although it would have been more logically added to the model for the known step inputs. The use of constant force instead of constant interval in the time–optimal model introduces only small errors if the inertia load is large; however, our step responses with velocity and acceleration measured, particularly Fig. 60, show a response similar to time–optimal (Fig. 79) for the predictable case, but different for the unpredictable case: Fig. 79 (center) indicates one possible type of action, but Fig. 79 (right) has to be considered as an equally likely possibility. As we shall discuss, physiological data up to now support either of these two modes of response, while we claim the time–optimal "bang–bang" model to be valid only for predictable steps.

INTERMITTENCY AND PSYCHOLOGY

Telford and later investigators [18, 34, 38, 60] have studied and described discontinuity in the operation of physiological systems which, like the hand control system, require a stimulus to be perceived before a response is elicited. A "refractory" period between responses has been observed and measured (0.2–0.5 sec), and exceptions to it, as well as effect of training, grouping of stimuli, etc., have been noted by these and other investigators. Most of their experiments referred to discrete stimuli, but Craik [18] and Bekey [8] have studied the extension to continuous stimuli. Ward, Bekey, and Lemay and Westcott [113, 8, 59] have proposed sampled–data models of the human tracker. Other investigators have questioned the presence of a sampler here. The hand sampler is not as clear as in the eye system both because of the greater use of prediction in hand tracking, and because of the considerable inertia of the hand which tends to smooth out neurological intermittency.

Our experiments suggest that intermittency is present for unpredictable inputs, but both our results and physiological considerations suggest it to be irregular and asynchronous. The experimental evidence such as the unpredictable ramp studies and the frequency responses such as those shown in Fig. 15 are prime sources of behavioral evidence for the intermittency. We have noted deterioration of performance such as is shown in Fig. 62f when the frequency of a repetitive signal exceeds 1.5 cycles/sec. A crucial experiment supporting the existence of intermittency is the refractory period of about 250 msec in short pulse responses. Open-loop and variable feedback responses also show intermittent corrections, which are generally faster, about 4/sec. The distribution of delays approaches that of an input–synchronized sampled system. Some of the frequency-response experiments (Fig. 75) show peaks in the gain and related phenomena in the phase around 1.5 cycles/sec, which can be interpreted as half a sampling frequency.

The impulse responses do not show intermittency. Indeed, the sampling rate is not constant, as is shown by the absence of peak for most of the spectra used by Bekey[8]. Another explanation might be that the presence of the proprioceptive loop obscures the phenomena. With regard to our models for unpredictable step responses (Fig. 79), the postulated intermittency means that corrective voluntary control can be applied only for small intervals of time separated by refractory periods. Reflex control is the only operative one between these periods of voluntary control.

With the postulated type of intermittency, the system is not equivalent to an error-sampled system either for the visual or proprioceptive loop. There is always a drive applied to the muscle dynamics. The signals applied to the muscle include new signals generated by the intermittent process, but they also contain signals from the direct input. This explains the diminishing but nonzero ability to track past 1.5 cycles/sec, although a dead zone could be another explanation for intermittency.

PHYSIOLOGY OF THE HAND CONTROL SYSTEM

A model of the motor coordination system, particularized to the motion of the supinator and pronator muscles of the forearm, is shown in Fig. 80. The visual loop and the external input are not included.

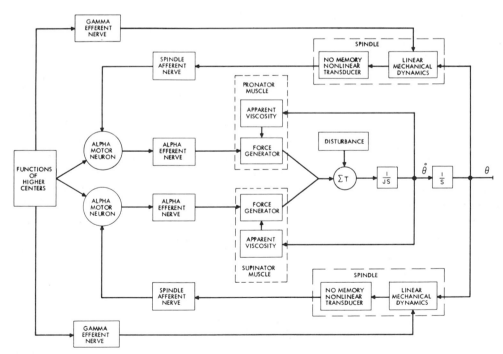

Fig. 80. Motor coordination system.

The pair of muscles, supinator and pronator, is required to rotate the wrist because of the inability of muscles to push; therefore only one muscle at a time actively aids movement (although the antagonist may be actively braking). The arm load is represented as an inertia, but some force is required to overcome the nonlinear apparent viscosity of the muscle.

The actuating elements are the force generators which control the contraction of the bundles of muscular fibers called motor units. For maximal excitation, all motor units (of the order of 10^4) are fired. We call excitation coefficient (α) that fraction of the muscular cross-sectional area that is activated. Thus, the applied torque becomes αT_{max}. The excitation of the muscle is determined by signals sent from the alpha motor neuron by means of the alpha efferent nerve. These signals originated in the cerebrum, cerebellum, and spinal cord. The alpha motor neuron also receives spindle signals transmitted by the afferent nerve; these signals depend on the length of the pertinent muscle. Other higher center signals come to the spindle through the gamma efferent nerve to control the myotatic or stretch reflex.

The dynamic characteristics of muscles have been studied at the Neurology Section at M.I.T. [48, 97]. The apparent viscosity has been found to be nonlinear [97], and higher for lengthening than for shortening, thus allowing for quicker damping than by a symmetrical element. The arm muscles can be represented as a linearized block:

$$\frac{\theta}{\alpha} = \frac{T_{max}}{s(\beta + Js)}$$

where β depends on α.

The spindle is a position feedback element connected in mechanical parallel to the muscle. The spindle afferent nerve is connected to a nuclear bag in the center of the spindle. The gamma efferent nerve can contract the spindle fibers, and acts as a level setter. Changes in length or in gamma bias, after being modified by the mechanical dynamics of the tendon, fibers, and nuclear bag, produce contractions in the nuclear bag fibers. These contractions then generate nervous signals which are transmitted to the alpha motor neuron by the spindle afferent nerve. The spindle is highly nonlinear, but the following approximate transfer function has been obtained by Houk for small signals [48]:

$$\frac{\alpha_s(s)}{\theta(s)} = \frac{ks\left(1 + \dfrac{s}{6.8}\right)}{\left(1 + \dfrac{s}{18}\right)\left(1 + \dfrac{s}{200}\right)} \qquad (14)$$

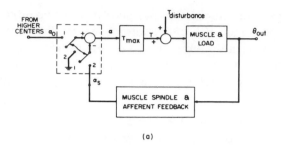

(a)

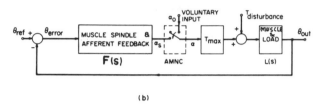

(b)

Fig. 81. Role of AMNC in tracking: (a) original model; (b) alternate version.

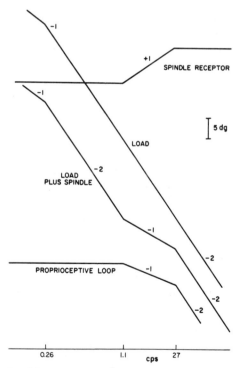

Fig. 82. Frequency characteristics of proprioceptive loop.

The differential behavior over a band of frequencies explains the role of the spindle in the damping of fast movements.

The stretch reflex operates as follows: When a limb is disturbed out of equilibrium, the spindle is stretched and the number of pulses per second sent to the alpha motor neuron is increased. The alpha efferent signal from the alpha motor neuron changes and causes the muscle to move the arm back to equilibrium.

Role of the Proprioceptive Loop

In the model presentation, alternation between voluntary and reflex control of the alpha efferent pathway — the final common pathway — was postulated. The behavior of the motor coordination system occurs as follows: Voluntary movements are preprogrammed in the higher centers and signaled as a whole. At the time of command release (Position 1, Figs. 56 and 81a) spindle afferent control is reduced and higher control of the alpha efferents is fully turned on. The system is operating proprioceptively open loop, characterized by high gain, low damping, and possible postural drift, during a short period of time of the order of 100 msec. Only apparent viscosity opposes motion.

After the motion has been substantially completed, the spindle afferent control of the alpha efferent pathway is fully turned on and the gamma efferent may be altered for a new set point. This corresponds to position 2 in Figs. 2 and 27. The arm may oscillate for a time and is finally clamped to a new position, lasting approximately 230 msec.

If there is still some error, a new correction is applied, and the process repeats itself at a rate of 3 corrections per second. Figure 81b shows an alternate model of action of the system, which shows the clamping operation to a new value of θ_{ref}. The following figures of merit can be defined:

$$F_{vol}(s) = \frac{\alpha_0(s)}{\theta_e(s)} = F(s) + \frac{1}{T_{max}L(s)} \tag{15}$$

$$F_{dist}(s) = \frac{T_{dist}(s)}{\theta_e(s)} = F(s)T_{max} + \frac{1}{L(s)} \tag{16}$$

High $F(s)$ means low gain and little stability. Low $L(s)$ means low gain.

This explanation is supported by considerable evidence. For example, the standard finger pointing and finger-nose-finger tests used by neurologists show the open-loop nature of rapid movements. Further, lack of antagonist tone during rapid voluntary "free-wheeling" movements favors a proprioceptive open-loop type of control [91].

Oscillations up to a few cycles per second are possible, and the frequency of oscillation is slower if the subject has to correct for disturbances and the proprioceptive loop is closed.

In Parkinson's disease, the spindle afferent path cannot be easily switched off, and the corticospinal input is thus weakened. Results are rigidity, inability to track, and low free-wheeling frequency [91, 98].

In experiments performed by Stark and Rushworth [104] the gamma input was blocked with procaine. It was found that the initial part of skillful movements (alpha controlled) was still correct, but the termination was affected; the subject lost control of his final arm position or of the termination of paper tracing because the proprioceptive feedback path was inoperative for clamping and damping.

Considerations of transient responses, particularly steps (Figs. 60 and 79), also support this dual mode behavior.

We must speak finally about frequency response for unpredictable inputs. On the basis of analytical and experimental work done at the Neurology Section at M.I.T., very approximate theoretical frequency response characteristics can be obtained. However, no simulation of the total model response has yet been done.

Using average values from typical tracking situations, we have computed $\alpha = 10^{-3}$ and

$$\frac{\theta}{T}(s) = \frac{10^4}{s(1 + 0.56s)}$$

The magnitude characteristics of the spindle and load are shown in Fig. 82. The gain of the proprioceptor loop is obtained by standard servo approximations. For example, we have closed the loop with a phase margin of only 30°. This is reasonable since we have already observed the low damping of the oscillations between corrective movements. In considering the visually open loop we have omitted the intermittency but added a peak at 1.5 cycles/sec, as shown in Fig. 83. The loop has then been closed with typical phase margins of 45° and 70°. The effect of the $(1 + k_2 s/k_1)$ term is included as a prefilter which is operative only from 0.5 to 2.5 cycles/sec, since at high frequency the operator has only the error to rely upon if he is to follow at all. This approximate prediction of total gain is in agreement with the experimental graphs of Fig. 75. However, simulation should be carried out, since an exact analysis of our model is too difficult. It would also be interesting to look at the impulse responses with reference to the above approximations.

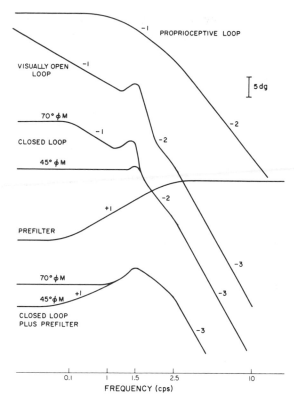

Fig. 83. Theoretical frequency characteristics of unpredictable tracking system.

SUMMARY

The major conclusions from our studies are summarized in the model presented. The loading impedance of the arm lies mainly in the inertia
of the limb (which we attempted to minimize with the wrist rotation movement studied), and in the apparent viscosity of the actively shortening and
lengthening agonist–antagonist muscle pair. On the input side, the predictor operator yields extraordinarily quick input adaptive recognition of repetitive signal characteristics and requires random or pseudo-random input
signals to control this factor. The dual control of hand movement requires
a continual switching process from postural to voluntary systems, which
thus determines which of the two modes has control of the final common
path, the alpha motor neuron. This phenomenon provides the intermittency or sampling type of behavior for the hand control system. Experiments
on reaction time distribution functions to random steps clearly show that

the sampling is input-synchronized, with some statistical spread, and not clock-synchronized is in usual engineering sampled-data systems.

The transient experiments demonstrate the refractory period in terms of the minimum-width pulse response to an impulse input. Also, the random ramp inputs show the intermittent steps of response, which vary in amplitude, but not in time interval, with variation in input ramp velocity. This latter experiment also demonstrates the position controller acting alone without a velocity controller, in contrast to the eye tracking system which has both position and velocity control under similar circumstances. Further evidence for the integrator, acting in the position controller, is seen in experiments under open-loop and variable feedback conditions. Here the response to a train of pulses is a sequence of steps. Again in contrast to eye movement tracking, the hand system can adapt to this bizarre change in the experimental arrangement.

The frequency-response experiments show different results for predictable and random signals, an illustration of the input adaptive characteristics. The deterministic signals show the hand to behave like a stable second-order system with an ideal predictor. Unpredictable signals show evidence for the sampled-data peak, providing, of course, that care is taken in selecting proper input spectra. The quantitative model in addition to intermittency, the integrator, time delays, a minor proprioceptive loop, and plant dynamics, has a lead-lag phase advance operator.

APPENDIX A

PERIPHERAL VS. CENTRAL ADAPTATION WITH AN EXPERIMENTAL TEST FOR THE HYPOTHESIS

This appendix is devoted to exploring possible decomposition of adaptive mechanisms into (a) those utilizing a variable state of gain or dynamics of the peripheral control system, and (b) those relying upon prefiltering of the preprogrammed ballistic intermittent commands which dominate the signal pathways during voluntary rapid movements.

The mechanical-impulse technique is put forward as an experimental means of testing the state of the peripheral control system during various types of adaptive changes. Those changes that alter the dynamical characteristics of the peripheral control system will be classed as peripheral adaptive changes, while those that do not so alter it are presumably central adaptive changes.

Adaptive control is an ill-defined phrase used to describe the ability of certain kinds of control systems to change their characteristics when the operating conditions of the system are altered so as to achieve behavior more appropriate, in some sense, than would be obtained if the system had remained unchanged. The human operator in a control situation is an example of a very versatile adaptive controller, one that can adapt to many different kinds of changes in operating conditions. Russell [83], Krendel [58], and Hall [33] have shown that in the steady state, at least, the human operator adjusts his own characteristics. Elkind [23, 24] has studied the manner in which human operator control characteristics depend upon the statistical properties of the input forcing function. Sheridan [88] has examined the response of the process of human adaptation to sudden and gradual changes in controlled element dynamics. Young et al. [121] have investigated in some detail the fine structure of the human operator's process of adaptation to sudden changes in systems dynamics and input forcing function.

Although admittedly tentative and incomplete, the model of human movement developed in the preceding chapters has clarified in a semi-quantitative way certain features of interest to anyone studying adaptation. A distinction is made between two systems: a lower level diffuse reflex postural system whose dynamic behavior is controlled via the muscle spindle, and a higher level specific control system which is open loop with respect to the proprioceptive feedback during important portions of its behavior.

The interaction between these two systems is complex and antagonistic in part. The voluntary system seems to be a discrete or discontinuous control which can switch off the opposing position servo for rapid movements. When it cannot, as in Parkinson's syndrome, pathological control behavior is evident. Prediction of input wave form and task nature seems to be a higher control function controlling the movement via the direct specific pathway. Adaptation of this type should operate independently of the gamma system. When plant dynamics change, different damping is required and this may well be a function of the diffuse postural system. It is known that mental set can change the stability margin of the postural system. Certain possible controls such as the gamma ratio would be ideal for this functional role.

As described in the preceding chapters, our studies have shown adaptation in the human operator to take several forms:

1. (a) Adaptation to a predictable signal involves recognition of the parameters of the input wave form in a carefully preprogrammed sequential set of signals to drive the muscles. There is some evidence that proprioceptive sensory feedback is necessary in a physiologically generated form, i.e., by active movement, before these "engrams" or adaptive matched filters can be formed. On the other hand, general sensory pattern clas-

sification of, for example, visual patterns, may well operate in this manner without requiring active motor participation by the subject.

(b) Adaptation to constraints on random signals has been demonstrated. The discontinuous or sampled-data position control system perhaps here operates in conjunction with higher control of the peripheral servo loop so that the antagonist muscle brought into action as part of the paired postural servo may act as a damping element.

2. Adaptation to changes in stick dynamics suggests that alteration of parameters in the postural servo is in fact the mechanism of adaptation. This would certainly be an economical method of adaptive control, requiring only an estimate of "plant" dynamics in the adaptive alteration of control system compensation. The alternative method of computing control signals in order to provide for compensation of subsequent distortion by the plant seems more complex. In fact, evidence indicating that only a restricted range of compensation can occur suggests indeed that the first scheme, naturally limited by the range of compensatory alterations, is most likely.

The postural servo is an ideal element for adaptive compensation. The body of the section has illustrated detailed quantitative analysis of its function. Such demonstrated refinements as the γ_1/γ_2 ratio seem even more designated for an adaptive role.

We now may make an explicit hypothesis: adaptation to stick dynamics is via the postural loop; adaptation to input wave form is via the direct proprioceptively open-loop path.

To test this hypothesis experimentally we need a method of measuring the gain and dynamic characteristics of the peripheral spindle feedback system. We propose that the mechanical impulse is such a test, since it results in a response wave form which defines the system transfer function. Some confusion may occur if the wrist is rotating at the moment of impact, but by monitoring the initial conditions — position, velocity, and acceleration, and by using positive and negative mechanical impulses, the difficulty may be circumvented.

The mechanical impulse experiments for evaluating the mechanical constants of the hand, or the gain and phase margin of the back-to-back position servos suggest themselves for testing alterations in the peripheral control system setting. Adaptation that takes place via the peripheral system should be reflected in changes in these parameters. Conversely, adaptation entirely in the form of changed input signals should leave the peripheral system unchanged.

Subjects for the experimental test of the hypothesis will be trained or adapted to three sets of conditions: 1, normal mode; 2, altered load, or stick dynamics; and 3, predictable wave form. These conditions must be matched for other parameters of the experimental situation. The hypothesis predicts that when the activity of the spindle feedback is now measured by testing as above, only condition 2 will be accompanied by a change in the postural feedback system parameters from their values in the normal mode.

While these experiments are being run, simulated runs on BIOSIM might also be carried out in order to test further the quantitative agreement between the results and the model.

REFERENCES AND FURTHER READING

1. Adams, J. A.: Human tracking behavior. Psych. Bull. 58: 55 (1961).
2. Adolph, A. R.: An application of feedback analysis and stochastic models to neurophysiology. Bull. Math. Biophys. 21: 195 (1959).
3. Arden, G. B., and Soderberg, U.: The transfer of optic information through the lateral geniculate body of the rabbit. In: Sensory Communication, W. A. Rosenblith, ed., Massachusetts Institute of Technology, Cambridge, Mass. (1962), p. 521.
4. Atwood, J., Elkind, J., Houk, J., King, M., Stark, L., and Willis, P.: Digital computer simulation of a neurological system. Quart. Prog. Rept. Research Laboratory of Electronics, M.I.T. 63: 215−217 (1961).
5. Atwood, J. G., Delaney, E., and Stark, L.: A transducer for testing motor coordination. Unpublished data (1956).
6. Barker, D., and Cope, M.: The innervation of individual intrafusal muscle fibers. In: Symposium on Muscle Receptors, D. Barker, ed., Hong Kong University Press (1962), pp. 263−269.
7. Barker, D., and Ip, M. C.: The primary and secondary endings of mammalian muscle spindle. J. Physiol. 153: 8 (1960).
8. Bekey, G. A.: The human operator as a sampled data system. IRE Trans. Human Factors Electron. 3: 43 (1962).
9. Bennett, C. A.: Sampled data tracking: Sampling of the operator's output. J. Exptl. Psych. 51: 429 (1956).
10. Bigland, B., and Lippold, O. C. J.: The relation between force velocity and integrated electrical activity in human muscles. J. Physiol. 123: 214 (1954).
11. Boshes, B., Wachs, H., Brumlik, J., Mier, M., and Petrovick, M.: Studies of tone, tremor and speech in normal persons and Parkinsonian patients. Neurol. 10: 805−813 (1960).
12. Boyd, I. A.: The nuclear−bag fibre and nuclear chain fibre systems in the muscle spindles of the cat. In: Symposium on Muscle Receptors, D. Barker, ed., Hong Kong University Press (1962), pp. 185−191.

13.   Boyle, D. L.: A neuron model which performs analog functions. San
      Diego Symposium for Biomedical Engineering, La Jolla, California
      (1962).

14.   Brindley, G. S.: Physiology of the Retina and the Visual Pathway,
      Edward Arnold, London (1960), Chap. III.

15.   Brown, A. C.: Analysis of myotatic reflex. Doctoral Dissertation,
      University of Washington, Seattle (1959).

16.   Chestnut, H., and Mayer, R. W.: Servomechanisms and Regulating
      System Design, Vol. 1, Wiley, New York (1959).

17.   Cortes, A.: The human servo: Arm control. Master's Thesis, Uni-
      versity of California, Engineering Library, Berkeley (1958).

18.   Craik, K. J. W.: Theory of the human operator in control systems.
      Brit. J. Psych. 38: 56−61 (1947) and 38: 142−148 (1947).

19.   Del Castillo, J.: The transmission of excitation from nerve to mus-
      cle. Neuromuscular Disorders Res. Publ. Assoc., N. M. Dis. 38:
      (1960).

20.   Eldred, E.: Posture and locomotion. In: Handbook of Neurophysiol-
      ogy, Vol. 2, H. W. Magoun, ed., Waverly Press, Baltimore (1960), p. 1067.

21.   Eldred, E., Granit, R., Holmgren, G., and Merton, P. A.: Proprioceptive
      control of muscular contraction and the cerebellum. J. Physiol. 123:
      46−47 (1953).

22.   Eldred, E., Granit, R., and Merton, P. A.: Supraspinal control of the
      muscle spindles and its significance. J. Physiol. 122: 498 (1953).

23.   Elkind, J. I.: Characteristics of simple manual control systems.
      Lincoln Laboratory Tech. Rept., M.I.T., Number III: 1−145 (1956).

24.   Elkind, J. I., and Green, D.: Measurement of time-varying and nonlinear
      dynamic characteristics of human pilots. Technical Report, Bolt,
      Beranek and Newman, Cambridge, Mass. (1961).

25.   Fenn, W. O., and Marsh, B. S.: Muscular force at different speeds of
      shortening. J. Physiol. 85: 277 (1935).

26.   Fulton, J. R., and Pi-Suner, J. A.: A note concerning the probable
      function of various afferent end-organs in skeletal muscle. Am. J.
      Physiol. 83: 554 (1928).

27.   Furman, G. G.: Neuromuscular control. Astia Report 270452 (1961).

28.   Graham, D., and McRuer, D.: Analysis of Nonlinear Control Systems,
      Wiley, New York (1961).

29.   Granit, R.: Receptors and Sensory Perception, Yale University
      Press, New Haven, Conn. (1955).

30.   Granit, R.: Neuromuscular interaction in postural tone of the cat's
      isometric soleus muscle. J. Physiol. 143: 387−402 (1958).

31.   Granit, R.: Some problems of muscle-spindle physiology. In: Sym-
      posium on Muscle Receptors, D. Barker, ed., Hong Kong University Press
      (1962), pp. 1−13.

32.   Granit, R., and Stark, L.: Unpublished experiments (1958).

33.   Hall, I. A. M.: Effect of controlled element on the human pilot.
      WADC Wright-Patterson AFB, Dayton, Ohio, TR 57-509 (1957).
34.   Halliday, A. M., Kerr, M., and Elithson, A.: Grouping of stimuli and
      apparent exceptions to the psychological refractory period. J. Exptl.
      Psych. 7: 72−89 (1960).
35.   Hammond, P. H.: An experimental study of servo action in human
      muscular control. Third International Conference on Medical Elec-
      tronics, London (1961).
36.   Hammond, P. H., Merton, P. A., and Sutton, G. G.: Nervous grada-
      tion of muscular contraction. Brit. Med. Bull. 12: 214 (1956).
37.   Harvey, R. J., and Matthews, P. B. C.: The response of deefferented
      muscle spindles in the cat's soleus to slow extensor of the muscle.
      J. Physiol. 157: 370 (1961).
38.   Hick, W. E.: Discontinuous functioning of the human operator in
      pursuit tasks. Quart. J. Exptl. Psych. 1: 36−57 (1948).
39.   Hill, A. V.: The heat of shortening and the dynamic constants of
      muscle. Proc. Roy. Soc. (London) B126: 136 (1938).
40.   Hill, A. V.: The dynamic constants of human muscle. Proc. Roy.
      Soc. (London) B128: 263 (1940).
41.   Hill, A. V.: A discussion on muscular contraction and relaxation:
      their physical and chemical basis. Proc. Roy. Soc. (London) B137:
      40 (1950).
42.   Hill, A. V.: The effect of series compliance on the tension developed
      in a muscle twitch. Proc. Roy. Soc. (London) B138: 325 (1951).
43.   Hill, A. V.: The series elastic component of muscle. Proc. Roy.
      Soc. (London) B137: 275 (1960).
44.   Hofmann, W.W.: Regulatory mechanisms in Parkinsonian tremor. J.
      Neurol. Neurosurg. Psychiat. 25: 109 (1962).
45.   Hofmann, W. W.: Observations on peripheral servomechanisms
      in Parkinsonian rigidity. J. Neurol. Neurosurg. Psychiat. 25: 203
      (1962).
46.   Houk, J., Okabe, Y., Rhodes, H. E., Stark, L., and Willis, P. A.:
      Transient response of the human motor coordination system. Quart.
      Prog. Rept., Research Laboratory of Electronics, M.I.T. 64: 315−
      326 (1962).
47.   Houk, J. C., Sanchez, V., and Wells, P.: Frequency response of a
      spindle receptor. Quart. Prog. Rept., Research Laboratory of Elec-
      tronics, M.I.T. 67: 223−227 (1962).
48.   Houk, J., and Stark, L.: An analytical model of a muscle spindle re-
      ceptor for simulation of motor coordination. Quart. Prog. Rept.,
      Research Laboratory of Electronics, M.I.T. 66: 384−389 (1962).
49.   Hunt, C. C., and Kuffler, S. W.: Further study of efferent small nerve
      fibres to mammalian muscle spindles: Multiple spindle innervation
      and activity during contraction. J. Physiol. 113: 283−297 (1951).

50.   Hunt, C. C., and Kuffler, S. W.: Stretch receptor discharges during
       muscle contraction. J. Physiol. 113: 298—315 (1951).
51.   Hyman, R.: Stimulus information as a determinant of reaction time.
       J. Exptl. Psych. 45: 188 (1953).
52.   Jansen, J. K. S., and Matthews, P. B. C.: The central control of the
       dynamic response of muscle spindle receptors. J. Physiol. 161: 57
       (1962).
53.   Jansen, J. K. S., and Matthews, P. B. C.: The effects of fusimotor
       activity on the static responsiveness of primary and secondary end-
       ings of muscle spindles in the decerebrate cat. Acta Physiol. Scand.
       55: 376 (1962).
54.   Johnson, A. R.: The servoanalysis of postural reflexes. Doctoral
       Dissertation, Dept. of Electrical Engineering, Massachusetts Insti-
       tute of Technology, Cambridge, Mass. (1959).
55.   Katz, B.: Depolarization of sensory terminals and the initiation of
       impulses in muscle spindle. J. Physiol. 111: 261 (1950).
56.   Katz, B.: The relation between force and speed in muscular contrac-
       tion. J. Physiol. 96: 45 (1939).
57.   Kipiniak, V. W.: Simulation of biological control systems. Quart.
       Prog. Rept., Research Laboratory of Electronics, M.I.T. 62: 284—
       286 (1961).
58.   Krendel, E. S., and McRuer, D. T.: A servomechanisms approach to
       skill development. J. Franklin Inst. 269: 24 (1960).
59.   Lemay, L. P., and Westcott, J. H.: The simulation of human operator
       tracking using an intermittent model. Presented at the International
       Congress on Human Factors in Electronics, Long Beach, California
       (1962).
60.   Licklider, J. C. R.: Quasi-linear operator models in the study of
       manual tracking. In: Developments in Mathematical Psychology,
       R. D. Luce, ed., Free Press, New York (1960).
61.   Lippold, O. C. J.: The relation between integrated action poten-
       tials in a human muscle and its isometric tension. J. Physiol.
       117: 492 (1952).
62.   Lippold, O. C. J., Nicholls, J. G., and Redfearn, J. W. T.: Electrical
       and mechanical factors in the adaptation of a mammalian muscle
       spindle. J. Physiol. 153: 207 (1960).
63.   Lippold, O. C. J., Redfearn, J. W. T., and Vuco, J.: The effect of
       sinusoidal stretching upon the activity of stretch receptors in volun-
       tary muscle and their reflex response. J. Physiol. 144: 373 (1958).
64.   Mason, S. J., and Zimmerman, H. J.: Electric Circuits, Signals and
       Systems, Wiley, New York (1960).
65.   McRuer, D. T., and Krendel, E. S.: Dynamic response of human op-
       erators. WADC Tech. Rept. 56: 524 (1957).

66. McRuer, D. T., and Krendel, E. S.: The human operator as a servo-system element. J. Franklin Inst. 267: 381−511 (1959).

67. McRuer, D. T., and Krendel, E. S.: The man-machine system concept. Proc. IRE 48: 117 (1962).

68. Merton, P. A.: Speculations on the servocontrol of movement. In: The Spinal Cord, a Ciba Foundation Symposium (1953), pp. 247−260.

69. Naves, F.: Sampling of quantatization in the human tracking system. Master's Thesis, Dept. of Electrical Engineering, Massachusetts Institute of Technology, Cambridge, Mass. (1963).

70. Naves, F., and Stark, L.: Sampling or intermittency in the hand control system. Biophys. J. 8: 252−302, 1968.

71. Noble, M., Fitts, P. M., and Warren, C. E.: The frequency response of skilled subjects in a pursuit tracking task. J. Exptl. Psych. 49: 249 (1955).

72. Okabe, Y., Rhodes, H. E., Stark, L., and Willis, P. A.: Further studies of the application of computer analysis to measurement of movement dynamic characteristics. Quart. Prog. Rept., Research Laboratory of Electronics, M.I.T. 62: 263−267 (1961).

73. Okabe, Y., Rhodes, H. E., Stark, L., and Willis, P. A.: Transient response of human motor coordination system. Quart. Prog. Rept., Research Laboratory of Electronics, M.I.T. 66: 389−395 (1962).

74. Okabe, Y., Rhodes, H. E., Stark, L., and Willis, P. A.: Simultaneous eye and hand tracking. Quart. Prog. Rept., Research Laboratory of Electronics, M.I.T. 66: 395−401 (1962).

75. Okabe, Y., Payne, R. C., Rhodes, H. E., Stark, L., and Willis, P. A.: Use of on-line digital computer for measurement of a neurological control system. Quart. Prog. Rept., Research Laboratory of Electronics, M.I.T. 61: 219−222 (1961).

76. Overmyer, R. G.: Is man necessary? Electron. Eng. 81: 174 (1962).

77. Patridge, L. D., and Glaser, G. D.: Adaptation in regulation of movement and posture: A study of stretch responses in spastic animals. J. Neurophys. 23: 257 (1960).

78. Pollock, L. J., and Davis, L.: Muscle tone in Parkinsonian states. Arch. Neurol. Psychiat. 23: 303−317 (1930).

79. Richter, H. R., and Pfatz, C. R.: A propos de l'electroculographie. Confinia Neurol. 16: 279 (1956).

80. Ritchie, J. M., and Wilkie, D. R.: The dynamics of muscular contraction. J. Physiol. 143: 104 (1958).

81. Roig, R.: Nonlinear adaptation in manual control systems. Electronic Systems Laboratory, M.I.T. ESL-7420-R-4: (1960).

82. Roberts, T. M.: Discussion of speculations on servo control of movement. In: The Spinal Cord, J. R. Malcom and J. A. B. Gray, eds., Churchill, London (1953), pp. 255−258.

83. Russell, L.: Characteristics of the human as a linear servoelement. Master's Thesis, Massachusetts Institute of Technology, Cambridge, Mass. (1951).

84. Ruston, W. A. H.: Peripheral coding in the nervous system. In: Sensory Communications, W. A. Rosenblith, ed., Massachusetts Institute of Technology, Cambridge, Mass. (1962), p. 169.

85. Schwab, R. S., England, A. C., and Peterson, E.: Akinesia in Parkinson's disease. Neurology 9: 65−72 (1959).

86. Sheridan, T. B.: Experimental analysis of time variation in the human operator transfer function. International Federation Automatic Control Congress, Moscow (1960), p. 1881.

87. Sheridan, T. B.: Time variable dynamics of human operator systems. Dynamic Analysis and Control Laboratory, M.I.T. PB−1471732: (1960).

88. Sheridan, T. B.: The human operator in control instrumentation. In: Progress in Control Engineering, Heywood & Co., London (1962), pp. 143−187.

89. Smith, O. J.: Nonlinear computations in the human controller. IRE Trans. Biomed. Electron. 11: 125−129 (1962).

90. Stark, L.: Stability, oscillations, and noise in the human pupil servomechanism. Proc. IRE 47: 1925 (1959).

91. Stark, L.: Neurological organization of the control system for movement. Quart. Prog. Rept., Research Laboratory of Electronics, M.I.T. 61: 234−238 (1961).

92. Stark, L.: Physiological models for human operator adaptive control. Memorandum Report to ASD, Wright−Patterson Air Force Base, Ohio (1963).

93. Stark, L.: Neurological feedback control systems: physiological models and behavioral characteristics of the adaptive motor coordination system. In: Advances in Bioengineering and Instrumentation, Frea Alt, ed., Plenum Press, New York (1966), Chapter 4, pp. 289−385.

94. Stark, L., and Baker, F.: Stability and oscillations in a neurological servomechanism. J. Neurophys. 22: 156 (1959).

95. Stark, L., Campbell, P. W., and Atwood, J.: Pupil unrest: an example of noise in a biological servomechanism. Nature 182: 867 (1958).

96. Stark, L., and Houk, J.: An analytical model of a muscle spindle receptor for simulations of motor coordination. Quart. Prog. Rept., Research Laboratory of Electronics, M.I.T. 66: 384 (1962).

97. Stark, L., Houk, J. C., Willis, P. A., and Elkind, J. I.: The dynamic characteristics of a muscle model used in digital−computer simulation of an agonist−antagonist muscle system in man. Quart. Prog. Rept., Research Laboratory of Electronics, M.I.T. 64: 309−315 (1962).

98.   Stark, L., and Iida, M.: Dynamical response of the movement coordi-
       nation system of patients with Parkinson syndrome. Quart. Prog.
       Rept., Research Laboratory of Electronics, M.I.T. 63: 204−213 (1961).
99.   Stark, L., Iida, M., and Willis, P. A.: Dynamic characteristics of the
       motor coordination system in man. Biophys. J. 1: 279 (1961).
100.   Stark, L., Okabe, Y., and Willis, P. A.: Sampled data properties of
       the human motor coordination system. Quart. Prog. Rept., Research
       Laboratory of Electronics, M.I.T. 67: 220−223 (1962).
101.   Stark, L., Okabe, Y., and Willis, P. A.: Transient response dynamics
       of the motor coordination system in man. (In preparation.)
102.   Stark, L., and Nelson, G. P.: Phototube glasses for measuring eye
       movement. Quart. Prog. Rept., Research Laboratory of Electronics,
       M.I.T. 67: 214−216 (1962).
103.   Stark, L., Payne, R., and Okabe, Y.: On−line digital computer for
       measurement of a neurological control system. CACM 5: 567−568
       (1962).
104.   Stark, L., and Rushworth, G.: Unpublished experiments (1958).
105.   Stark, L., and Sandberg, A.: Analog simulation of the human pupil
       system. Quart. Prog. Rept., Research Laboratory of Electronics,
       M.I.T. 66: 420−428 (1962).
106.   Stark, L., and Sherman, P. M.: A servoanalytic study of the consen-
       sual pupil reflex to light. Neurophys. 20: 17 (1957).
107.   Stark, L., Vossius, G., and Young, L.: Predictive eye movement
       control. IRE Trans. Human Factors Electron. HFE−3: 52−57 (1962).
108.   Telford, C. O.: The refractory phase of voluntary and associative
       responses. Exptl. Psych. 14: 1 (1931).
109.   Truxall, J. G.: Control System Synthesis, McGraw−Hill, New York
       (1955).
110.   Tustin, A.: The nature of the operator's response in manual con-
       trol and its implications for controller design. J. Inst. Electronic
       Engrs. (London) 94: 190 (1947).
111.   Vince, M. A.: The intermittency of control and the psychological re-
       fractory period. Brit. J. Psych. 38: 249 (1948).
112.   Walsh, F. M. R.: Muscular rigidity of paralysis agitans. Brain 47:
       159 (1924).
113.   Ward, J. R.: The dynamics of a human operator in a control system.
       Doctoral Dissertation, Department of Aeronautics, University of
       Sydney, Sydney, Australia (1958).
114.   Webster, D. D.: Dynamic measurement of rigidity, strength, and
       tremor in Parkinson patients before and after destruction of mesial
       globus pallidus. Neurol. 10: 157−163 (1960).
115.   Welford, A. T.: The psychological refractory period and the timing
       of high speed performance: A review and a theory. Brit. J. Psych.
       43: 2 (1952).

116. Wiener, N.: Cybernetics, Wiley, New York (1958).
117. Wilkie, D. R.: The relation between force and velocity in human muscle. J. Physiol. 110: 249−280 (1950).
118. Wilkie, D. R.: Facts and theories about muscle. Prog. Biophys. Biophys. Chem. 4: 288 (1954).
119. Willis, P. A.: Mechanical output impedance of motor coordination transducer. Unpublished data (1960).
120. Young, L. R.: A sampled data model for eye movements. Doctoral Dissertation, Dept. of Aeronautics, Massachusetts Institute of Technology, Cambridge, Mass. (1962).
121. Young, L. R., Green, D. M., Elkind, J., and Kelly, J. A.: The adaptive dynamics response characteristics of the human operator in simple manual control. NASA TN D-2255: (1964).
122. Young, L., and Stark, L.: A discontinuous biological system−eye movement control. Presented at the 15th Annual Conference on Engineering Tech. in Medicine and Biology, Chicago (1962).
123. Young, L., and Stark, L.: A sampled data model for eye tracking movement. Quart. Prog. Rept., Research Laboratory, M.I.T. 66: 371−384 (1962).

The Author's Published Papers
Excluding Abstracts

1952—1959

1. Jose del Castillo-Nicolau and Lawrence Stark, The effect of calcium ions on the motor end-plate potentials, J. Physiol. 116: 507—515 (1952).

2. Jose del Castillo-Nicolau and Lawrence Stark, Local responses in single medullated nerve fibres, J. Physiol. 118 (2): 207—215 (March 1952).

3. Gilbert H. Glaser, Lawrence Stark, and Mary O. Godenne, The diagnosis of atypical poliovirus infection, Trans. Am. Neurol. Assoc. 81: 18—22 (1956).

4. Lawrence Stark and Philip M. Sherman, A servoanalytic study of consensual pupil reflex to light, J. Neurophysiol. 20: 17—26 (1957).

5. Lawrence Stark and Tom N. Cornsweet, Testing a servoanalytic hypothesis for pupil oscillations, Science 127: 588 (1958).

6. Gilbert H. Glaser and Lawrence Stark, Excitability in experimental myopathy: I. Measurement of refractory period; quinidine effect; cortisone myopathy, Neurology 8: 640—644 (August 1958).

7. Gilbert H. Glaser and Lawrence Stark, Excitability in experimental myopathy: II. Potassium deficiency: an initial study, Neurology 8: 708—711 (September 1958).

8. Lawrence Stark, Fergus W. Campbell, and John Atwood, Pupil unrest: An example of noise in a biological servomechanism, Nature 182: 857—858 (September 1958).

9. Lawrence Stark and Frank Baker, Stability and oscillations in a neurological servomechanism, J. Neurophysiol. 22: 156—164 (1959).

10. Lawrence Stark, Transfer function of a biological photoreceptor, Air Force Tech. Rep. WACD TR 59-311: 1—22 (August 1959).

11.   Lawrence Stark, Stability, oscillations, and noise in the human pupil
       servomechanism, Proc. IRE 47 (11): 1925–1939 (November 1959).
12.   Lawrence Stark, Vision: Servoanalysis of pupil reflex to light, in:
       Medical Physics, Vol. 3, Otto Glasser (ed.), Year Book Publishers,
       Inc., Chicago (1959), pp. 701–719.

1961–1962

13.   Lawrence Stark, Mitsuo Iida, and Paul A. Willis, Dynamic charac-
       teristics of the motor coordination system in man, Biophys. J. 1:
       279–300 (1961).
14.   Lawrence Stark and Howard T. Hermann, Transfer function of a
       biological photoreceptor, Nature 191: 1173–1174 (September 1961).
15.   Lawrence Stark and Howard T. Hermann, The transfer function of a
       photoreceptor organ, Kybernetik 1: 124–129 (December 1961).
16.   Lawrence Stark, Environmental clamping of biological systems:
       Pupil servomechanism, J. Opt. Soc. Am. 52 (8): 925–930 (August
       1962).
17.   Lawrence Stark, Gerhard Vossius, and Laurence R. Young, Predic-
       tive control of eye tracking movements, IRE Trans. Human Factors
       Electron. HFE-3: 52–57 (September 1962).
18.   Lawrence Stark, Mitsuharu Okajima, and Gerald H. Whipple, Com-
       puter pattern recognition techniques: Electrocardiographic diagnosis,
       Commun. Assoc. Computing Machinery 5: 527–532 (October 1962).
19.   Lawrence Stark, Biological rhythms, noise, and asymmetry in the
       pupil–retinal control system, Ann. N.Y. Acad. Sci. 98: 1096–1108
       (October 30, 1962).
20.   Lawrence Stark, Henk van Der Tweel, and Julia Redhead, Pulse re-
       sponse of the pupil, Acta Physiol. Pharmacol. Neerl. 11: 235–239
       (1962).
21.   Howard T. Hermann, Lawrence Stark, and Paul A. Willis, Instrumen-
       tation for processing neural signals, J. Electroencephalog. Clin.
       Neurophysiol. 14: 557–560 (1962).
22.   Lawrence Stark, Robert Payne, and Yutaka Okabe, On-line digital
       computer for measurement of a neurological control system, Com-
       mun. Assoc. Computing Machinery 5: 567–568 (November 1962).

1963–1964

23.   Lawrence Stark, Physiological models for human operator adaptive
       control, Aeronautical Systems Division, Air Force Systems Com-
       mand, U.S. Air Force, Wright-Patterson Air Force Base, Ohio,
       ASRMCM-3, Memorandum Rep. 10124-1, Job No. 11116 (February
       15, 1963).

24.   Laurence R. Young and Lawrence Stark, A discrete model for eye
       tracking movements, IEEE Trans. Military Electron. MIL-7 (2-3):
       113–115 (April–July 1963).
25.   Mitsuhara Okajima, Lawrence Stark, Gerald Whipple, and Shoji Yasui,
       Computer pattern recognition techniques: Some results with real
       electrocardiographic data, IEEE Trans. Biomed. Electron. 10: 106–
       114 (July 1963).
26.   Lawrence Stark, Stability, oscillation, and noise in the human pupil
       servomechanism, Bol. Inst. Estud. Med. Biol. (Mex.) 21 (2): 201–222
       (August 1963).
27.   Laurence R. Young and Lawrence Stark, Variable feedback experi-
       ments testing a sampled data model for eye tracking movements,
       IEEE Trans. Human Factors Electron. (Special Manual Control Is-
       sue) HFE-4 (1): 38–51 (September 1963).
28.   Lawrence Stark and Howard Hermann, Prerequisites for a photo-
       receptor structure in the crayfish tail ganglion, Anat. Rec. 147 (2):
       209–217 (October 1963).
29.   Lawrence Stark and Howard T. Hermann, Single-unit responses in a
       primitive photoreceptor organ, J. Neurophysiol. 26: 215–228 (1963).
30.   Lawrence Stark, Allen A. Sandberg, Saul Stanten, Paul A. Willis, and
       James F. Dickson, On-line digital computer used in biological ex-
       periments and modeling, Ann. N.Y. Acad. Sci. 115 (2): 738–762 (July
       31, 1964).
31.   Shoji Yasui, Gerald H. Whipple, and Lawrence Stark, Comparison of
       human and computer electrocardiographic wave-form classification
       and identification, Am. Heart J. 68 (2): 236–242 (August 1964).
32.   Lawrence Stark and Laurence R. Young, Defining biological feedback
       control systems, Ann. N.Y. Acad. Sci. 117 (1): 426–442 (September
       10, 1964).
33.   Arne Troelstra, Bert L. Zuber, David Miller, and Lawrence Stark,
       Accommodative tracking: A trial-and-error function, Vision Res.
       4: 585–594 (1964).

1965

34.   Gerald H. Whipple, James F. Dickson, Hiroshi Horibe, and Lawrence
       Stark, Remote on-line, real-time computer diagnosis of the clinical
       electrocardiogram, Commun. Assoc. Computing Machinery 8: 49–52
       (January 1965).
35.   Lawrence Stark and Yoshizo Takahashi, Absence of an odd-error
       signal mechanism in human accommodation, IEEE Trans. Biomed.
       Eng. BME-12: 138–146 (July–October 1965). Also Symposium on
       Automatic Control, Systems Science, Cybernetics and Human Fac-
       tors, the IEEE International Convention (March 1965).

36.   Laurence R. Young and Lawrence Stark, Biological control systems —
      A critical review and evaluation: Development in manual control,
      NASA Contractor Rep. CR-190: 1—221 (March 1965).
37.   Lawrence Stark and James F. Dickson, Mathematical concepts of
      central nervous system function, Neurosciences Research Program
      Bull. 3: 1—72 (May 14, 1965). Neurosciences Research Symposium
      Summaries, Vol. 1, F. O. Schmitt and T. Melnechuk (eds.), M.I.T.
      Press, Cambridge, Mass. (1966), pp. 109—178.
38.   Lawrence Stark, Classical and statistical mathematical models for a
      neurological feedback system, Neurosciences Research Program
      Bull. 3: 55—60 (May 1965). Neurosciences Research Symposium
      Summaries, Vol. 1, F. O. Schmitt and T. Melnechuk (eds.), M.I.T.
      Press, Cambridge, Mass. (1966), pp. 164—169.
39.   Lawrence Stark, Carl Kupfer, and Laurence R. Young, Physiology of
      the visual control system, NASA Contractor Rep. CR-238: 1—88 (June
      1965).
40.   Lewis G. Bishop and Lawrence Stark, Pupillary response of the
      screech owl, Otus asio, Science 40: 1750—1752 (June 1965).
41.   Lawrence Stark and James F. Dickson, Remote computerized medi-
      cal diagnostic systems, Computers and Automation 14: 18–21
      (July 1965).
42.   Lawrence Stark, James F. Dickson, Gerald H. Whipple, and Hiroshi
      Horibe, Remote real-time diagnosis of clinical electrocardiograms
      by a digital computer system, Ann. N. Y. Acad. Sci. 126: 851–872 (1965).
43.   Martin Lorber, Bert L. Zuber, and Lawrence Stark, Suppression of
      the pupillary light reflex in binocular rivalry and saccadic suppres-
      sion, Nature 208: 558—560 (November 6, 1965).
44.   Lawrence Stark, Yoshizo Takahashi, and George Zames, Nonlinear
      servoanalysis of human lens accommodation, IEEE Trans. Systems
      Sci. Cybernetics SSC-1: 75—83 (1965).
45.   Bert L. Zuber and Lawrence Stark, Microsaccades and the velocity—
      amplitude relationship for saccadic eye movement, Science 150:
      1459—1460 (December 1965).

1966

46.   Lawrence Stark, Bioengineering: A definition by example of the inter-
      action between the engineering sciences and biology and medicine,
      Presbyterian-St. Luke's Hosp. Med. Bull. 5: 42—52 (1966). Also
      Medical Electronics News, J. N. Martin (ed.), Library of Congress
      Cat. No. 66-30546.
47.   Bert L. Zuber, Lawrence Stark, and Martin Lorber, Saccadic sup-
      pression of the pupillary light reflex, Exp. Neurol. 14: 351—370
      (March 1966).

48. Allen Sandberg and Lawrence Stark, Functional analysis of pupil
     nonlinearities, Presbyterian-St. Luke's Hosp. Med. Bull. 5: 89—105
     (1966).
49. Lawrence Stark, Neurological feedback control systems, in: Advances
     in Bioengineering and Instrumentation, Fred Alt (ed.), Plenum Press,
     New York (1966), Chapter 4, pp. 289—385.
50. James F. Dickson and Lawrence Stark, Remote real-time computer
     system for medical research and diagnosis, J. Am. Med. Assoc. 196:
     149—154 (June 13, 1966).
51. Saul F. Stanten and Lawrence Stark, A statistical analysis of pupil
     noise, IEEE Trans. Biomed. Eng. BME—13: 140—152 (July 1966).
52. Laurence R. Young, Bert L. Zuber, and Lawrence Stark, Visual and
     control aspects of saccadic eye movements, NASA Contractor Rept.
     CR-564: 1—138 (September 1966).
53. Lawrence Stark, Laurence R. Young, Robert Taub, Arthur Taub, and
     Peter G. Katona, Biological control systems: A critical review and
     evaluation, NASA Contractor Rept. CR-577: 1—391 (September 1966).
54. Bert L. Zuber and Lawrence Stark, Saccadic suppression: Elevation
     of visual threshold associated with saccadic eye movements, Exptl.
     Neurol. 16: 65—79 (September 1966).
55. Gerald Cook, Lawrence Stark, and Bert L. Zuber, Horizontal eye
     movements studied with the on-line computer, Arch. Ophthalmol. 76:
     589—595 (October 1966).
56. James Houk, Ronald W. Cornew, and Lawrence Stark, A model of
     adaptation in amphibian spindle receptors, J. Theoret. Biol. 12: 196—
     215 (1966).
57. Joel A. Michael and Lawrence Stark, Interactions between eye move-
     ments and the visually evoked responses in the cat, Electroencepha-
     log. Clin. Neurophysiol. 21: 478—488 (1966).
58. Guillermina N. Yankelevich, Jose Negrete-Martinez, George Theo-
     doridis, and Lawrence Stark, Analisis del flujo de informacion en la
     respuesta de marcha del cambarino orconectes virilis a la estimula-
     cion fotica abdominal, Bol. Inst. Estud. Med. Biol. (Mex.) 24: 23—52
     (1966).

1967

59. Joel A. Michael and Lawrence Stark, Electrophysiological correlates
     of saccadic suppression, Exptl. Neurol. 17: 233—246 (1967).
60. Gerald Cook and Lawrence Stark, Derivation of a model for the hu-
     man eye-positioning mechanism, Bull. Math. Biophys. 29: 153—174
     (1967).
61. Lawrence Stark, Y. Takahashi, and George Zames, Biological control
     mechanisms: Human accommodation as an example of a neurological
     servomechanism, Theoret. Exptl. Biophys. 1: 129—397 (1967).

62. Jerald Brodkey and Lawrence Stark, Feedback control analysis of accommodative convergence, Am. J. Surg. 114: 150−158 (July 1967).

63. Jerald Brodkey and Lawrence Stark, Accommodative convergence: An adaptive nonlinear control system, IEEE Trans. Systems Sci. Cybernetics SSC-3: 121−133 (1967). Also Presbyterian-St. Luke's Hosp. Med. Bull. 6: 30−43 (1967).

64. Ronald W. Cornew, James C. Houk, and Lawrence Stark, Fine control in the human temperature regulation system, J. Theoret. Biol. 16: 406−426 (1967).

65. Lawrence Stark, Information rate in patterns of nerve impulses to the crayfish photoreceptor−walking movement system, in: Information and Control Processes in Living Systems, D. M. Ramsey (ed.), New York Academy of Sciences, New York (1967), pp. 213−222.

66. Lawrence Stark, Pattern recognition as a model for neurophysiology, in: Information and Control Processes in Living Systems, D. M. Ramsey (ed.), New York Academy of Sciences, New York (1967), pp. 85−103.

1968

67. Fernando Navas and Lawrence Stark, Sampling or intermittency in the hand control system, Biophys. J. 8: 252−302 (1968).

68. Gerald Cook and Lawrence Stark, Dynamic behavior of the human eye-positioning mechanism, Commun. Behavioral Biol. 1: 197−204 (March 1968). Also Presbyterian-St. Luke's Hosp. Med. Bull. 6: 44−51 (1967).

69. Gerald Cook and Lawrence Stark, The human eye-movement mechanism: Experiments, modeling, and model testing, Arch. Ophthalmol. 79: 428−436 (April 1968). Also SRL-0005, Office of Aerospace Research United States Air Force (June 1967).

70. Lawrence Stark, Robert Arzbaecher, Gyan Agarwal, Jerald Brodkey, Derek Hendry, and Willian O'Neill, Status of research in biomedical engineering: A Report for the Engineering for Biology and Medicine Training Committee of the National Institute of General Medical Sciences, IEEE Trans. Biomed. Eng. BME−15 (July 1968, in press).

71. William D. O'Neill and Lawrence Stark, Triple Function Ocular Monitor, J. Opt. Soc. Am. 58: 570−573 (April 1968).

72. Jerald S. Brodkey and Lawrence Stark, New Direct Evidence Against Intermittency or Sampling in Human Smooth Pursuit Eye Movements, Nature 218: 273−275 (April 1968).

73. Lawrence Stark, The Pupillary Control System: Its Nonlinear Adaptive and Stochastic Engineering Design Characteristics (to appear in Proceedings of the International Federation of Automatic Control Symposium on "Technical and Biological Problems in Cybernetics," September 24−28, 1968, Yerevan, USSR).

74. Bert L. Zuber and Lawrence Stark, Dynamical Characteristics of the
     Fusional Vergence Eye-Movement System, IEEE Trans. Systems
     Sci. Cybernetics SSC-4: 72—79 (March 1968).
75. Lawrence Stark, John Semmlow, and Joseph F. Terdiman, Anatomical
     Transfer Function, Mathematical Biosciences (in press).
76. Gyan Agarwal, Gerald Gottlieb, and Lawrence Stark, Models of Muscle
     Proprioceptive Receptors, 4th Annual NASA Univ. Conference on
     Manual Control, March 1968 (in press).
77. Gyan Agarwal, Bradley Berman, Michael Hogins, Peter Lohnberg,
     and Lawrence Stark, Effect of External Loading on Human Motor Re-
     flexes, 4th Annual NASA Univ. Conference on Manual Control, March
     1968 (in press).

Index